D0378246

TECHNICAL WRITING

SITUATIONS AND STRATEGIES

Second Edition

MICHAEL H. MARKEL *Drexel University*

ST. MARTIN'S PRESS *New York*

TECHNICAL WRITING

SITUATIONS AND STRATEGIES

Second Edition

Library of Congress Catalog Card Number: 87-060517

Copyright © 1988 by St. Martin's Press, Inc.
All rights reserved.
Manufactured in the United States of America.
21098
fed

For information, write St. Martin's Press, Inc.
175 Fifth Avenue, New York, NY 10010

Cover design: Darby Downey
Cover photo: © Susan Lapides 1980/Design Conceptions.

ISBN: 0-312-00269-6

ACKNOWLEDGMENTS

Flow chart on solar energy from SOLAR ENERGY UTILIZATION by Tim Michels. Copyright © 1979 by Van Nostrand Reinhold. All rights reserved.

"How to Use Pollution Abstracts" reprinted with the permission of the publisher. © 1986 Cambridge Scientific Abstracts. All rights reserved.

Passage from UNTIL THE SUN DIES by Robert Jastrow reprinted by permission of W. W. Norton & Company, Inc. Copyright © 1977 by Robert Jastrow.

Diagrams from COMPLETE DO-IT-YOURSELF MANUAL © 1973 by The Reader's Digest Association, Inc. Reprinted by permission of the publisher.

Excerpt from the user's guide for Microsoft® Multiplan® Electronic Worksheet Program for Apple®Macintosh (1984). Reprinted by permission of the Microsoft Corporation.

Instructions for installing a sliding door for a shower. Copyright 1986. NoviAmerican, Inc. Novi, Michigan.

Instructions on how to apply panels to stud walls. Reprinted with the permission of the National Retail Hardware Association.

Explanatory page from the HOSPITAL LITERATURE INDEX. Reprinted with permission of the American Hospital Association, © 1985.

Acknowledgments continue on page 590, which constitutes an extension of the copyright page.

To My Parents

PREFACE

The second edition of *Technical Writing: Situations and Strategies* is intended to provide a clear and comprehensive teaching tool for technical writing instructors while at the same time preparing students to approach the working world's various writing situations. The idea behind the book is that the best way to learn to write is to write and rewrite. Accordingly, the text contains not only numerous samples of actual technical writing but also dozens of writing and revising exercises that enable students to apply what they have learned in realistic technical writing situations.

The first two chapters explain the concept on which the rest of the book is based: that the difficult part of technical writing is not learning to solve the complicated little problems (such as how to decide when to write out numbers) but to handle successfully the simple, overriding problem involved in every kind of factual communication. This problem could be called "How to Write About Your Subject, Not About Yourself." Or it might be called "How to Write to Your Readers, Not to Yourself." But by any name, the common problem in technical writing is that most writers simply fail to consider *why* and *for whom* they are writing. Too often, they write as if they were still in school and their readers were their instructors. The results are, of course, all too predictable: long, complicated documents that few people have the time, desire, or knowledge to read.

If *Technical Writing* has a single goal, it is to impress upon students the need to analyze the writing situation—the audience and purpose of the communication—and how this situation affects the structure, development, and style of the document. In this second edition, coverage of audience and purpose has been expanded substantially, as has attention to the technical writing process in general. The important details of the writing process are addressed in Part One, with Chapter 2 devoted to the assessment of audience and purpose, including heuristics to help students analyze an audience. Chapter 3 examines prewriting, drafting, and revising, followed by material on finding and using information in Chapter 4. The part concludes with a discussion of technical writing style.

The techniques of technical writing are the subject of Part Two. The five chapters in this part discuss the basic rhetorical techniques used in technical writing—definition, analysis, and description—as well as graphic aids, which are such an integral part of effective technical communication. New to this edition is a chapter on instructions and manuals, including a guide to writing a manual.

Part Three concentrates on the technical report, beginning with a chapter on format and moving on to types of reports. The treatment of the memo in this book is substantially different from that in most other technical writing texts. Here it is discussed not as a brief in-house news item, but rather as a short internal report. The chapters on proposals, progress reports, and completion reports have a dual purpose: first, to describe how these documents function in the world of business and industry, and second, to provide logical and realistic models for student writing in the course. Many technical writing instructors have found that these three types of reports, considered in sequence, help their students understand the basic future-present-past movement of most on-the-job writing:

☐ the proposal: what I want to investigate, and why

☐ the progress report: how the investigation is proceeding

☐ the completion report: what the investigation revealed, and what the findings mean

Part Four covers other technical writing applications: technical articles, oral presentations, correspondence, and the job-application procedure. Three appendixes provide additional information for students. They consist of a handbook of style, punctuation, and mechanics; a guide to documentation of sources; and a selected bibliography.

For the second edition of this text I have tried to retain what worked well in the first while adding topics new and important to the field of technical writing. Two topics presented in this edition for the first time should be mentioned in particular: computers and collaborative writing. Throughout the text are numerous discussions of ways to use computers and word processors—from prewriting to compiling a bibliography to generating computerized graphics. And the student-written sample reports that run through Chapters 13, 14, and 15 study the purchase of a personal computer. Collaborative writing, basic to most businesses and increasingly common in college classrooms, is discussed in Chapter 3; in addition, suggestions on how to use this text collaboratively can be found in the instructor's manual.

The instructor's manual, like that of the first edition, includes sample course plans and chapter-by-chapter guides. New to this edition, however, are additional readings that might be photocopied and given out to your students, including both good examples that can serve as models and flawed ones that might be used as exercises in revision.

All the examples in this book—from single sentences to complete reports—are real. Some were written by my students at Clarkson College in Potsdam, New York. Others were written by Drexel University students, either while they were in school in Philadelphia or while they were off campus working as co-op students. The bulk of the writing, however, was done by engineers, scientists, and managers with whom I have worked as a consultant over the last fourteen years. Because most of the information in the writing is proprietary, I have changed brand names, company names, and other identifying nouns. I would like to thank the dozens of people—students and professionals alike—who have graciously allowed me to print their writing.

In working on the second edition of *Technical Writing* I have benefited greatly from the perceptive observations and helpful suggestions of my fellow teachers throughout the country: Jeanne Vasterling Allen, University of Missouri at Kansas City; Valarie Arms, Drexel University; Eugene Baer, Wisconsin Lutheran College; Barry Batorsky, Bridgewater State College; David F. Beer, University of Texas at Austin; Roberta Boss, University of Maryland at College Park; Sara Brown, Tulsa Junior College; Cynthia S. Butler, Florida Institute of Technology; Nancy Coppola, New Jersey Institute of Technology; B. J. Craig, Del Mar College; Virginia Cyrus, Rider College; Everett Devine, Purdue University; Sandra Donaldson, University of North Dakota; T. Jeff Evans, University of Maine; Edwin M. Falk, Mesa Community College; Charlene Fletcher, University of Nebraska at Omaha; Ruth Foreman, South Dakota State University; Roger Freling, Central

Michigan University; Julia M. Gergits, University of Minnesota; Steve Gerson, Johnson County Community College; Barry Greer, Oregon State University; William J. Griese, DeAnza College; Kendall Hane, Indiana University Northwest; P. C. Hironymous, University of Maryland-Baltimore County; Penny L. Hirsch, Northwestern University; Sharon Irvin, Florida Institute of Technology; Bruce L. Janoff, University of Pittsburgh; Laura B. Kennelly, North Texas State University, Jeanne Halgren Kilde, University of Minnesota; Carol Kremer, William Woods College; David R. Lampe, M.I.T.; Carmen Lanning, Texas A & M University; Jean Hoskinson Leach, University of Maine at Orono; Donna Lenhoff, Butte College; Christine Levenduski, University at Minnesota; Beverly Levy, University of Maine at Orono; Frank K. Lewis, Texas State Technical Institute; Richard Louth, Southeastern Louisiana University; Ann Marlowe, Missouri Southern State College; Cornelius P. McAdorey, SUNY-Farmingdale; David B. Merrell, Abilene Christian University; Dana Nelson, Kent State University; Elinore H. Partridge, California State University at San Bernardino; Priscilla Pratt, University of Minnesota; Esther N. Rauch, University of Maine at Orono; John S. Robertson, Iowa Central Community College; Simeon J. Round, Johns Hopkins University; Michael Rose, Lane Community College; Christopher Sawyer-Laucanno, M.I.T.; Robert A. Schwegler, University of Rhode Island; Nancy Shoemaker, University of Minnesota; Elizabeth Slocum, University of Minnesota; Timothy Sweet, University of Minnesota; Jacqueline Trace, SUNY-Fredonia; Eugene Vasilew, SUNY-Binghamton; David K. Vaughan, University of Maine at Orono; Damon Vincent, The School of the Ozarks; Margaret Anne Walterhouse, Florida Institute of Technology; and Gayle H. Woolley, University of the Pacific.

In particular I would like to thank Virginia A. Book, University of Nebraska at Lincoln; Stephen C. Brennan, Tulane University; Michael Connaughton, St. Cloud State University; Rick Eden; Dixie Elise Hickman, University of Southern Mississippi; Robert T. Knighton, University of the Pacific; and Delia C. Temes, Syracuse University. My thanks go, too, to Susan Anker, Marilyn Moller, Anne McCoy, and Christine Pearson of St. Martin's Press, for their many good ideas and great assistance throughout this project.

My greatest debt, however, is to my wife, Rita, who over the course of many months and two editions has helped me to say what I mean. To her should go praise for the good things in this book. For not always following her advice, I take responsibility.

Michael H. Markel

CONTENTS

TECHNICAL WRITING

SITUATIONS AND STRATEGIES

Second Edition

CHAPTER ONE

THE ROLE OF TECHNICAL WRITING IN BUSINESS AND INDUSTRY

CHARACTERISTICS OF EFFECTIVE TECHNICAL WRITING

CLARITY

ACCURACY

COMPREHENSIVENESS

ACCESSIBILITY

CONCISENESS

CORRECTNESS

INTRODUCTION TO TECHNICAL WRITING

Technical writing can be defined as writing about a technical subject, intended to convey specific information to a specific audience for a specific purpose. Much of what you read every day is technical writing—textbooks (including this one), the owner's manual for your car, cookbooks. The words and graphic aids of technical writing are meant to be practical: that is, to communicate a body of factual information that will help an audience understand a subject or carry out a task. For example, an introductory biology text helps students understand the fundamentals of plant and animal biology and carry out basic experiments. An automobile owner's manual describes how to operate and maintain that particular car.

THE ROLE OF TECHNICAL WRITING IN BUSINESS AND INDUSTRY

The working world depends on written communication. Virtually every action taken in a modern organization is documented in writing (whether on paper or in a computer's memory). The average company has dozens of different forms to be filled out for routine activities, ranging from purchasing office supplies to taking inventory. If you need some filing cabinets at a total cost of $300, for instance, you can purchase them after an office manager signs a simple form. But expenditures over a specified amount—such as $2,000 or $10,000—require a brief report showing that the purchase is necessary and that the supplier you want to buy from is offering a better deal (in terms of price, quality, service, etc.) than the competition. Similarly, most technicians record their day-to-day activities on relatively simple forms. But when these same technicians travel to a client's plant to teach a class on a new procedure, they have to write up a detailed memo or report when they return. That report describes the purpose of the trip, analyzes any problems that arose, and offers suggestions that would make future trips more successful.

When a major project is being contemplated within an organization, a proposal must be written to persuade management that the project would be not only feasible but also in the best interests of the organization. Once the project is approved and under way, progress reports must be submitted periodically to inform the supervisors and managers about the current status of the project and about any unexpected developments. If, for instance, the project is not going to be completed on schedule or is going to cost more than anticipated, the upper-level managers need to be able to reassess the project in the broader context of the organization's goals. Often, projects are radically altered on the basis of the information contained in progress reports. When the project has ended, a completion report must be written to document the work and enable the organization to implement any recommendations.

In addition to all this in-house writing, every organization must communicate with other organizations and the public. The letter is the basic format for this purpose. Inquiry letters, sales letters, goodwill letters, and claim and adjustment letters are just some of the types of daily correspondence. And if a company performs contract work for other companies, then proposals, progress reports, and completion reports are called for again. The world of business and industry is a world of communication.

If you are taking a technical writing course now, the chances are that it was created in response to the needs of the working world. You have undoubtedly seen articles in newspapers and magazines about the importance of writing and other communication skills in a professional's career. The first step in *getting* a professional-level position, in fact, is to write an application letter and a résumé. If there are other candidates as well qualified as you—and in the job market today there almost certainly are—your writing skills will help determine whether an organization decides to interview you. At the interview, your oral communication skills will be examined along with your other qualifications. Once you start work, you will write memos, letters, and short reports, and you might be asked to contribute to larger projects, such as proposals. During your first few months on the job, your supervisors will be looking at both your technical abilities and your ability to communicate information. The facts of corporate life today are simple. If you can communicate well, you are valuable to your organization; if you cannot, you are much less valuable. The tremendous growth in continuing-education courses in technical and business writing reflects this situation. Organizations are paying to send their professionals back to the classroom to improve their writing skills. New employees who can write and speak effectively bring invaluable skills to their positions.

CHARACTERISTICS OF EFFECTIVE TECHNICAL WRITING

Technical writing is meant to get a job done. Everything else is secondary. If the writing style is interesting, so much the better. But keep in mind six basic characteristics of technical writing:

1. clarity
2. accuracy
3. comprehensiveness
4. accessibility
5. conciseness
6. correctness

CLARITY   The primary characteristic of effective technical writing is clarity. That is, the written document must convey a single meaning that the reader can understand easily.

Unclear technical writing is expensive. The cost of a typical business letter today, for instance, counting the time of the writer and the typist and the cost of equipment, stationery, and postage, is more than eight dollars. Every time an unclear letter is sent out, the reader has to write or phone the writer and ask for clarification. The letter then has to be rewritten and sent again to confirm the correct information. On a larger scale, clarity in technical writing is essential because of the cooperative nature of most projects in business and industry. One employee is assigned to investigate one aspect of the problem, while other employees work on other aspects of it. The vital communication link among the various employees is usually the report. If one link in the communication chain is weak, the entire project could be jeopardized. For example, consider the case of an electronics manufacturing company that is trying to decide whether to develop a new product. The marketing department is in charge of determining the size of the market. If that report recommends against producing the product but is misinterpreted, the company could lose a substantial amount of money by beginning research and development. Such a breakdown could represent a huge financial loss for the organization.

More important, unclear technical writing can be dangerous. Poorly written warnings on bottles of medication are a common example, as are unclear instructions on how to operate machinery safely. A carelessly drafted building code tempts contractors to save money by using inferior materials or techniques. The events that led to the 1986 space-shuttle tragedy might have been prevented had the officials responsible for deciding whether to launch received clear reports of the safety risks involved that day.

ACCURACY   Inaccurate writing causes as many problems as unclear writing. Accuracy is a simple concept in one sense: you must record your facts carefully. If you mean to write *4,000*, don't write *40,000*. If you mean to refer to Figure 3-1, don't refer to Figure 3-2. The slightest error in accuracy will at least confuse and annoy your readers. Inaccurate instructions, naturally, can be dangerous.

In another sense, however, accuracy is a more difficult concept. Technical writing must be as objective and free of bias as a well-conducted scientific experiment. If the readers suspect that you are slanting information—by overstating the significance of a particular fact or by omitting reference to an important point—they have every right to doubt the validity of the entire document. Technical writing must be effective by virtue of its clarity and organization, but it must also be reasonable, fair, and honest.

COMPREHEN-   A comprehensive technical document provides all the information
SIVENESS   the readers will need. It describes the background so that readers who are unfamiliar with the project will be able to understand the problem or opportunity that led to the project. It includes a clear description of the methods the writer used to carry out the project, as well as a complete statement of the principal findings—the results and any conclusions and recommendations.

Comprehensiveness is crucial for two reasons. First, the people who will act on the document need a complete, self-contained discussion so that they can apply the information effectively, efficiently, and safely. Second, the document will be the official company record of the project, from its inception to its completion.

For example, a scientific article reporting on an experiment comparing the reaction of a new strain of bacterium to two different compounds will not be considered for publication unless the writer has described fully the methods used in the experiment. Because other scientists want to be able to replicate the researcher's methods, every detail is included, including the names of the companies from which the researcher obtained all the materials.

Or consider a report recommending that a company network its computers. The company will probably want to have the recommendations analyzed in detail before committing itself to such an expensive and important project. The team charged with studying the report needs all the details. If the recommendations are implemented, the company needs a single, complete source of information in case changes have to be made several months or years later.

ACCESSIBILITY

Accessibility refers to the ease with which readers can locate the information they seek. One of the major differences between technical writing and other kinds of nonfiction is that most technical documents are made up of small, independent sections. Some readers are interested in only one or several sections; other readers might want to read many or most of the sections. Because relatively few people will pick up the document and start reading from the first page all the way through, the writer's job is to make the various parts of the document accessible. That is, the readers should not be forced to flip through the pages to find the appropriate section.

To increase the accessibility of the writing, use headings and lists (see Chapter 5). For reports, include a detailed table of contents (see Chapter 11).

CONCISENESS

Technical writing is meant to be useful. The longer a document is, the more difficult it is to use, for the obvious reason that it takes more of the reader's time. In a sense, conciseness works against clarity and comprehensiveness. If a technical explanation is to be absolutely clear, you must describe every aspect of the subject in great detail. The solution to this conflict is to balance the claims of clarity, conciseness, and comprehensiveness. The document must be just long enough to be clear—given the audience, purpose, and subject—but not a word longer. You can shorten most writing by

10–20 percent simply by eliminating unnecessary phrases, choosing short words rather than long ones, and using economical grammatical forms.

The battle for concise writing, however, is often more a matter of psychology than of grammar. A utility company, for instance, recently experienced a fairly serious management problem caused by its employees' writing. Each branch manager of the company was required to submit a semiannual status report, informing the corporate managers about any technical problems that had occurred during the previous six months. The length of the report was clearly specified as three pages, maximum.

As it turned out, the corporate headquarters was paralyzed for almost a month after the status reports came in. They averaged 17 pages. The branch managers had read the "three pages, maximum" directive; they just didn't believe it. They wanted to impress their supervisors. So each put a good many weeks into creating reports that would cover everything. No detail was too small to be included. In their attempt to produce the ultimate status report, they forgot the simple fact that somebody had to read it. From the reader's point of view, few things are more frustrating than having to read 17 pages that say, in effect, that no problems occurred. The corporate managers had to spend hours chopping the 17-page reports down to 3 pages so that they could be filed for future reference. Justifiably enough, these managers felt they were doing the branch managers' jobs.

The utility company hired a writing consultant to make one point to the branch managers: not only is it acceptable to write three-page status reports, but more than three pages is less than good. Once this point was communicated, the rest of the time was spent discussing ways to get all the necessary information into three pages.

CORRECTNESS Good technical writing is correct: it observes the conventions of grammar, punctuation, and usage, as well as any appropriate format standards.

Many of the "rules" of correctness are clearly important. If you mean to write, "The three inspectors—Bill, Harry, and I—attended the session," but you use commas instead of dashes, your readers might think six people (not three) attended. If you write, "While feeding on the algae, the researchers captured the fish," some of your readers might have a little trouble following you—at least for a few moments.

Most of the rules, however, make a difference primarily because readers will form an impression of your writing on the basis of how it looks and sounds. If your report is sloppy or contains grammar errors, your readers will doubt the accuracy of your information, or will at least lose their concentration. You will still be communicating, but the message won't be the one you had intended. As a result, the document will not achieve its purpose.

Technical writing is meant to fulfill a mission: to convey information to a particular audience or to persuade that audience to a particular point of view. To accomplish these goals, it must be clear, yet accurate, and complete, yet easy to access. It must be economical and correct. The writer must be invisible. The only evidence of his or her hard work is a document that works—without the writer's being there to explain it.

The rest of this book describes ways to help you say what you want to say.

PART ONE

THE TECHNICAL WRITING PROCESS

CHAPTER
TWO

THE TWO COMPONENTS OF THE WRITING SITUATION

AUDIENCE

ANALYZING YOUR AUDIENCE

THE TECHNICAL PERSON

THE MANAGER

THE GENERAL READER

ACCOMMODATING THE MULTIPLE AUDIENCE

UNDERSTANDING THE PERSONAL ELEMENT

PURPOSE

STRATEGY

AUDIENCE AND PURPOSE CHECKLIST

EXERCISES

ASSESSING THE
AUDIENCE AND PURPOSE

Chapter 1 defines technical writing as writing about a technical subject that conveys specific information to a specific audience for a specific purpose. In other words, the content and form of any technical document are determined by the situation that calls for that document: your audience and your purpose. Understanding the writing situation helps you devise a strategy to meet your readers' needs—and your own.

THE TWO COMPONENTS
OF THE WRITING SITUATION

The concepts of audience and purpose are hardly unique to technical writing: apart from idle chit-chat, most everyday communication is the product of the same two-part environment. When you write, of course, you must be more precise than you are in conversation: your audience cannot stop you to ask for clarification of a complex concept, nor can you rely on body language and tone of voice to convey your purpose.

Typical everyday examples of the two-part writing situation abound. For example, if you have recently moved to a new apartment and wish to invite a group of friends over, you might duplicate a set of directions explaining how to get there from some well-known landmark. In this case, your writing situation would be clearly identifiable:

AUDIENCE

the group of friends

PURPOSE

to direct them to your apartment

Given your comprehension of this writing situation, you would include in your directions the information that each member of the group might need in order to get to your apartment and arrange this information in a useful sequence.

When a classified advertisement describes a job opening for prospective applicants, the writing situation of the advertiser can be broken down similarly:

AUDIENCE

prospective applicants

PURPOSE

to describe the job opening

When a notice is posted on a departmental bulletin board advising students interested in enrolling in a popular course to sign up on a waiting list, the department's writing situation also conforms to the two-part model:

AUDIENCE

students interested in enrolling in the course

PURPOSE

to advise signing up on the waiting list

Once you have established the two basic elements of your writing situation, you must analyze each before deciding upon the content that your document should include and the form that it should take. Effective communication satisfies the demands of both elements of the situation. Although you might assume that purpose would be your primary consideration, an examination of audience is in fact more helpful as a first step: an exploration of its characteristics often influences the course you take in refining your initial, general conception of purpose. The separation of audience and purpose is to some extent artificial—you cannot realistically banish one from your mind as you contemplate the other—

but it is nonetheless useful. The more thoroughly you can isolate first audience and then purpose, the better grasp you will have of the writing situation and the more exactly you will be able to tailor your document to the situation.

AUDIENCE

Identifying and analyzing the audience can be difficult. You have to put yourself in the position of people you might not know reading something you haven't yet started to write. Most writers want to concentrate first on content—the nuts and bolts of what they have to say. Their knowledge and opinions are understandably what they feel most comfortable with. They want to write something first and shape it later.

Resisting this temptation is especially important. Having written mostly for teachers, students do not automatically think much about the audience. Students in most cases have some idea of what their teachers want to read. Further, teachers establish guidelines and expectations for their assignments and usually know what students are trying to say. The typical teacher is a known quantity. In business and industry, however, you will often have to write to different audiences, some of whom you will know almost nothing about and some of whom will have very limited technical knowledge of your field.

In addition, these different audiences might well have very different purposes in reading what you have written. A primary audience, for example, would consist of people who have to act on your recommendations by making important decisions. An executive who decides whether to authorize building a new production facility is a primary reader. So is the treasurer who has to plan for paying for it. A secondary audience would consist of people who need to know what is being planned, such as salespeople who want to know where the facility will be located, what products it will produce, and when it will be operational.

ANALYZING YOUR AUDIENCE
How do you analyze your audience? Sometimes it's difficult because you don't know who the readers will be. You might be writing a report for your supervisor, who is likely to distribute it—but you don't know to whom. Or you might be addressing an audience of several hundred or even several thousand. Although these many readers share some characteristics, they are not a unified group.

Despite these difficulties, however, you can often analyze an audience accurately. First, try to determine whether one person or several will be reading. For every reader whom you can identify, ask yourself such questions as the following:

1. What is the person's name and job title?
2. What are the person's chief responsibilities on the job?
3. What is the person's educational background?
4. What is the person's professional background (previous positions or work experience)?
5. What will the person do with the document: file it, skim it, read only a portion of it, study it carefully, modify it and submit it to another reader, attempt to implement its recommendations?
6. What are the person's likes and dislikes that might affect his or her reaction to the document?

A profile of your audience is a good idea because it forces you to be as specific as possible. Following are a couple of examples.

You are writing a monthly status report on the activities of your mechanical engineering group. Your reader is Bill Harpole, a 42-year-old manager of engineering operations, who supervises the operations of four different groups. Bill is trained as a civil engineer, with a graduate degree in business administration. He is reasonably knowledgeable in your area: for six years he worked as a project manager and interacted with civil engineers daily.

Bill will take the information from the four status reports, summarize it, and submit it to his supervisor. You know very little about Bill's supervisor, Karen Stahl, Director of Operations, except that she is trained in operations research, the study of systems. Bill has told you not to consider her as you write to him.

Because Bill meets with you and the other group leaders several times each week, he already has a good idea of what will be included in each status report. For this reason, what he really wants is information on which to base the report he has to write to Karen. In fact, he has given the four groups a checklist with questions to guide them in filling out the status report. Bill likes concise reports that avoid unnecessary technical terms.

Later, this chapter will discuss how to use an audience profile in devising a strategy for the document.

Here is another example of an audience profile. As a systems analyst for a manufacturing company, you have written a software program that helps you keep track of production. As finished goods are manufactured and shipped, technicians enter data into the computer, which then creates reports on inventory, productivity, and so forth. You will be writing the users' manual for the system.

Your readers will consist of high-school graduates with a clerical background, as well as some persons with trade-school certificates in data processing. Some readers have never operated any kind of computer. Most of the readers are in their mid-twenties. Their jobs will be fairly repetitive: they will take numbers from written forms and enter them on different lines on the screen. They will need about a week to learn the basics of the system. Some of the functions, however, will be used rarely; therefore, the manual will also have to function as a quick reference.

As these two audience profiles suggest, virtually every writing situation is unique. However, many writers find it useful to classify their readers according to their general backgrounds, expectations, and needs. The writer can then modify this basic guide to accommodate the special needs of the particular audience.

Three basic types of audience can be identified:

1. the technical person
2. the manager
3. the general reader

THE
TECHNICAL
PERSON

The term *technical person* is used here to cover a fairly broad range of reader, from the expert who carries out original research and writes articles for technical journals to the technician who operates equipment. Roughly in the middle of this range is the technically trained professional—the engineer, the biologist, the accountant—who analyzes and solves problems as they arise.

Technical people devise and implement computer programs and fix the computer when it isn't working properly. They carry out audits to determine if proper accounting procedures are being followed. They analyze water samples to determine if the ecology of a stream is threatened. Yet all these people share one characteristic—a technical understanding of a subject area and an interest in that area. There are differences, of course. The expert understands the theory to a much greater degree than either the middle-level professional or the technician. The middle-level professional is well equipped to carry out experiments to measure productivity or quality on a production line but might not be able to conduct basic research. The technician identifies a problem in a piece of equipment, improvises a temporary solution until the defective part can be replaced, and installs the replacement part when it arrives.

When you write to a technical person, keep in mind his or her needs. The expert feels quite at home with technical vocabulary and formulas. Thus you can get to the details of the technical subject right away, without sketching in the background; the expert already knows it. The middle-level technical person, such as the engineer, might need a brief orientation to the subject; unlike the expert, the engineer isn't always familiar with theoretical background. The technician, however, needs schematic diagrams, parts lists, and step-by-step instructions to apply to a concrete task.

Following is a passage, from an article in a professional journal (April and Newton 1985: 380), that illustrates the needs and interests of the middle-level professional audience. The subject of the article is a study of the effects of acid rain on lakes. The researchers are explaining one reason that two nearby lakes—Panther and Woods—have significantly different levels of acidification.

The three basic flowpaths by which water enters a lake are: (1) direct precipitation on the lake surface, (2) overland and shallow interflow, and (3) baseflow of ground water. We have calculated the amount of water flowing along each of these flowpaths in Woods and Panther lake-watersheds using the hydrologic simulation module of the ILWAS model (see Gherini *et al.*, 1985). Surface runoff and shallow interflow predominate in Woods Lake basin because of its thin and less permeable soils (Figure 5). Hydrologically, the watershed is a surface-water and shallow interflow-dominated system producing rapid runoff (see Peters, 1985). In contrast, Panther Lake basin is a ground-water dominated-system in which a significant portion of incident precipitation moves through the ground water system as baseflow (Figure 5). The thick and relatively permeable upper-till deposits in Panther provide for extensive ground water storage and produce a uniform baseflow.

The longer residence time of water in the ground water system of Panther Lake allows the rate-limited neutralization reactions such as primary mineral weathering to proceed further toward equilibrium. Sufficient acid neutralization takes place to keep the waters of Panther Lake near a pH of 7 for most of the year. Only during the annual springmelt, when a large portion of water enters the lake through overland flow and shallow interflow, does the pH of Panther Lake surface water (upper 1 m) drop to approximately 5.

The water of Woods Lake is acidic because most incident acidic precipitation reaches the lake by surface and shallow interflow. The water has a short residence time in the thin soils and glacial till comprising the surficial deposits of the basin. Consequently, neutralization of acidic waters is incomplete. Woods Lake maintains a pH that, for the most part, ranges between 4.4 to 4.9 yr round.

This paragraph exhibits several characteristics of technical writing addressed to the middle-level professional. First, it assumes that the readers are familiar with the concepts, terminology, and symbols of the technical field. Second, it contains a lot of technical data. And third, it refers to other technical sources that the readers are likely to want to consult.

THE MANAGER   The manager is harder to define than the technical person, for the word *manager* describes what a person does more than what a person is or knows. A manager makes sure the organization operates smoothly and efficiently. The manager of the procurement department at a manufacturing plant is responsible for seeing that raw materials are purchased and delivered on time so that production will not be interrupted. The manager of the sales department of that same organization is responsible for seeing that salespeople are out in the field, creating interest in the products and following up leads. In other words, managers coordinate and supervise the day-to-day activities of the organization.

Upper-level managers, known as executives, are responsible for longer-range concerns. They have to foresee problems years ahead. Is a technology that is currently being used at the company becoming obsolete? What are the newest technologies? How expensive are they? How much would they disrupt operations if they were adopted? What other plans would have to be postponed or dropped altogether? When would the conversion start to pay for itself? What has been the experience of other companies that have adopted the new technologies? Executives are concerned with these and dozens of other questions that go beyond the day-to-day managerial concerns.

Management is a popular subject in the business curriculum, and many managers today have a background in general business, psychology, and sociology. Often, however, managers are trained in the technical areas their organizations work in. An experienced chemical engineer, for instance, might manage the engineering division of a consulting company. Although he has a solid background in chemical engineering, he earned his managerial position because of his broad knowledge of engineering and his ability to deal well with colleagues. In his daily work he might have little opportunity to use his specialized engineering skills.

Although generalizing about the average manager's background is difficult, identifying the manager's needs is an easier task. Managers are mainly interested in the "bottom line." They have to get a job done on schedule; they don't have time to study and admire a theory the way an expert does. Rather, managers have to juggle constraints—money, data, and organizational priorities—and make logical and reasonable decisions quickly.

When you write to a manager, try to determine his or her technical background and then choose an appropriate vocabulary and sentence length. Regardless of the individual's background, however, focus on the practical information the manager will need to carry out his or her job. For example, an engineer who is describing to the sales manager a new product line that the research-and-development department has created will want to provide some theoretical background on the product so that the sales representatives can communicate effectively with potential clients. For the most part, however, the description will concentrate on the product's capabilities and its advantages over the competition.

Following is a statement addressed to a managerial audience. The subject is industry's attempts to understand and confront the problem of acid rain (Wantuck 1984: 30).

A solution to the problem of acid rain has been a high priority for industry. For example, the Electric Power Research Institute, research arm of the electric utility industry, has spent more than $25 million over the past six years studying acid rain and has earmarked $75 million for additional research. Other monies are targeted toward a search for new technologies that can reduce emissions.

EPRI and organizations like the Hudson Institute, a think tank for public policy research, suggest that industrial pollutants in rain contribute only slightly, compared with natural acidity, to high acidity in lakes and streams.

William Brown, director of technological studies at the Hudson Institute, notes that only a small fraction of rain falls directly into lakes. Most of it first passes through a series of acidic or alkaline natural filters in the watershed. For example, the humus—plant or animal matter in the soil—filter can inject far more acid into the rainwater percolating through it than any amount of anticipated industrial pollution, Brown says. In fact, estimates are that the humus may contain as much as 1,000 times the acid that falls from the sky in a year.

The forests themselves are thus natural acid creators.

Fortunately, Brown says, the other filters—often several feet thick—through which the rainwater passes generally contain generous amounts of limestone and other alkaline substances to neutralize the acid. If they do not, the water remains acidic.

René Malès, vice president of the Electric Power Research Institute, says his research supports the theory that acid rain has a minimal effect. "We did testing in the Adirondacks to try to determine why lakes have different levels of acidity," he says. "We found that while [acid] deposition plays a role, much more important are the soil characteristics and the land's natural systems—the biology around the lakes."

This example is characteristic of technical writing addressed to managers. It discusses the problem of acid rain as a technical dilemma that will require studies and remedies that cost money. The writing uses quotations from reputable individuals. Rather than focusing on the technical details of acid rain, it generalizes. Terms that might not be understood, such as *humus*, are defined parenthetically.

THE GENERAL
READER

Occasionally you will have to address the general reader, sometimes called the layperson. In such cases, you must avoid technical language and concepts and translate jargon—however acceptable it might be in your specialized field—into standard English idiom. A nuclear scientist reading about economics is a general reader, as is a homemaker reading about new drugs used to treat arthritis.

The layperson reads out of curiosity or self-interest. The average article in the magazine supplement of the Sunday paper—on attempts to increase the populations of endangered species in zoos, for example—will attract the general reader's attention if it seems interesting and well written. The general reader may also seek specific information in an unfamiliar field: someone interested in buying a house might read articles on new methods of alternative financing.

In writing for a general audience, use a simple vocabulary and relatively short sentences when you are discussing areas that might confuse the layperson. Use analogies and examples to clarify your discussion. Sketch in any special background—historical or ethical, for example—so that your reader can follow your discussion easily. Concentrate on the implications for the general reader. For example, in discussing a new substance that removes graffiti from buildings, focus on its effectiveness and cost, not on its chemical composition.

Following is an example of a statement addressed to the general reader ("Dumping" 1984: 18). The subject is the Environmental Protection Agency's ruling on a request by three eastern states to force several midwestern states to reduce their sulfur-dioxide emissions.

Last week the EPA finally responded. The agency said, in a proposed ruling that presumably will become final after a mandatory 30-day period for any public objections to be heard, that it intended to reject the petition. The ruling contends that the Clean Air Act can be invoked only against the interstate transmission of specific pollutants cited in the law and that acid rain is not one of them.

The agency argued that the scientific link between sources of sulfur dioxide and the impact of acid rain on the three states had not yet been demonstrated to its satisfaction. This reasoning is in line with the claim by EPA Administrator William Ruckelshaus that numerous studies (including one prepared for the White House) were not persuasive in concluding that this form of pollution actually causes the damage that has been observed in Northeastern forests and lakes. He has asked for yet more research before committing his agency to ordering the polluting states to reduce their sulfur-dioxide emissions substantially.

Officials of the suing states were distressed by the EPA position. Thornburgh suggested that it continued a pattern of "discriminatory enforcement." New York's Democratic Governor Mario Cuomo claimed that his state has "the most comprehensive program in the nation to reduce acid rain, but 90% of the acid rain killing our lakes originates in other states. The Administration is leaving us all but defenseless." As for Maine's Democratic Governor Joseph Brennan, he angrily accused the Administration of "saying, in effect, it's O.K. to dump your garbage on your neighbor's lawn."

These paragraphs are typical of writing addressed to a general audience. The writer avoids the technical issues of acid rain completely, portraying the controversy almost as a personal feud: one group of governors against the EPA. The passage ends with a colorful quotation, a suggestion that the dispute has degenerated into a name-calling contest, with all of the participants more interested in protecting their own interests than in working together to understand and solve the problem.

ACCOMMODATING THE MULTIPLE AUDIENCE

Sometimes you will write a technical letter to one reader. Much more frequently, however, you will be addressing more than one person. And in many cases, these different readers will represent two or even three of the basic categories. A common situation, for instance, is for a technical memo to be addressed to a technical reader, with copies going to technical readers outside the writer's field, as well as to managers.

Twenty years ago, the need to address a multiple audience placed no special demands on the writer, because most managers were technical people who had risen from the technical staff. Today, however, the knowledge explosion and the increasing complexity of business operations have opened a fairly wide gap between the knowledge of the technical staff and that of most managers. Whereas two decades ago the manager of a paint manufacturing company almost certainly would have known all about paint, today that manager might have worked most recently for a brokerage house or a bakery.

Another complicating factor is an innocent-looking piece of hardware—the photocopy machine. In the days of carbon paper, a secretary could make only three or four legible copies of a typed page. Thus the number of readers—and the likelihood of having different kinds of readers—was relatively limited. But today, running off a few dozen copies is so easy and inexpensive that more people (and more kinds of people) see almost everything that is written on company stationery. In many companies the manager of an engineering division will routinely receive copies of *everything* written by its 100 engineers. And, with the increasingly standard electronic office, writing has become dispersed even more widely.

Because of this increase in the number and types of readers, you have to assess the full range of your audience carefully before you write. If you think you might have a multiple audience, structure the document accordingly. For memos and reports, include a preliminary section addressed to the manager and a detailed section addressed to the technical reader. (See Chapter 11 for a further discussion.) Clearly distinguish the two sections so that the managers know exactly where their discussion begins and ends. In each section, use a vocabulary and a sentence length and structure appropriate to those who will read it.

UNDERSTANDING THE PERSONAL ELEMENT

Identifying your readers' backgrounds helps you understand who they are and why they are reading your document. Don't forget, however, that your readers are individuals. If you know your readers, consider their character traits—their biases and eccentricities as well as their particular interests—which will probably tell you more about them as readers than their backgrounds do. When you don't know your readers, remember as you try to "put yourself in their shoes" that biases exist and that you want to avoid at all costs offending delicate sensibilities.

Personal preferences range from the important to the trivial. Some readers appreciate an introductory summary of the basic problem and the results of the project. Others want the summary in a letter of transmittal (see Chapter 11). Some readers are annoyed by the use of the pronoun *I* in a technical document; others appreciate it as a means of avoiding the passive voice. Some readers are distracted by the use of computer jargon, such as *feedback* and *input*; others find it a useful shorthand.

Common sense suggests that you should accommodate as many of the readers' preferences as possible. Sometimes, of course, such general accommodation is impossible, either because several of the preferences are contradictory or because of some special demands of the subject. In any case, never forget that technical writing is intended to get a job done; it is not a personal statement. Whenever possible, try to avoid alienating or distracting your readers.

PURPOSE

Once you have identified and analyzed your audience, reexamine your general purpose in writing. Ask yourself this simple question: "What do I want this document to accomplish?" When your readers have finished reading what you have written, what do you want them to *know*, or *believe*? Think of your writing not as an attempt to say something about the subject but as a way to help others understand it or act on it.

To come to a clearer definition of your purpose, think in terms of verbs. Try to isolate a single verb that represents what you are trying to do and keep it in mind throughout the writing process. Almost without exception, you will find that to some extent your purpose can be expressed in a simple English verb. (Of course, in some cases a technical document has several purposes, and therefore you might want to choose several verbs.) Here are a few examples of verbs that indicate typical purposes you might be trying to accomplish in technical documents. The list has been divided into two categories: verbs used when you primarily want to communicate information to your readers, and verbs used when you want to convince them to accept a particular point of view.

"COMMUNICATING" VERBS	"CONVINCING" VERBS
to explain	to assess
to inform	to request
to illustrate	to propose
to review	to recommend
to outline	to forecast
to authorize	
to describe	

This classification of verbs is not absolute. For example, "to review" could in some cases be a "convincing" verb rather than a "communicating" verb: one writer's review of a complicated situation might be very different from another's review of the same situation. Following are a few examples of how these verbs can be used in clarifying the purpose of the document. (The verbs are italicized.)

1. This report *describes* the research project to determine the effectiveness of the new waste-treatment filter.
2. This report *reviews* the progress in the first six months of the heat-dissipation study.

3. This letter *authorizes* the purchase of six new word processors for the Jenkintown facility.

4. This memo *proposes* that we study new ways to distribute specification revisions to our sales staff.

As you devise your purpose statement, remember that your real purpose might differ from your expressed purpose. For instance, if your real purpose is to persuade your reader to lease a new computer system rather than purchase it, you might phrase the purpose this way: "to explain the advantages of leasing over purchasing." Many readers don't want to be "persuaded"; they want to "learn the facts."

Make sure, as you define your purpose in writing, that you understand your audience's needs. Most technical writing is used either to document a project—so that the organization will have a record of what it did and why—or to communicate some information necessary for carrying out a current project. If the document is intended to initiate or facilitate a technical task, the structure must help the audience to understand why and how to begin. If the document is intended as a reference, other structures—such as "problem-methods-solutions" (see Chapter 4)—might be most appropriate. The important point to remember is that your purpose in writing should answer your audience's needs. Accordingly, a thorough understanding of audience is necessary before purpose can be defined precisely.

STRATEGY

Once you have analyzed your audience and purpose, you can devise a strategy for the document. Strategy refers to what information will be included and how it will be expressed.

For example, you can determine the scope—the boundaries—of the document. If you have some information on the history of computers, but your audience needs an instruction manual that explains how to operate a particular system, you might well decide that the historical information is not relevant.

More important, a good sense of audience and purpose will remind you that some kinds of information *must* be included. For example, if you are proposing that your company purchase some computer equipment, you know that you must define the need for the equipment, survey the available equipment, and make the case that buying it is less expensive than not buying it. The following example shows how an understanding of the audience and purpose would determine the content and form of such a proposal.

Your principal reader is Harry Becker, a middle manager who screens all proposals for capital expenditures. He is trained in management and has a good understanding of the kinds of equipment the different departments need to function effectively. Although he is not an expert on the kind of equipment that you want to buy, he isn't hostile to the idea of new high-technology equipment.

He knows that your department is productive and that you have never requested unnecessary or inappropriate purchases. On the other hand, he did authorize a large, unwise expenditure just three months ago that nobody has forgotten.

His standard procedure for handling a request that he approves is simply to attach a covering memo and send it to his supervisor.

This profile tells you that you have to accommodate the needs of both *your* reader and *his* reader. You have to be direct, straightforward, and objective. You must avoid excessive technical details and vocabulary. You should focus clearly on the practical advantages the new equipment would provide the company. You should include a brief summary for the convenience of your two readers.

Next, you define your purpose in writing. In this case, the purpose is clear: to convince Harry Becker that your proposal is reasonable, so that he will endorse it and pass it on to his supervisor, who in turn will authorize the purchase of the equipment.

What do your audience and purpose tell you about what to include in the document? Because your readers will be particularly careful about large capital expenditures, you must clearly show that the type of equipment you want is necessary and that you have recommended the most effective and efficient model. Your proposal will have to answer a number of questions that will be going through your readers' minds:

1. What system is being used now?
2. What is wrong with that system, or how would the new equipment improve our operations?
3. On what basis should we evaluate the different kinds of available equipment?
4. What is the best piece of equipment for the job?
5. How much will it cost to purchase (or lease), maintain, and operate the equipment?
6. Is the cost worth it? At what point will the equipment pay for itself?
7. What benefits and problems have other purchasers of the equipment experienced?

8. How long would it take to have the equipment in place and working?

9. How would we go about getting it?

A careful definition of the writing situation is the first step in planning any technical writing assignment. If you know whom you are writing to, and why, you can decide on your strategy—what to include and how to shape it. The result will be a document that works.

AUDIENCE AND PURPOSE CHECKLIST

Following is a checklist of questions you should ask yourself when you profile your audience and purpose.

1. To whom are you writing? Is your audience one person, several people, a large group, or several groups with various needs?

2. What do you know about your reader's (readers') job responsibilities and professional background?

3. What do you know about your reader's (readers') personal preferences that will help you plan the writing?

4. What will your readers be doing with the document? Why will they be reading it?

5. What is your purpose in writing? What is the document intended to accomplish?

6. Is your purpose consistent with your audience's needs?

7. How does your understanding of your audience and of your purpose determine your strategy: the scope, structure, organization, tone, and vocabulary of the document?

EXERCISES

1. Choose two articles on the same subject, one from a general-audience periodical, such as *Reader's Digest* or *Newsweek*, and one from a more technical journal, such as *Scientific American* or *Forbes*. Write an essay comparing and contrasting the two articles from the point of view of the authors' assessment of the writing situation: the audience and the purpose. As you plan the essay, keep in mind the following questions:

   a. What is the background of each article's audience likely to be? Does either article require that the reader have specialized knowledge?

   b. What is the author's purpose in each article? In other words, what is each article intended to accomplish?

   c. How do the differences in audience and purpose affect the following elements in the two articles?
      (1) scope and organization
      (2) sentence length and structure

    (3) vocabulary

    (4) number and type of graphic aids

    (5) references within the articles and at the end

2. Choose a 200-word passage from a technical article addressed to an expert audience. Rewrite the passage so that it is clear and interesting to the general reader.

3. Audience is the primary consideration in many types of nontechnical writing. Choose a magazine advertisement for an economy car, such as a Subaru, and one for a luxury car, such as a Mercedes. In an essay, contrast the audiences for the two ads by age, sex, economic means,

---

GREENLAWN   April 5, 19--

Dear Mrs. Smith,

    Thank you for inquiring about the safety of the Greenlawn program. The materials purchased and used by professional landscape companies are effective, nonpersistent products that have been extensively researched by the Environmental Protection Agency. Scientific tests have shown that dilute tank-mix solutions sprayed on customers' lawns are rated "practically nontoxic," which means that they have a toxicity rating equal to or lower than such common household products as cooking oils, modeling clays, and some baby creams. Greenlawn applications present little health risk to children and pets. A child would have to ingest almost 10 cupsful of treated lawn clippings to equal the toxicity of one baby aspirin. Research published in the American Journal of Veterinary Research in February 1984 demonstrated that a dog could not consume enough grass treated at the normal rate of application to ingest the amount of spray material required to produce toxic symptoms. The dog's stomach simply is not large enough.

    A check at your local hardware or garden store will show that numerous lawn, ornamental, and tree care pesticides are available for purchase by homeowners either as a concentrate or combined with fertilizers as part of a weed and feed mix. Label information on these counter products contain generally the same pesticides as those programmed for use by professional lawn-care companies,

lifestyle, etc. In contrasting the two audiences, consider the explicit information in the ads—the writing—as well as the implicit information—hidden persuaders such as background scenery, color, lighting, angles, and the situation portrayed by any people photographed.

4. The letter to Mrs. Smith was written by a branch manager of a lawn service company to a homeowner who inquired about the safety of the lawn care service. The homeowner specifically mentioned in her letter that she has an infant that likes to play outside on the lawn. Write an essay evaluating the way the letter responds to the concerns of its reader.

---

but contain higher concentrations of these pesticides than found in the dilute tank-mix solutions applied to lawns and shrubs. By using a professional service, homeowners can eliminate the need to store pesticide concentrates and avoid the problems of improper overapplication and illegal disposal of leftover products in sewers or household trash containers.

Greenlawn Services Corporation has a commitment to safety and to the protection of the environment. We have developed a modern delivery system using large droplets for lawn-care applications. Our applicators are trained professionals and are licensed by the state in the proper handling and use of pesticides. Our selection of materials is based on effectiveness and safety. We only apply materials that can be used safely, with all applications made according to label instructions.

On the basis of these facts, I am sure that you will be pleased to know that the Greenlawn program is a safe and effective way to protect your valued home landscape. I have also enclosed some additional safety information. I encourage you to contact me directly should you have any questions.

Sincerely,

Helen Lewis

Helen Lewis
Branch Manager

REFERENCES   April, R., and R. Newton. 1985. Influence of geology on lake acidification in the ILWAS watersheds. *Water, Air, and Soil Pollution* 26 (December): 380.

"Dumping garbage on neighbors." 1984. *Time* (10 September): 18.

Wantuck, M. 1984. Little is plain about acid rain. *Nation's Business* 17 (November): 30.

CHAPTER
THREE

PREWRITING, DRAFTING, AND REVISING

The highly complex mental functions involved in writing are not completely understood. One thing, however, is certain: just as no two persons think alike, no two persons use the same process in writing. And the same person might well use different techniques on different occasions. In one sense, therefore, the term *writing process* is misleading. There are really an infinite number of writing processes.

Yet research conducted over the last decade has shown that, despite considerable variations, most persons do their best work when they treat writing as a three-part process:

1. prewriting
2. drafting
3. revising

Many technical people dislike writing and try to get it over with as quickly as possible. They treat writing as a two-part process: writing and typing. They start with a clean sheet of paper, write "I. Introduction" at the top, and hope the rest will follow. It doesn't. No wonder they dislike writing. A few geniuses can write like that, but the rest of us find that it doesn't work.

We end up staring at the blank sheet for half an hour or so, unable to think of what to say. After another half hour we have created a miserable little paragraph. We then spend the rest of the morning trying to turn that paragraph into coherent English. We change a little here and there, clean up the grammar and punctuation, and consult the thesaurus. Finally, it's time to go to lunch. We return, happy that we have at least created a good paragraph. Then we reread what we have written. We realize that all we have done is wasted a morning. We start to panic, which makes it all the more difficult to continue writing.

Some writers use a somewhat more effective and efficient process. They realize they need some sort of outline—a plan to follow. But they do only a half-hearted job on the outline, scribbling a few words on a pad. Outlining has bad associations for most people. All of us at one time or another have had to hand in outlines to English teachers or history teachers who seemed more interested in correcting the format—the Roman numerals and the indention—than in suggesting ways to help write the papers. And sometimes the outline assignment was a punishment for having written a poorly organized paper. "Take this home and outline it. Then you'll see what a mess it is!"

The writing process described in the following section makes writing easier because it breaks it down into smaller, more manageable tasks. Writing becomes less mysterious and more like the other kinds of technical tasks that we carry out all the time.

PREWRITING

Prewriting refers to the various processes that occur before you actually start to write sentences and paragraphs. For experienced writers, prewriting is a crucial step. In fact, when they have completed prewriting, most writers consider themselves halfway home.

Prewriting consists of three steps:

1. analyzing audience and purpose
2. brainstorming
3. outlining

ANALYZING AUDIENCE AND PURPOSE   The first step in any writing assignment is to analyze your audience and purpose. Chapter 2 discusses this concept in detail. Briefly, it involves profiling your readers—their background, experience, responsibilities, personal factors, and reasons for reading—and identifying your purpose in writing—to provide information, change your readers' attitudes, or motivate action.

Although it is important to keep audience and purpose in mind throughout the writing process, you must be absolutely sure at this stage that you have a clear understanding of the writing situation. You are committing yourself to a plan of attack—a direction, a scope, a sequence, a particular tone and level of difficulty, and so on. You must know what you are doing. Otherwise you will eventually have to rewrite substantially—or throw out—long sections of your first draft.

One way to be sure you understand the writing situation is to define it in a complete sentence. Following are several examples:

1. I want to explain to management how a catalog describing the different generators we sell would help our sales staff.

2. I want to describe to the technical staff a new method for removing miscible organic pollutants from streams and rivers.

3. I want to persuade my supervisor that we ought to consider developing a contingency plan for dealing with a total loss of our computer facilities.

BRAINSTORMING   Brainstorming is a way to generate ideas for your document. The process is to spend 15 or 20 minutes listing ideas about your subject. You list the ideas as quickly as you can, using short phrases, not sentences. For instance, if you are describing the Wankel automobile engine, the first five items in your brainstorming list might look like this:

```
greater efficiency

advantages of Wankel engine

bad EPA ratings

principle of operation

no spark plugs
```

As the word *brainstorming* suggests, you are not trying to impose any order on your thinking. You will probably skip around from one idea to another. Some of the ideas you jot down will be subsets of larger ideas. Don't worry about it; you will straighten that out later.

Brainstorming has one purpose: to free your mind, within the limits of your situation, so that you can think of as many items as possible that relate to your subject. When you get to the next stage in prewriting—constructing an outline—you will probably find that some of the items you listed do not in fact belong in the document. Just toss them out at that stage. The advantage of brainstorming is that for most people it is the most effective and efficient way to catalogue those things that *might* be important to the document. A more structured way of generating ideas, ironically, would miss more of them.

Brainstorming is actually a way of classifying everything you know into two categories: (1) information that probably belongs in the document and (2) information that does not belong. Once you have decided what material probably belongs, you can start to write the outline.

Following is a brainstorming list created by a writer who is reporting to management on the feasibility of starting a health-promotion program at her company. Management authorized her project, and she is trying to persuade them that a health-promotion program would be a cost-effective solution to a serious problem. She has already gathered and skimmed all the information she will need from the professional journals. (Chapter 4 discusses library research and the technique of skimming.)

```
costs of replacing sick employees

necessary equipment

payback period?

history of health-promotion plans

how to measure interest?

increased productivity of workers?

how to measure success?

necessary publicity

converting space

start-up costs

make it mandatory?

do interest surveys?

maintenance costs

national trends?

how to maintain interest

who would staff it?
```

The next phase of prewriting, constructing an outline, involves refining this brainstorming list into a clear and organized plan.

OUTLINING   Creating an outline involves three main tasks:

1. placing similar items together in a group

2. sequencing the items in the groups

3. sequencing the groups

PLACING SIMILAR ITEMS TOGETHER IN A GROUP Start by looking at the first item on your brainstorming list. For the writer working on the feasibility study of health-promotion programs, that item is "costs of replacing sick employees." The writer decides that this item should be part of the background for her report. She scans the brainstorming list, looking for another background item. Finally she comes upon "history of health-promotion plans" and "national trends."

The writer rewrites the items as follows:

```
Background
    costs of replacing sick employees
    increased productivity of workers?
    history of health-promotion programs
    national trends
```

Looking at this list, the writer realizes that she has omitted an important item: a definition of health-promotion programs. She adds it to the list.

The writer then goes to the second item on the brainstorming list: "necessary equipment." She scans the list, looking for other items related to costs. She writes this list:

```
Costs
    necessary equipment
    necessary publicity
    converting space
    start-up costs
    maintenance costs
    who would staff it?
```

The writer realizes that this list contains two items that embrace the others: start-up costs and maintenance costs. She revises the list accordingly:

```
Costs
    start-up costs
        necessary equipment
```

```
    necessary publicity
    converting space
 maintenance costs
    who would staff it?
```

Looking at the maintenance costs, she realizes that she has forgotten insurance, so she adds it to the list.

The writer follows the same procedure as she goes through her original brainstorming list: grouping related items and adding new items that she had omitted.

Finally, she has created a series of small lists:

```
Background
    costs of replacing sick employees
    increased productivity of workers?
    history of health-promotion programs
    national trends
    definition of health-promotion programs

Costs and Benefits
    Costs
        start-up costs
            necessary equipment
            necessary publicity
            converting space
        maintenance costs
            who would staff it?
            how to maintain interest
            insurance
    Payback period?
    Benefits

How to measure interest?
    do interest surveys?

How to measure success?
```

The writer realizes that these lists are very rough. Some of the groups, such as the last one, have no subgroups. And the writer re-

alizes that one of the items from her original brainstorming list—
"make it mandatory?"—will be covered in the definition of
health-promotion programs.

At this point writers face a decision. Some writers start to draft
the document at this point, even though the outline is incomplete.
They like to see what they have to say, and the only sure way to do
that is to start to write. Once they have started drafting, they feel
they are better able to decide how to sequence items. They juggle
paragraphs instead of juggling outline items.

Other writers feel more comfortable working from a more re-
fined outline. For these writers, the next step is to sequence the
items within each group.

SEQUENCING THE ITEMS IN THE GROUPS   The writer looks at the
background section of her outline:

```
Background
    costs of replacing sick employees
    increased productivity of workers?
    history of health-promotion programs
    national trends
    definition of health-promotion programs
```

She realizes that any report should begin with a definition of important terms. Next, she decides, should be a brief history of the
programs. Two other items—"costs of replacing sick employees"
and "increased productivity"—she decides could be worked into
the history section, because they were the principal reasons to start
the programs in the first place. However, she also realizes that
they could be a part of the definition of health-promotion plans.
She makes a note on the outline to check later to see where the information works best. The remaining item—"national trends"—
follows naturally from the history section.

The tentative sequence, then, looks like this:

```
Background
    definition of health-promotion programs
    history of health-promotion programs
      costs of replacing sick employees
      increased productivity of workers?
    national trends
```

Next, the writer turns to the extensive Costs and Benefits section of the tentative outline:

```
Costs and Benefits
    Costs
        start-up costs
            necessary equipment
            necessary publicity
            converting space
        maintenance costs
            who would staff it?
            how to maintain interest
            insurance
    Payback period?
    Benefits
```

She decides to work from major items to minor items, although some writers prefer to work in the opposite direction. Because "costs and benefits" is a common term, she chooses the following sequence:

```
costs
benefits
payback period
```

Now she starts to sequence subunits. Within the costs section, "start-up costs" should precede "maintenance costs" because the chronological sequence seems most appropriate here.

```
Costs
    start-up costs
    maintenance costs
```

Within the "start-up costs" section, the writer faces a decision. She can use a chronological pattern, which might call for publicity followed by space and equipment. Or she can use a pattern that calls for the more important items followed by the less important ones. Because she is describing costs, she determines that her readers

will want to read about the major costs first. Therefore, she decides on the following sequence:

```
Start-up costs
   converting space
   equipment
   publicity
```

Turning to the maintenance costs, the writer decides to continue
with the "more important to less important" pattern. She refines
her headings as she rewrites them. In addition, the writer realizes
she should add "equipment maintenance" to the list:

```
Maintenance costs
   insurance
   staffing
   equipment maintenance
   continued publicity
```

The writer decides to leave the "payback period" item where it is:
at the end of the costs section. The complete costs section now
looks like this:

```
Costs
   start-up costs
      converting space
      equipment
      publicity
   maintenance costs
      insurance
      staffing
      equipment maintenance
      continued publicity
   payback period
```

The writer now turns to the two remaining groups in her outline:
"How to measure interest?" and "How to measure success?" She
cannot think of any ways to measure interest except through sur-

veys. Realizing that a unit should not be followed by a single sub-unit, she decides to revise the item as follows:

```
Using surveys to measure interest
```

Turning to the problem that her "How to measure success" group contains no subgroups, she turns back to the journal literature and identifies several accepted techniques used in industry:

```
How to measure success
   participation rates
   absenteeism rates
   long-term disability rates
   changes in productivity
```

The writer has started with the factors easiest to measure and moved toward the more difficult to measure because she feels her readers will be most interested in knowing first how to gather the most specific evidence.

SEQUENCING THE GROUPS Sequencing the groups involves the same logical processes as sequencing the items within each group. The first step is to determine the needs of the audience. In the outline being discussed here, the readers are expecting a traditional report structure: an executive summary and introduction followed by the background, the discussion, and the findings (appropriate results, conclusions, and recommendations). The writer decides to sequence the groups she has outlined, and then add the other elements. "Background," of course, comes first. "Measuring interest" is a logical second group, because unless the employees are interested, there is no point in proceeding with the analysis. "Costs and benefits" is a logical next group, followed by "measuring success." With the traditional report elements added, the writer's sequence for the major items looks like this:

```
executive summary
introduction
background
discussion
   using surveys to measure interest
   costs and benefits
   how to measure success
```

```
results
conclusions
recommendations
```

As this discussion shows, outlines are created and constantly re-fined as the writer tries to organize the material. Some writers prefer to create only a rough outline and then start drafting, stop-ping occasionally to revise and expand the outline. However, even those writers who like to create a very specific outline often end up changing it as they draft the document. Whatever technique helps you meet your audience's needs and achieve your purpose is the best technique for you.

AVOIDING COMMON LOGICAL PROBLEMS IN OUTLINING As you create and refine your outline, keep in mind two common logical problems: faulty coordination and faulty subordination.

Faulty coordination involves equating items that are not of equal value. Look, for example, at the following listing:

```
common household tools
    screwdriver
    drill
    claw hammer
    ball-peen hammer
```

Although there are claw hammers and ball-peen hammers, this listing is not parallel, because the other items are not similarly dif-ferentiated. For instance, there are electric drills and manual drills, as well as different kinds of screwdrivers. The listing should read:

```
common household tools
    screwdriver
    drill
    hammer
```

A related problem of coordination involves overlapping items, that is, creating more than one heading under which an item might be placed:

```
accounting services
    Automated Data Services
```

```
Data Ready
   The Accountants
The Tax Service, Inc.
```

Because "The Tax Service, Inc." is an accounting service, it should be subordinated to "accounting services." However, if you wish to devote more attention to The Tax Service, Inc., than to the other companies, an acceptable solution would be to revise the listing as follows:

```
three accounting services
   Automated Data Services
   Data Ready
   The Accountants
The Tax Service, Inc.
```

Adding the word *three* before "accounting services" prevents the overlapping. Another solution would be to revise the listing like this:

```
The Tax Service, Inc.
other accounting services
   Automated Data Services
   Data Ready
   The Accountants
```

Adding the word *other* and reversing the order of the two groups eliminates the problem of overlapping.

Faulty subordination occurs when an item is made a subunit of a unit to which it does not belong:

```
power sources for lawnmowers
   manual
   gasoline
   electric
   riding mowers
```

In this excerpt, "riding mowers" is out of place because it is not a power source of a lawnmower. Whether it belongs in the outline at all is another question, but it certainly doesn't belong here.

A second kind of faulty subordination occurs when only one subunit is listed under a unit:

```
types of sound reproduction systems
   records
      phonograph records
   tapes
      cassette
      eight-track
      open reel
   compact discs
```

It is illogical to list "phonograph records" if there are no other kinds of records. To solve the problem of faulty subordination, incorporate the single subunit into the unit:

```
types of sound reproduction systems
   phonograph records
   tapes
      cassette
      eight-track
      open reel
   compact discs
```

USING COMMON PATTERNS IN DEVELOPING IDEAS As the discussion of the outlining process for the paper on health-promotion programs shows, you as a writer must constantly choose a pattern of development to use in developing your ideas. The analysis of your audience and your purpose is your best guide. At one point in your outline, you might think that a chronological approach would best suit your purposes; at another point, that a spatial pattern might work best. There is no single approach to developing ideas.

However, as you work on your outline, keep in mind that there are some standard patterns that usually work well in particular situations. Understanding these different patterns—and how to combine them to meet your specific needs—gives you more options as you put together your document.

The following five patterns of development can be used effectively in technical writing:

1. chronological (time)
2. spatial
3. general to specific
4. more important to less important
5. problems-methods-solution

1. *Chronological Pattern.* The chronological pattern works well when you want to describe a process, give instructions on how to perform a task, or explain how something happens.

The following outline is structured chronologically. The writer is outlining a pamphlet describing to a general audience the geological history of St. John's, United States Virgin Islands.

    I.   Introduction
   II.   Stage One: Submarine Volcanism
         A.   Logical Evidence: The Pressure Theory
         B.   Empirical Evidence: Ram Head Drillings
  III.   Stage Two: Above-Water Volcanism
         A.   Logical Evidence
              1.   The Role of the Lowered Water Level
              2.   The Role of Continental Drift
         B.   Empirical Evidence: Modern Geological Data
   IV.   Stage Three: Depositing of Organic Marine Sediments
         A.   The Role of the Ice Age
              1.   Water Level
              2.   Water Temperature
         B.   The Role of "Drowned" Reefs
    V.   Conclusion

Notice that on the A and B levels of this outline the reader continues the chronological pattern. "Logical Evidence," the hypothetical explanations offered by earlier scientists, precedes the "Empirical Evidence," the measurable data gathered by modern scientists. In level IV, the Ice Age preceded and caused the drowned reefs, so it is discussed first.

2. *Spatial Pattern.* In the spatial pattern, items are organized according to their physical relationships to one another. The spa-

tial pattern is useful in structuring a description of a physical object.

A spatially organized outline follows. The writer is a heating contractor who is writing a proposal to be delivered to a rural fire department. He is describing the heating system he recommends for the firehouse.

I.   System for Offices, Dining Area, and Kitchen

    A.   Description of Floor-mounted Fan Heater

    B.   Principal Advantages

        1.   Individual Thermostats

        2.   Quick Recovery Time

II.   System for Bathrooms

    A.   Description of Baseboard Heaters

    B.   Principal Advantages

        1.   Individual Thermostats

        2.   Small Size

III.   System for Garage

    A.   Description of Ceiling-mounted Fan Heater

    B.   Principal Advantages

        1.   High Power

        2.   High Efficiency

The basic organizing principle behind this outline is the physical layout of the firehouse. This principle makes sense because the readers—in this case, the firefighters—are familiar with the firehouse. Other spatial patterns could be developed here (such as top-to-bottom: ceiling fans, baseboards, floor fans), but since the readers are unfamiliar with heating systems, they would have a harder time following the discussion. Other patterns—from most expensive to least expensive, from general principles to specific recommendations, and so forth—could be used, but the writer has determined that his readers want simple and straightforward information that they can understand as they read the proposal and study their firehouse.

3. *General-to-Specific Pattern.* The movement from general to more specific information is common in sales literature and reports intended for a multiple audience. The general information enables the manager to understand the basics of the discussion without having to read all the details. The technical reader, too,

appreciates the general information, which serves as an introductory overview.

Following is an outline for a sales brochure promoting a plastic coating used on institutional and industrial floors. The readers will be maintenance managers, who might not have heard of the product because it is fairly new. The purpose of the brochure is to persuade the readers to request a visit from the company's sales representative.

I.   The Story of EDS Plastic Coating
    A.   What Is EDS Plastic Coating?
    B.   Why Is EDS the Industry Leader?
       1.   Strength and Durability
       2.   Long-lasting Results
       3.   Ease of Application
       4.   Low Cost
    C.   How to Find Out More About EDS Plastic Coating
II.   How EDS Works
    A.   How EDS Bonds on Different Floorings
       1.   Wood
       2.   Tile
          a.   Flat
          b.   Embossed
       3.   Linoleum
          a.   Flat
          b.   Embossed
    B.   How EDS Dries on the Floor
III.   How to Apply EDS Plastic Coating
    A.   Preparing the Surface
    B.   Preparing the EDS Mixture
    C.   Spreading EDS Plastic Coating
       1.   Wetting the Applicators
          a.   Spreading Pads
          b.   Paint Brushes
       2.   Determining Where to Begin
          a.   Tiled Flooring
          b.   Continuous Flooring

   3. Coordinating Several Workers

   4. Stopping for the Day

  D. Cleaning Up

   1. Spreading Pads

   2. Paint Brushes

Parts I and II of this outline are addressed to the person who would decide whether to seek more information about this product. The general information in Part I answers the three basic questions such a reader might have:

1. What is the product?
2. What are its important advantages?
3. How can I find out more about it?

Part II explains how the product works. Part III of the outline, which contains more specific information about how to use the product, is intended for the reader who actually has to supervise the use of the product. The pattern within this part is chronological, which is the obvious choice for a description of a process.

  4. *More-Important-to-Less-Important Pattern.* Another basic pattern is the movement from more-important to less-important information. This pattern is effective even in describing events or processes that would seem to call for a chronological pattern. For example, suppose you were ready to write a report to a client after having performed an eight-step maintenance procedure on a piece of electronic equipment. A chronological pattern—focusing on what you did—would answer the following question: "What did I do, and what did I find?" A more-important-to-less-important pattern—focusing first on the problem areas and then on the no-problem areas—would answer the following question: "What were the most important findings of the procedure?" Most readers would probably be more interested, first, in knowing *what* you found than in *how* you found it.

  The outline that follows is an example of the more-important-to-less-important pattern. The writer manages the engine-repair division of a company that provides various services to owners of small aircraft. The purpose of the report is to explain clearly what work was done and why.

  I. Introduction

   A. Customer's Statement of Engine Irregularities

   B. Explanation of Test Procedures

II.   Problem Areas Revealed by Tests

    A.   Turbine Bearing Support

        1.   Oversized Seal Ring Housing Defective

        2.   Compressor Shaft Rusted

    B.   Intermediate Case Assembly

        1.   LP Compressor Packed with Carbon

        2.   Oversized LP Compressor Rear Seal Ring Defective

III.   Components That Tested Satisfactorily

    A.   LP Turbine

    B.   HP Turbine

    C.   HP Compressor

In most technical writing, bad news is more important than good news, for the simple reason that breakdowns or problems have to be fixed. Therefore, the writer of this outline has described first the components that are not working properly.

5. *Problem-Methods-Solution Pattern.* This is a basic pattern for outlining a complete project: begin with the problem, discuss the methods you followed, and then finish with the results, conclusions, and recommendations.

The following outline was written by an electrical engineer working for a company that makes portable space heaters. He has been asked by his supervisor to determine the best way to improve the efficiency of the temperature controls used in their space heaters. After explaining the reasons for wanting to improve the efficiency, he describes the technology the company currently uses. Next he describes his research on the different methods of controlling temperature, concluding with the method he recommends.

I.   The Need for Greater Efficiency in Our Heaters

    A.   Financial Aspects

    B.   Ethical Aspects

II.   The Current Method of Temperature Control: The Thermostat

    A.   Principle of Operation

    B.   Advantages

    C.   Disadvantages

III.   Alternative Methods of Temperature Control

    A.   Rheostat

        1.   Principle of Operation

        2.   Advantages

        3.   Disadvantages

    B.   Zero-Voltage Control

        1.   Principle of Operation

        2.   Advantages

        3.   Disadvantages

IV.   Recommendation: Zero-Voltage Control

    A.   Projected Developments in Semiconductor Technology

    B.   Availability of Components

    C.   Preliminary Design of Zero-Voltage Control System

    D.   Schedule for Test Analysis

V.   References

Notice that the writer begins with a discussion of the problem (Part I), followed by a discussion of the existing type of control used on the company's products.

In Part III of this outline the writer has placed the discussion of the recommended alternative—zero-voltage control—after the discussions of the two other alternatives. The recommended alternative does not *have* to come last, but most readers will expect to see it in that position. This sequence *appears* to be the most logical—as if the writer had discarded the unsatisfactory alternatives until he finally hit upon the best one. In fact, we cannot know in what sequence he studied the different alternatives, or even whether he studied only one at a time. But the sequence of the outline gives the discussion a sense of forward momentum.

This pattern of development is particularly popular when the writer's company has been hired by the reader's company. In consulting arrangements of this sort, the reader expects to see a pattern that emphasizes the amount of work the writer performed.

CHOOSING AN OUTLINE FORMAT   All the outlines included so far in this chapter are topic outlines: that is, they consist only of phrases that suggest the topic to be discussed. You should know, however, about another kind of outline, the sentence outline, in which all of the items are complete sentences. If the outline concerning temperature-control components in heaters were a sentence outline, here is how Part IIIA would appear:

A.   Temperature could be controlled by a rheostat.

    1.   A rheostat operates according to the principle of variable

         resistance: adjustments to the rheostat would allow more or

```
less current to flow in the heater by decreasing or increas-
ing the amount of resistance in the line.
```
2.   The principal advantages of rheostats are that they are sim-
     ple and reliable.
3.   The principal disadvantage of rheostats is their size: a
     rheostat capable of controlling the current in our 1,500-
     watt heater would have to be as big as the heater itself.

Few writers use sentence outlines, because writing out all the sentences is time-consuming. In addition, most of the sentences will
need to be revised later anyway. The main advantage of the sentence outline is that it keeps you honest: if you would be content to
list a topic heading—such as "Advantages of the Proposed System"—without knowing what you're going to say, a sentence outline would force you to find out the necessary information. However, most people will happily cut corners and use a topic outline
unless another person has specifically requested a sentence outline.

The other aspect of outlines that most writers aren't overly
concerned about is the mechanical system for signifying the various levels of headings. Many writers, especially in private outlines, simply indent to set off subunits:

first-level heading
  second-level heading
    third-level heading

In the traditional notation system for formal outlines, numbers
and letters go in front of the headings:

I.
  A.
    1.
    2.
  B.
    1.
      a.
      b.
      c.
        (1)
        (2)
          (a)
          (b)
    2.
  C. etc.

A popular variation is the decimal style, which is used by the military and often by scientists:

1.0
  1.1
    1.1.1
    1.1.2
  1.2
2.0 etc.

An advantage of the decimal system is that it is simple to use and to understand. You don't have to remember how to represent the next heading level. In the same way, the readers can easily determine what level they are reading: a three-digit number is a third-level heading, and so forth.

Whichever system helps you stay organized—or is required by your organization—is the one to use.

DRAFTING

A rough draft is a preliminary version of the final document. Some rough drafts are rougher than others. An experienced writer might be able to write a draft, fix a comma here and change a word there, and have a perfectly serviceable document. On the other hand, every writer remembers the stubborn report that refused to make any sense at all after several drafts.

Many writers devise their own technique for drafting, but the key is to write quickly. Leave the revision to the next step.

If you are working on paper, write on one side only so you can cut and paste later. Triple space so that you can add material easily. When you start to draft, remember that there is no reason to begin at the beginning of the document. Most writers like to begin with a section from the middle of the document, usually a technical section they are comfortable with. Postponing the introductory elements of the document makes sense for another reason: because the introductory elements are based on the body—and have to flow smoothly into the body—you have to know what the body will say before you can introduce it. Until you have written the body, you cannot be sure what it is going to say.

To make sure they keep writing as fast as they can, many writers force themselves to draft for a specified period, such as an hour, without stopping. Writing the rough draft is closer in spirit to brainstorming than to outlining. When you make up the outline, you are concentrating hard, trying to figure out precise relationships. When you write the draft, you are just trying to turn your outline into paragraphs as fast as you can.

Your goal is to get beyond "writer's block," the mental paralysis that can set in when you look down at a blank piece of paper. Write *something*: it will be easy to revise later. Once you get rolling, you will be able to see if your outline works. If you really don't know what to say about an item, you will probably want to stop and think it out. Perhaps you will revise your outline at this point. But by not stopping to worry about your writing, you don't lose your momentum. Some writers are so intent on keeping the rough draft flowing that they refuse to stop even when they can't figure out an item on their outlines; they just pick up a new piece of paper and start with the next item that *does* make sense.

REVISING

Some writers can simply read their drafts and instantly recognize all the problems in them; others devise comprehensive checklists so they will not forget to ask themselves important questions.

Yet all writers agree on one thing: you should not revise your draft right after you have finished writing it. You can identify immediately some of the smaller writing problems—such as errors in spelling or grammar—but you cannot accurately assess the quality of what you have said: whether the information is clear, comprehensive, and coherent.

To be able to revise your draft effectively, you have to set it aside for a time before looking at it again. Give yourself at least a night's rest. If possible, work on something else for a few days. This will give you time to "forget" the draft and approach it more as your readers will. In a few minutes, you will see problems that would have escaped your attention even if you had spent hours revising it right after completing the draft.

How exactly do you revise a draft? Naturally, there is no single way; you should try to develop a technique that works for you.

In your first pass through the document, concentrate on the largest issues: *content* and *organization*. Recreate your outline by writing all your headings in a list. Perhaps in writing the draft you omitted a heading from your original outline. Check to see if all the material is included. Having completed the first draft, you might now have a somewhat different perspective on the subject. In the haste of drafting you might have added material that now looks irrelevant. If so, mark it. You might want to move it or omit it altogether. Look at the headings to see that the sequence is clear and logical: remember that you are trying to meet your audience's needs. If you now think that a different organizational pattern will work better, make the changes.

Looking at your headings will also give you a sense of the *emphasis* you have given to the different topics in the document. If a relatively minor topic seems to be treated at great length, check the draft itself. The problem might be merely that you created more headings at that point than you did in treating some of the other topics. But if your treatment is in fact excessive, mark passages to consider condensing.

Reread your draft also for *accuracy*. Have you provided all the necessary data? Are the data correct? Check them against your notes.

Once you are satisfied that you have included the right information in the right order, revise for *style*. Have you used an appropriate level of vocabulary for your audience? Have you used consistent terminology throughout and provided a glossary—a list of definitions—if any of your readers will need it?

Have you varied your sentences appropriately? Are the sentences grammatically correct? Have you avoided awkward constructions? Are all the words spelled correctly? See Chapter 5 for a discussion of technical writing style.

After you have revised your draft, give it to someone else to read. This will provide a more objective assessment of your writing. Choose a person who comes close to the profile of your eventual readers; someone more knowledgeable about the subject than your intended audience will understand the document even if it isn't clear and hence will not be sufficiently critical. After your colleague has read the document, find out what he or she thinks about it: strong points, unclear passages, sections that need to be added, deleted, or revised, and so forth. If possible, have your colleague read it as your eventual readers will: if the document is a set of instructions, see if he or she can perform the task.

WORD PROCESSING AND THE WRITING PROCESS

Most people who have never used a word processor are skeptical when they hear stories about how terrific it is. How can a machine increase a writer's productivity by 100 percent, 200 percent, or more? How can a machine eliminate misspellings and mistakes in grammar? How can a machine make it fun to write?

These claims, and dozens more, are made every day. Some of

the claims are merely advertising talk. Computer products and services are a multibillion-dollar industry: there is a lot of hardware and software to be sold, and many thousands of people earn their salaries helping you make a wise purchase and teaching you how to use the products.

Another explanation for the sometimes exaggerated claims is that word processing makes the writing process sound more objective, easier to perform. Sometimes people get carried away with the idea of making a complicated and idiosyncratic process more rational and orderly.

The truth is that most of the claims about word processing contain a good deal of truth. You can write faster and much more easily. You can make fewer errors. You can produce neater and more professional-looking writing. You can collaborate more easily with other writers to produce longer, more ambitious documents.

There is only one catch: you still have to do the writing. The word processor won't tell you what to say. But if you think of it as a sophisticated tool that enables you to write better, you will find it almost as terrific as people claim.

Of course, there are some drawbacks. The instruction manuals for the hardware and software can be frustrating. And most users have lost some text because they forgot to "file" it on a disk—or because the system "crashed."

In the following discussion, the term *word processor* will be used to refer to dedicated word processors (computers that perform only word processing), microcomputers used with word-processing programs, and larger computers that the user accesses through remote terminals.

Word processors offer many advantages over pen and paper at every stage of the writing process—prewriting, drafting, and revising.

PREWRITING   A number of idea-generating programs exist that appear to help many students create topics to write about. These programs ask you to respond to a series of questions about a subject area of your choice. Such questions force you to think about the audience and purpose of the report and to indicate what kinds of information you will need to write it. There is no scientific proof at this point that using these programs is more effective than simply answering the same questions on paper. However, some professors report that as their students use idea-generating programs several times, they learn how to ask the right questions and therefore write better reports.

A second use of word processors is in taking notes. If you can bring your source material to the word processor, you can save time and trouble later. Rather than take notes on index cards, type them into the word processor. When you want to use that information in your document, you can put it there easily and revise it any way you want. Making your bibliography will be much easier, too, and you won't introduce as many errors, because you'll simply move the entries into position without having to type them in from handwritten notes.

Once you sit down to brainstorm and outline, the word processor will help you save time. Because even slow typists can work quickly on a word processor, you can create a full brainstorming list quickly and easily—and everything you write will be easy to read. Moving items from one position to another on the screen is almost effortless; therefore, you can easily try out different alternatives as you classify items into groups. The same holds true when you sequence the groups. Creating and sequencing the groups on paper can be cumbersome.

DRAFTING   The word processor is a useful tool during the drafting stage because it helps you write faster. Knowing that you can easily move information from one place in the text to another encourages you to begin writing in the middle: on a technical point you are familiar with. When you find that you can't think of what to write on one subject, just skip a few lines and go to the next subject on your outline. Shuffling your text later takes only a few seconds.

With a word processor, you can easily write your draft right on your outline. The advantage is that you are less likely to lose sight of your overall plan. Before you start to draft, make a copy of your outline. You'll be able to use it later to create a table of contents. On the other copy of the outline, start to draft. As you write, the portion of the outline below will scroll downward to accommodate your draft. You don't have a separate outline and a separate draft: the outline becomes the draft.

Because word processors are relatively quiet and easy to type with, you will probably find that you can generate much more writing in a given period of time. You don't have the physical effort involved in writing by hand, and you don't have to return the carriage at the end of the line as you do on a typewriter. Producing a lot of writing quickly is exactly the point of drafting: you want to have material to revise later.

Another factor that encourages productivity is that you don't worry about the quality of the writing or about typographical errors. You can concentrate on what you are trying to say because making changes, large or small, requires so little effort.

Some writers find that the word processor provides an effective way of breaking the habit of revising what they have just typed. The screen has a contrast knob, just as a television set does. Turning the contrast knob all the way makes the screen black. This technique, called invisible writing, encourages writers to close their eyes or to look at their outline or the keyboard. The result is that they don't stop typing so often.

A common feature on word processors, the global search-and-replace function, also increases writing speed during the drafting stage. The search-and-replace function lets you find a phrase, word, or group of characters and replace it with something else throughout the document. For example, if you will need to use the word *potentiometer* a number of times, you can simply type in *po* each time. Then, during the revising stage, you can instruct the word processor to change every *po* to *potentiometer*. This use of the search-and-replace function also reduces the chances of misspelling, for you have to spell the word correctly only once.

REVISING   Word processors make every kind of revising easier. Most obviously your writing is legible, so you see what you've done without being distracted by sloppy handwriting. And because your writing is typed neatly, you have a more objective perspective on your work. You are looking at it as others will.

If the obvious typographical errors distract you, fix them so that you can concentrate on substantive changes. All word processors have easy-to-use add and delete functions so that you can change *hte* to *the* in a second.

You can make major revisions to the structure and organization of the document easily. All word processors let you move text simply and quickly—anything from a single letter to whole paragraphs. Therefore, you can try out different versions of the document without the bother of cutting and pasting pieces of paper. Most word processors also have a copy function, which lets you copy text—such as an introductory paragraph—and move it to some other location without moving the original text. With the copy function you can in effect create two different versions of the document simultaneously and decide which one works better.

A number of different editing programs are available that help you identify problem areas that need to be fixed.

One common editing program is called a spelling checker, which compares what you have typed with a dictionary, usually of 20,000 to 90,000 words. The program can usually check approximately five thousand to ten thousand words per minute. The program will alert you when it sees a word that isn't in its dictionary. Although that word might be misspelled, it might be a correctly spelled word that isn't in the dictionary. You can add the word to your dictionary so that in the future it will know that the word is not misspelled. If the word is misspelled, you have to look it up, a process that might help you learn the correct spelling. Without the spelling checker to point out the error, you might not have known the word was misspelled.

One limitation of any spelling-checker program is that it cannot tell whether you have used the correct word; it can tell you only whether the word you have used is in its dictionary. Therefore, if you have typed, "We need too dozen test tubes," the spelling checker will not see a problem.

A related program is a thesaurus, which lists words similar in meaning to many common words. A thesaurus program has the same strengths and weaknesses as a printed thesaurus: if you know the word you are looking for but can't quite think of it, the thesaurus will help you to remember it. But the terms listed might not be related closely enough to the key term to function as synonyms. Unless you are aware of the shades of difference, you might be tempted to substitute an inappropriate word. For example, the word *journal* is followed in *Roget's College Thesaurus* by the word *diary*. A personal journal is a diary, but a professional periodical certainly isn't.

A word-usage program measures the frequency of particular words and the length of words. We all overuse some words, but without the word processor we have a difficult time in determining which ones. Word length is a useful factor to know, for technical terms frequently are long words. After we analyze our audience and purpose, we need to consider the amount of technical terminology to include in the document. Readability formulas, which rely on word length and sentence length, are discussed in Chapter 5.

Finally, there are style programs, many of which perform several functions. For instance, they count such factors as sentence length, number of passive voice constructions and expletives, and types of sentence (see Chapter 5). Many style programs identify abstract words and suggest more specific ones. Many point out sexist terms and provide nonsexist alternatives. Many point out fancy words, such as *demonstrate*, and suggest substitutes, such as *show*.

Keep in mind just what these different programs can do and what they cannot do. They can point out your use of the passive voice, but they cannot tell you whether the passive voice is preferable to the active voice in a particular sentence. In a way these style programs make your job as a writer more challenging: by pointing out potential problems they force you to make decisions about issues that you might not have noticed without the word processor. But the payoff is that wise use of the programs will give you a better document.

Many of the functions performed by these style programs can be performed with the search function. For example, if you know that you overuse expletives ("It is . . . ," "there is . . . ," "there are . . ."), you can search for words such as *is* and *there*. Some of the uses of these words, of course, will not be expletives, but you can revise those that are inappropriate. Or if you realize you overuse nominalizations (noun forms of verbs, such as *installation* for *install*), you can search for the common suffixes (such as *-tion*, *-ance*, and *-ment*) used in nominalizations. These stylistic problems are discussed in Chapter 5.

Although word processors can help you do much of the work involved in revision, they cannot replace a careful reading by another person. Revision programs will calculate sentence length more accurately than a person can, but they cannot identify unclear explanations, contradictions, inaccurate data, inappropriate choice of vocabulary, and so forth. Use the revision programs, revise your document yourself, and then get help from someone you trust.

Three other aspects of word processing will be discussed later in this text. Chapter 4, "Finding and Using Information," discusses data-base searching, a technique for accessing information stored electronically. Although data bases can be accessed through a personal computer, most students find it simpler and more effective to seek the assistance of the professional librarian at the college or university library. Chapter 10, "Graphic Aids," refers to some of the common types of graphics that can be produced with graphics software on many kinds of computers. Chapter 11, "Elements of a Report," discusses some of the formatting techniques that word processors make convenient.

COLLABORATIVE WRITING

This chapter has described writing as essentially a solitary activity. Most small documents—such as letters, memos, and brief reports—are in fact conceived and written by a single person. However, much writing on the job is collaborative. In addition, collaborative writing is becoming much more popular in universities for big documents such as senior design proposals and reports.

Collaborative writing can be defined as any writing in which more than one person participates in at least one of the stages of the writing process: prewriting, drafting, or revising. Following are some common examples of collaborative writing in industry.

1. A manufacturer of computer software has completed a program and wants to write a user's manual. The manual begins with notes written by the systems analyst who devised the program. In conversations between the systems analyst and the writer, the manual is fleshed out. An artist creates the illustrations. Then the writer drafts the manual and solicits comments and revisions from the systems analyst and the marketing department. Finally, a third person edits the manual.

2. An electronics company wants to respond to a government request to design a piece of high-technology equipment. The project manager creates a proposal-writing team consisting of several engineers in different areas encompassed by the product to be designed, a graphic artist, several writers and editors, and a contracts specialist.

3. Three biologists who have conducted a set of experiments on the reaction of a strain of bacteria to a genetically engineered molecule decide to write up their work for publication in a professional journal. Each biologist writes a different section after a meeting to decide the best strategy for the article. Finally, one of the three revises the whole article.

As these examples suggest, collaborative writing can involve any number of people, can address any kind of audience, and can lead to any kind of document.

Still, some basic guidelines can make almost any kind of collaborative writing project effective and efficient:

1. *The participants must agree to a strategy and a set of procedures.* All members of the group must agree in their analysis of audience and purpose. On the basis of this analysis, the group must establish a strategy covering such matters as length, writing style, level of vocabulary, and the like. In addition, the group must formally agree to their operating procedures, including such matters as schedules for meetings and other communication among members.

2. *Each of the participants must have a specific role in the collaborative process.* For example, one person accepts responsibility for investigating and writing up a particular section of the document. Another person is responsible for gathering and collecting all the technical data.

3. *The group must have a coordinator.* One person must be the leader. That person must be responsible for scheduling and chairing meetings

of the group, maintaining effective communications, providing motivation and technical support, and helping the group reach consensus
when differences of opinion occur.

Although people can work together at all stages of the writing process, the greatest amount of collaboration occurs during
the prewriting and revising stages.

In fact, the term *brainstorming* was coined almost half a century ago to refer to planning sessions attended by as many as a
dozen people. At these brainstorming sessions one person was the
leader, responsible for keeping the ideas flowing. He or she would
encourage the less outspoken participants and keep the more vocal
ones from dominating the session. In any kind of brainstorming
session, a group of people can create a much longer and more useful list of topics than a solitary person can. One person's idea gives
another person an idea. A skillful leader can motivate everyone to
contribute.

During the drafting stage, most people work alone. This does
not mean that they cannot talk with other people or call for assistance. But individual sentences and paragraphs are created most
efficiently by a person working alone.

During the revision stage, collaboration is again useful. The
collaboration can be formal or relatively informal. Informal collaboration might consist of one person asking another person's
opinion on a particular passage in the document. Formal collaboration is more structured. The group members exchange their
work, write comments and suggestions, and return the work to
each writer. The group meets to talk through any problems. The
process can be repeated any number of times, depending on the
length and complexity of the document. Eventually, time constraints force the group members to collect all the work and put it
together in a single document. At this point, one person usually
oversees the creation of the unified document. The process involves making sure the different segments are consistent in terms
of physical format, writing style, pagination, and so forth. The
goal is to make the document look as if it were written by one person, not by a group.

The computer has become an invaluable tool in the collaborative writing process. Word processors allow the group members to
revise their writing easily and effectively. If the group members
are linked through a central computer, the different pieces of
writing can be transmitted electronically and then revised. Even if
the writers are working on separate word processors, the writing
can be transmitted through telephone lines, or, if their systems are
compatible, the group members can merely exchange disks.

Word processors save a lot of time and effort when it comes to putting the document together. Obviously, word processors make it unnecessary to retype the different sections. But in addition, the principal writer can use functions such as search and replace to make sure that technical vocabulary is being used consistently and to create a list of key words or an index when necessary.

EXERCISES

1. Choose a topic you are familiar with and interested in (such as some aspect of your academic study, a neighborhood concern, or some issue of public policy). Brainstorm for 10 minutes, listing as many items as you can that might be relevant in a report on the topic. Finally, after analyzing your list, arrange the items in an outline, discarding irrelevant items and adding necessary ones that occur to you.

2. The following brainstorming list was created by a student interning as a chemist with a large chemical company.

   The subject is trihalomethanes (THMs), a series of chemicals that pollute drinking water. The Environmental Protection Agency has ruled that within one year all municipalities must reduce the percentage of THMs in drinking water to less than 0.05 percent. The chemical company for which the writer works currently uses four chemicals for removing THMs: chlorine dioxide, polymers, phosphates, and activated carbon. Each chemical has advantages and disadvantages.

   Turn this brainstorming list into an outline for a report to management of the chemical company, recommending that the company create an aggressive marketing document that can be distributed to municipal governments. Where necessary, note information that has to be added to the outline.

   Then, turn the brainstorming list into an outline for a report addressed to the municipal government of a town for which the activated-carbon method of reducing THMs is the most effective. Where necessary, note information that has to be added to the outline.

   EPA guideline
   activated carbon advantages and disadvantages
   publicity through environmental groups on THM danger
   regions of country where THM is biggest problem
   what are THMs
   how to determine the best method
   15 states currently analyzing water for THMs
   start-up costs for activated carbon high
   polymers already used for other applications in water supplies
   chlorine dioxide can boost other chemicals above limits
   chemical structures of common THMs

safety record of our company excellent
phosphates are somewhat dangerous to handle
activated carbon can be reused; therefore cheap to use
financial penalties for noncompliance
relationship between temperature and THM incidence
describe our free water analysis program
schedule for other states to start analyzing water
no product development or testing cost to us: 4 methods EPA ok'd
chlorine dioxide working already in one plant
chlorine dioxide an effective broad-range biocide
phosphates also inhibit future THM growth
activated carbon requires little operator attention

3. The following brainstorming list was created by a manager in the personnel department of a city government.

   A recent study conducted by the department shows that turnover and absenteeism are high and that productivity is low. Of the many factors that contribute to these problems, an ineffective method for evaluating job performance appears to be the most important.

   Turn this brainstorming list into an outline for a report to the city manager, recommending that the city investigate alternative methods of job-performance evaluation. Where necessary, note information that has to be added to the outline.

   Then, turn the brainstorming list into an outline for a memo addressed to the city workers, explaining that alternative methods of job-performance evaluation will be studied and that the best one will be implemented. Where necessary, note information that has to be added to the outline.

   characteristics of current method of evaluation
   goal: decrease turnover
   ethical reasons to improve system
   current evaluations not performed on any schedule
   purposes of an evaluation system
   evaluations should be performed on a regular schedule
   economic reasons to improve system
   system should be useful to mgt and workers
   characteristics of a fair system
   federal govt prevents discrimination
   objectives of the system should be clear
   number of suits against city has increased
   goal: decrease absenteeism
   criteria of our system are subjective

why institute a fair system?

legal reasons to improve system

criteria should be objective

goal: reward good performance, discourage bad performance

4. The following portions of outlines contain logical flaws. In a sentence each, explain the flaws.

   a. I. Advantages of collegiate football
        A. Fosters school spirit
        B. Teaches sportsmanship
     II. Increases revenue for college
        A. From alumni gifts
        B. From media coverage

   b. A. Effects of new draft law
        1. On Army personnel
        2. On males
        3. On females

   c. I. Components of a Personal Computer
        A. Central Processing Unit
        B. External Storage Device
        C. Keyboard
        D. Magnetic Tape
        E. Disks
        F. Diskettes

   d. A. Types of Common Screwdrivers
        1. Standard
        2. Phillips
        3. Ratchet-type
        4. Screw-holding tip
        5. Jeweler's
        6. Short-handled

   e. A. Types of Health Care Facilities
        1. Hospitals
        2. Nursing Homes
        3. Care at Home
        4. Hospices

CHAPTER
FOUR

CHOOSING A TOPIC

FINDING THE INFORMATION

USING THE LIBRARY
Reference Librarians / *Card Catalogs* / *Reference Books* /
Periodicals Indexes / *Abstract Journals* / *Government*
Publications / *Guides to Business and Industry* / *Computerized*
Information Retrieval

PERSONAL INTERVIEWS
Choosing a Respondent / *Preparing for the Interview* /
Conducting the Interview / *After the Interview*

LETTERS OF INQUIRY

QUESTIONNAIRES
Creating Effective Questions / *Sending the Questionnaire*

USING THE INFORMATION

EVALUATING THE SOURCES

SKIMMING

TAKING NOTES
Paraphrasing / *Quoting*

EXERCISES

FINDING AND USING INFORMATION

W e live in what is called the information age. The mark of a true professional in every technical field is the ability to find and use effectively the massive amounts of information available. This chapter could not hope to provide a comprehensive look at how to conduct research: there are literally thousands of different kinds of research tools and techniques ranging from familiar reference books to computerized mathematical models. Rather, this discussion will introduce the basic methods of finding and using printed information. It will also discuss the techniques used in questionnaires and interviews.

CHOOSING A TOPIC

Whereas very few professionals have the opportunity to choose their topics, you as a student are likely, at least on occasion, to have this freedom. Like most other freedoms, this one is a mixed blessing. An instructor's request that you "come up with an appropriate topic" is for many people frustratingly vague; they would rather be told what to write about—how the Soviet Union's Sputnik changed the American policy on space exploration, for instance—even though the topic assigned might not interest them.

Provided you don't spend weeks agonizing over the decision, the freedom to choose your own topic and approach is a real advantage: if you are interested in your topic, you'll be more likely to want to read and write about it. Therefore, you will do a better job.

Start by forgetting topics such as the legal drinking age, the draft, and abortion. These topics are unrelated to most practical writing situations. Ask yourself what you are *really* interested in: perhaps something you are studying at school, some aspect of your part-time job, something you do during your free time. Browse through three or four recent issues of *Time* or *Newsweek*, and you will find dozens of articles that suggest interesting, practical topics involving technical information. The April 28, 1986, *Newsweek*, for example, contained the following articles:

SUBJECT OF THE ARTICLE	POSSIBLE SUBJECT AREAS FOR A REPORT
U.S.-built picture tubes used in Japanese TV sets	Japanese dominance in the electronics market
	foreign manufacturers that produce goods in the United States
	Congress and protectionism
U.S. air attack on Libya	the quality of U.S. military hardware
	the quality of U.S. intelligence operations
	airport security systems
	business relationships between Europe and Libya and between the United States and Libya
the Titan missile explosion	"O-ring" technology
	spy-in-the-sky satellites
video format of *Playboy* centerfold	new technologies for old products
buying mortgages	the growth of a high-risk business venture
new designs for restaurants	changing dining patterns of Americans
Vladimir Horowitz plays in Moscow	cultural exchange programs
Vancouver's Expo 86	the economics of expositions
men's studies as an academic discipline at colleges and universities	the history of women's studies as a subject
Excuse Booth: a phone booth that plays prerecorded background noises	high-technology toys

Many of these topics are not sufficiently focused for a research report of 10 to 15 pages, but they are good starting points. The subject of airport security systems, for example, could be narrowed to a topic on new screening techniques being developed to combat the use of plastic explosives that cannot be detected by X-rays. The subject of Americans' changing dining patterns could be narrowed to a topic on the restaurant industry's response to the simultaneous increase in the number of meals we eat away from home and our new interest in lower-fat nutrition. The point is that good topics are all around. If you are interested in music, you might write about what happened to eight-track tapes or quadrophonic recording, or the use of new materials in the manufacture of instruments, or the Japanese dominance in audio components. If you are interested in health care, you might write about hospices, or birthing rooms, or the nursing shortage, or the regulation of so-called quack medicine.

Once you have chosen a tentative topic for your research report, the next step is to see if you have the resources necessary to do the job well. The three resources you must consider are:

1. your own knowledge of the subject
2. other information on the subject
3. time

First, determine how much you already know about the subject by brainstorming. Keep in mind that if you don't know much about it, you will have a much more difficult time doing your research. You will have to do a lot of background reading in general sources such as encyclopedias and handbooks before you will understand even the basics. Sometimes, of course, the subject will be so new that there are no convenient sources for background reading. If, however, you already know the basics of the subject, you know what kind of information you will need, and you will have an easier time finding that information.

Most of this chapter discusses how to find information, both in the library and through personal interviews and questionnaires. Actually, the job of finding information involves two separate tasks: determining if the information exists, and trying to obtain it. Everyone at one time or another makes the mistake of assuming that if a book is listed in the card catalog, it can be found on the shelf. But fully 10 percent of a library's holdings might be missing: miscatalogued, misplaced, stolen, or checked out indefinitely by a faculty member.

Time is the third crucial resource. You might not have the two or three weeks needed to get a book through interlibrary loan, or the month required to send out a questionnaire. You might not have the time for extensive background reading. And even if you have all your information in hand, it might not be possible for you to write a good report in time. Before you start to do any serious research, skim through your sources to determine if you will have enough time. Changing topics is much easier *before* you've put in a lot of hours.

Some writers like to have an absolutely precise topic in mind before they begin to search the literature. A few examples of precise topics follow:

1. the effect that energy-efficiency tax credits have had on the home-improvement industry
2. the effect of salt-water encroachment on the aquifer in southern New Jersey
3. the market, during the coming decade, for videodisc instructional systems designed to be used in elementary schools

The advantage of a precise topic is that you can begin your research quickly, because you know what you're looking for.

Other writers prefer to begin their research with only a general topic in mind. For example, you might know that you want to write about computers but be unsure about what aspect of computers to focus on. The basic research techniques described in this chapter will enable you to discover—quickly and easily—what aspects of computers will make the most promising topics. Perhaps you didn't know about the large amount of research being carried out to help companies reduce the theft of computerized data. That topic might appeal to you. Or you might want to research new techniques used in computer graphics. The advantage of being flexible about your topic is obvious: in 15 minutes you can discover dozens of possible topics from which to choose.

FINDING THE INFORMATION

The best place to find information about most topics is the library. Learning to use the library is essential. However, some topics require that you interview a person or persons or that you send questionnaires.

USING THE
LIBRARY
You should become familiar with the different kinds of libraries. The local public library in a small city is a good source for basic reference works, such as general encyclopedias, but it is unlikely

to have more than a few specialized reference works. Most college libraries have substantially larger reference collections and receive the major professional journals. Large universities, of course, have comprehensive library collections. Many large universities have specialized libraries that complement selected graduate programs, such as those in zoology or architecture. Large cities often have special scientific or business libraries that you can use as well.

REFERENCE LIBRARIANS The most important information sources at any library are the reference librarians. Although as a college student you are expected to know how to use the library, reference librarians are there to help you solve special problems. They are invariably willing to suggest new ways to search for what you need—specialized directories, bibliographies, or collections that you didn't know existed. Perhaps most important, they will tell you if the library *doesn't* have the information you need and suggest other libraries to try. Reference librarians can save you a lot of time, effort, and frustration. Don't be afraid to ask them questions when you run into problems.

CARD CATALOGS With few exceptions, every general library maintains a main card catalog, which lists almost all the library's books, microforms, films, phonograph records, tapes, and other materials. Some libraries list periodicals (newspapers and journals) in the main catalog; others keep separate "serials" catalogs. The library's nonfiction books and pamphlets, as well as many nonliterary resources, are usually entered on three types of cards (author, title, and subject) in the main catalog. Periodicals, whether in the main catalog or in the serials catalog, are catalogued by title; to find an article, you must know the name and date of the journal in which it appears (see the discussion of periodicals indexes below).

Accordingly, if you know the title or author of the nonserial item for which you are looking, you can determine easily whether the library has it. Your search is more difficult if you have no specific work in mind and are simply looking for a discussion of a topic by consulting the subject cards. (Some libraries separate their subject cards from their author and title cards.) In such cases you must determine the likeliest subject headings under which relevant publications might be classified. If you have trouble finding appropriate materials, a reference librarian should be able to suggest other subject headings to look under. The broader the subject, the greater the number of cards there will be that you will have to go through. However, broad subjects are often subdivided: *biol-*

ogy, for example, might be subdivided into *cytology*, *histology*, *anatomy*, *physiology*, and *embryology*, following the range of general *biology* entries. Also, at the end of a range of subject cards, a "see also" card often suggests other subject headings under which to look.

Most libraries post a guide to their classification system, indicating the locations of the books in all the different subject areas. This guide is useful if you want to make sure you are using the correct terminology in searching for a subject area.

All three types of catalog cards provide the same basic information about an item, as Figure 4-1 shows. In addition to the basic bibliographic information—author, title, place of publication, publisher, number of pages, International Standard Book Number (ISBN), and cataloguing codes—each card also lists the call number (the number that indicates where on the shelves the item is located) in the upper left-hand corner. Call numbers are based on one of two major classification systems. The older, the Dewey Decimal system, uses numbers (such as 519.402462) to designate subject areas. The newer system, the Library of Congress classification, uses combinations of letters and numbers (such as TA330.H68). Some libraries have converted completely from Dewey Decimal to Library of Congress; others are still in the process. In Dewey Decimal, for example, 519 is the category for engineering mathematics; in Library of Congress, engineering mathematics is TA. Also posted in every library is a map that will direct you to the area where the books are shelved (if the stacks are open).

Card catalogs are changing in format. Some libraries are transferring their catalogs to microfiche or to computer-generated printed format. And some libraries, such as the New York Public Library, have converted completely to on-line catalogs. Instead of walking around the card catalog area to find the appropriate tray of cards, you simply sit before a terminal and enter commands that call up the appropriate entries on the screen. There are a number of different on-line catalog systems, but they all involve the standard accessing categories of author, title, and subject.

REFERENCE BOOKS   Some of the books in the card catalog will have call numbers preceded by the abbreviation *Ref.* These books are part of the reference collection, a separate grouping of books that normally may not be checked out of the library.

In the reference collection are the general dictionaries and encyclopedias, biographic dictionaries (*International Who's Who*), almanacs (*Facts on File*), atlases (*Rand McNally Commercial At-*

AUTHOR CARD

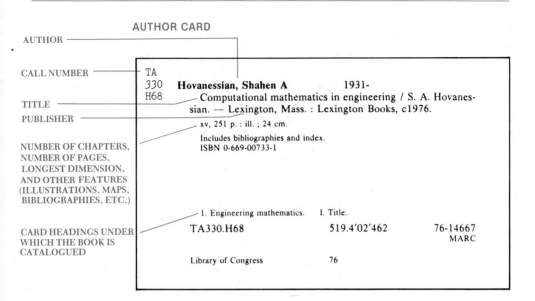

AUTHOR

CALL NUMBER

TITLE

PUBLISHER

NUMBER OF CHAPTERS,
NUMBER OF PAGES,
LONGEST DIMENSION,
AND OTHER FEATURES
(ILLUSTRATIONS, MAPS,
BIBLIOGRAPHIES, ETC.)

CARD HEADINGS UNDER
WHICH THE BOOK IS
CATALOGUED

TA
330
H68   **Hovanessian, Shahen A**   1931-
       Computational mathematics in engineering / S. A. Hovanes-
       sian. — Lexington, Mass. : Lexington Books, c1976.

       xv, 251 p. : ill. ; 24 cm.

       Includes bibliographies and index.
       ISBN 0-669-00733-1

       1. Engineering mathematics.   I. Title.

       TA330.H68   519.4'02'462   76-14667
                                                                   MARC

       Library of Congress   76

TITLE CARD

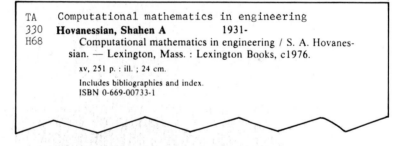

TA   Computational mathematics in engineering
330   **Hovanessian, Shahen A**   1931-
H68   Computational mathematics in engineering / S. A. Hovanes-
          sian. — Lexington, Mass. : Lexington Books, c1976.

          xv, 251 p. : ill. ; 24 cm.

          Includes bibliographies and index.
          ISBN 0-669-00733-1

SUBJECT CARD

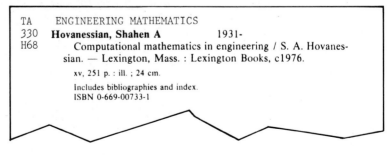

TA   ENGINEERING MATHEMATICS
330   **Hovanessian, Shahen A**   1931-
H68   Computational mathematics in engineering / S. A. Hovanes-
          sian. — Lexington, Mass. : Lexington Books, c1976.

          xv, 251 p. : ill. ; 24 cm.

          Includes bibliographies and index.
          ISBN 0-669-00733-1

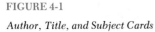

FIGURE 4-1

Author, Title, and Subject Cards

las and Marketing Guide), and dozens of other general research tools. In addition, the reference collection contains subject encyclopedias (*Encyclopedia of Banking and Finance*), dictionaries (*Psychiatric Dictionary*), and handbooks (*Biology Data Book*). These specialized reference books are especially useful when you begin a writing project, for they can provide an overview of the subject and often list the major works in the field.

How do you know if there is a dictionary of the terms used in a given field, such as nutrition? The answer, as you might have guessed, can be found in the reference collection. It would be impossible to list in this chapter even the major reference books in science, engineering, and business, but you should be familiar with the reference books that list the available reference books. Among these guides-to-the-guides are the following:

Downs, R. B., and C. D. Keller. 1975. *How to do library research.* 2d ed. Urbana, Ill.: University of Illinois Press.

Guide to reference books. 1976. 9th ed., ed. E. P. Sheehy. Chicago: American Library Association.

Guide to reference material. 1973. 3d ed. 3 vols., ed. A. J. Walford. London: Library Association.

Look under *Reference* in the subject catalog to see which guides the library has. The most comprehensive is Sheehy's *Guide to Reference Books* (also see its 1980 and 1982 *Supplements*), an indispensable resource that lists ·bibliographies, indexes, abstracting journals, dictionaries, directories, handbooks, encyclopedias, and many other sources. The items are classified according to specialty (for example, *organic chemistry*) and annotated. One of the most useful features of Sheehy's book is that it directs you to other guides (such as Henry M. Woodburn's *Using the Chemical Literature: A Practical Guide*) geared to your own specialty. Read the prefatory materials in Sheehy; you can save yourself many frustrating hours.

PERIODICALS INDEXES Periodicals are the best source of information for most research projects, because they offer recent discussions of subjects whose coverage is often otherwise limited. The hardest aspect of using periodicals is identifying and locating the dozens of pertinent articles that are published each month. Although there may be only half a dozen major journals that concentrate on your field, a useful article might appear in one of a hundred other publications. A periodical index, which is simply a listing of articles classified according to title, subject, and author, can help you determine which journals you want to locate.

After using a periodical index you might want to find out more information about the journals containing the articles. *Ulrich's International Periodicals Directory,* which is updated every two years, listed 69,000 journals in 557 subject areas in its 1985 edition. *Ulrich's* is indexed by subject area and by title of journal and lists the circulation and publisher for each entry.

Figure 4-2, an explanatory page from the *Hospital Literature Index,* demonstrates how most indexes work.

Some periodical indexes are more useful than others. A number of indexes—such as *Engineering Index* and *Business Periodicals Index*—are very comprehensive. However, if you are going to rely on a narrower, more specialized index when compiling a preliminary bibliography for your report, you should determine if it is accurate and comprehensive. The prefatory material—the publisher's statement and the list of journals indexed—will supply answers to the following questions:

1. Is the index compiled by a reputable organization? Many of the better indexes, such as *The Readers' Guide to Periodical Literature,* are published by the H. W. Wilson Company. Almost all the reputable indexes are sponsored by well-known professional societies or associations.

2. What is the scope of the index? An index is not very useful unless it includes all the pertinent journals. Scan the list of journals that the index picks up. Also, determine if the index includes materials other than articles, such as annual reports or proceedings of annual conferences. Does the index include articles in foreign languages? In foreign journals? Ask the reference librarian if you have questions.

3. Are the listings clear? Does the index define the abbreviations, and do the listings provide enough bibliographic information to enable you to locate the article?

4. How current is the index? How recent are the articles listed, and how frequently is the index published? Don't restrict your search to the bound annual compilations; most good indexes are updated monthly or quarterly.

5. How does the index arrange its listings? Most of the better indexes are heavily cross-indexed: that is, their listings are arranged by subject, author, and (where appropriate) formula and patent.

The better subject indexes include *Applied Science and Technology Index* (New York: H. W. Wilson Company, 1913 to date), *Science Citation Index* (Philadelphia: Institute for Scientific Infor-

SUBJECT SECTION

Many articles in *Hospital Literature Index* appear under more than one subject heading. Under each subject heading, the articles are arranged alphabetically by journal title abbreviation. Bibliographic information is given in the following order: title, first author (if more than one), journal title abbreviation, date of issue, volume number, issue or part number, inclusive pagination. For example:

In the case of review articles, the number of references appears at the end of the citation.

AUTHOR SECTION

References in the Author Section cite a maximum of three authors, with the complete bibliographic citation appearing only under the name of the first author. Names of the second and third authors appear as cross-references to the full citation under the first author's name. Authors' names are followed by other bibliographic elements as they appear in the Subject Section described above. For example:

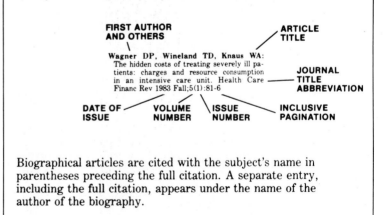

Biographical articles are cited with the subject's name in parentheses preceding the full citation. A separate entry, including the full citation, appears under the name of the author of the biography.

mation, 1961 to date), and *Index Medicus* (Washington, D.C.: National Library of Medicine, 1960 to date). The *New York Times* and *Wall Street Journal* also publish indexes of their own articles.

ABSTRACT JOURNALS Abstract journals not only contain bibliographic information for articles listed, but also provide abstracts—brief summaries of the articles' important results and conclusions. (See Chapter 11 for a discussion of abstracts.) The advantage of having the abstract is that in most cases it will enable you to decide whether to search out the full article. The title of an article, alone, is often a misleading indicator of its contents.

Whereas some abstract journals, such as *Chemical Abstracts*, cover a broad field, many are specialized rather than general. *Adverse Reaction Titles*, for instance, covers research on the subject of adverse reactions to drugs. *Applied Mechanics Review*, too, is relatively narrow in scope. A major drawback of abstract journals is that they take longer to compile than indexes (about a year, as opposed to about two months); therefore, the articles listed are not so recent. In many fields, however, this disadvantage is more than offset by the usefulness of the abstracts.

In evaluating the quality of an abstract journal, apply the same criteria that you apply to indexes, but add one other: how clear and informative is the abstract? To be useful, abstracts have to be brief (usually no more than 250 words), and summarizing a long and complex article in so few words requires great skill. Poorly written abstracts can be very confusing. Most abstract journals reserve the right to revise or rewrite the abstracts provided by authors. Read a few of the abstracts to see if they make sense.

Figure 4-3 shows a guide included in *Pollution Abstracts* that provides an excellent explanation of how to use that journal. (Notice, too, how well written the abstract is.) Most abstract journals are organized in this way.

GOVERNMENT PUBLICATIONS The United States government publishes about twenty thousand documents annually. In researching any field of science, engineering, or business, you are likely to find that a government agency has produced a relevant brochure, report, or book.

Government publications usually are not listed in the indexes and abstract journals. The *Monthly Catalog of United States Government Publications*, put out by the U.S. superintendent of documents, provides extensive access to these materials. The *Monthly Catalog* is indexed by author, title, and subject. If you are doing any research that requires information published by the govern-

FIGURE 4-3

*Guide to an Abstract
Journal*

HOW TO USE
pollution abstracts®

TO RESEARCH A SPECIFIC SUBJECT: Determine which terms best describe the subject you are researching such as scrubbers, wastewater outfalls, pyrolysis, etc. Look up the terms in the Subject Index at the back of the issue. The subject index lists alphabetically all controlled terms assigned to the abstract: each entry refers you by accession number to the corresponding abstract. If you cannot find a listing for a certain term, try a synonym or related term. A complete listing of the controlled terms used in indexing *Pollution Abstracts* may be obtained by writing to the CSA Editorial Department.

EXAMPLE: You are seeking literature about noise control in industry. You look up **Noise Reduction** in the subject index and find the following entry:

Noise reduction, Data bases, Pollution control,
Occupational health, Environmental protection 08646

Citation 83-08646 would also be listed under the controlled terms: Data bases, Pollution control, Occupational health, Environmental protection. Free terms may further describe the material in the abstract but are not listed alphabetically in the index.

You then turn to the abstract numbered 83-08646:

┌──── *Text language*
Original title ┌─*Summary language*

Citation number **83-08646P Data Bank of noise control case histories: The reasons and the experience in Tuscany, Italy. [En;en] Presented at Brit. Occup. Hyg. Soc., Annl. Conf. 1982, Univ.**
Author **London, London, UK, 29 Mar-1 Apr 1982.** by S. Silvestri (Serv. Prev., Ig Sicurezza Luoghi Lavoro, Unit. Sanit. Locale 10/D, vis della
Publication Cupola 64, Firenze, Italy) ANN. OCCUP. HYG., vol. 27, no. 2, 1983, pp. 209-211.
This paper describes the way in which a data bank of noise control case histories has been
Abstract built up by the Occupational Hygiene and Safety Prevention Services in Tuscany, Italy. The data bank aims to become a source of information on noise control techniques for trade unions, factory inspectors, and companies. The paper argues that such steps are essential for the effective control of noise in industry.

Language abbreviations
text summaries

Af af Afrikaans	Ee ee Estonian	Ja ja Japanese	Ro ro Romanian
Ar ar Arabic	En en English	Ko ko Korean	Ru ru Russian
Be be Belorussian	Es es Spanish	Li li Lithuanian	Sh sh Serbo-croat
Bg bg Bulgarian	Fi fi Finnish	Lv lv Latvian	Sk sk Slovak
Ch ch Chinese	Fr fr French	Ma ma Macedonian	Sn sn Slovenian
Cs cs Czech	Gr gr Greek	Nl nl Dutch	Sv sv Swedish
Da da Danish	He he Hebrew	No no Norwegian	Tr tr Turkish
De de German	Hu hu Hungarian	Pl pl Polish	Uk uk Ukrainian
	It it Italian	Pt pt Portuguese	

ment before 1970, you should know about the *Cumulative Subject Index to the Monthly Catalog, 1900–1971* (Washington, D.C.: Carrollton Press, 1973–76). This 15-volume index eliminates the need to search through the early *Monthly Catalogs.*

Government publications are usually catalogued and shelved separately from the other publications. They are classified according to the Superintendent of Documents system, not the Dewey Decimal or the Library of Congress system. See the reference librarian or the government documents specialist for information about finding government publications in your library.

State government publications are indexed in the *Monthly Checklist of State Publications* (1910 to date), published by the Library of Congress.

If you would like more information on government publications, consult the following four guides:

Lesko, M. 1983. *Information U.S.A.* New York: Viking.

Morehead, J. 1978. *Introduction to United States public documents.* 2d ed. Littleton, Col.: Libraries Unlimited.

Schmeckebier, L. F., and R. B. Eastin. 1969. *Government publications and their use.* Washington, D.C.: Brookings Institution.

Schwarzkopf, L. C. 1984. *Government reference books 82/83.* Littleton, Col.: Libraries Unlimited. A biennial publication.

GUIDES TO BUSINESS AND INDUSTRY Many kinds of technical research require access to information about regional, national, and international business and industry. If, for example, you are researching a new product for a report, you may need to contact its manufacturer and distributors. For information such as where to write to discover the range of the product's applications, you could consult one of the following two directories:

Thomas' register of American manufacturers. New York: Thomas. 1905 to date.

Poor's register of corporations, directors, and executives. New York: Standard and Poor's. 1928 to date.

These annual directories provide information on products and services and on corporate officers and executives. In addition to these national directories, many local and regional directories are published by state governments, local chambers of commerce, and business organizations. These resources are particularly useful for students and professionals who wish to communicate with potential vendors or clients in their own area.

Also valuable as resources are investment services—publications that provide balance sheets, earnings data, and market prices for corporations. Two major investment services exist:

Moody's investor's service. New York: Moody's.

Standard and Poor's corporation records. New York: Standard and Poor's.

Two government periodicals contain valuable information about national business trends:

The Federal Reserve bulletin. Washington, D.C.: U.S. Board of Governors of the Federal Reserve System. A monthly publication that outlines banking and monetary statistics.

Survey of current business. Washington, D.C.: U.S. Office of Business Economics. Details general business indicators.

For more information on how to use business reference materials, consult the following guides:

Brownstone, D. M., and G. Carruth. 1979. *Where to find business information.* New York: Wiley.

Daniells, L. M. 1976. *Business information sources.* Berkeley: University of California Press.

Washington Researchers. 1981. *Where to find information about companies.* 2d ed. Washington, D.C.: Washington Researchers.

COMPUTERIZED INFORMATION RETRIEVAL Most college and university libraries—and even some public libraries—have facilities for computerized information retrieval. Technological improvements in computer science, along with continuing growth in the amount of scholarly literature produced, ensure that in the foreseeable future computerized information retrieval will be the primary means of gaining access to information.

Even today, when computerized information retrieval is in its infancy, more than 150 million items are contained in the hundreds of different data bases. A data base is a machine-readable file of information from which a computer retrieves specific data. Libraries today lease access to different data-base services, with which they communicate by computer. The largest data-base service is DIALOG Information Services, which by 1986 offered over

250 different data bases containing some 100 million bibliographic entries in a wide range of fields. For example, DIALOG carries Agricola, an agriculture index; Biosis, a biology index; Insurance Abstracts; Pharmaceutical News Index; Commerce and Business Daily; Chemical Industry Notes; Aptic, an air pollution index; World Textiles; Mathfile; Historical Abstracts; and World Affairs Report.

Computerized literature searches are now standard in many fields, especially in science and technology. Although most computerized literature searches require the assistance of a trained librarian, the concept is relatively simple. You decide what key words (or phrases) to use in your search, and what limitations you wish to place on it. For example, you might wish to limit your search to articles, excluding other types of literature, or you might wish to retrieve only those articles written over a limited period. When you enter a key word, the screen will show you how many items in the data base are filed according to that key word. You can ask the computer to combine several key words and show the number of items that include all the key words. For example, the key "solar energy" might elicit 735 entries. The key "home heating" might elicit 1,438. Combined, the two might elicit 459 entries—that is, 459 items dealing with solar energy *and* home heating.

By trying out various combinations of key words, you can come up with an effective search strategy. You can then ask the computer to print the bibliography—or in some cases the actual abstracts—at the terminal or at the computer site.

Computerized information retrieval offers several advantages over manual searching:

1. It is faster. The search can take as little as a few minutes.

2. It is more comprehensive. The computer doesn't overlook any items: it follows your instructions to the letter.

3. It is more flexible. You can devise your strategy at the terminal, modifying it effortlessly.

4. It is more up-to-date. Data bases are produced and updated more quickly than printed indexes or abstract services.

5. It is more accurate. The printout of items contains no mechanical errors or indecipherable handwriting.

However, computerized information retrieval also has its disadvantages:

1. It is expensive. An average search can cost $10 to $20 in computer time. Students usually have to assume at least part of this expense.

2. It can produce a lot of irrelevant information. Unless you tell it not to, the computer will print out items you hadn't anticipated—such as speeches or films. In addition, even a well-planned search strategy can yield items that are not useful. For instance, an article filed under the key words "Japan" and "whale hunting" might relate to whale hunting by the Japanese—as you had hoped—but deal almost exclusively with Japanese consumption of whale products.

3. It is inconvenient. Your library might not have access to the data base you need, and in most cases you cannot perform the search without assistance.

For more information on data-base services, see the *Directory of Online Databases* (New York: Cuadra/Elsevier), a quarterly publication that describes nearly three thousand data bases. You can access the directory by five indexes, including data-base service, subject area, and producer. For more information on how to use data bases, see *Getting On-Line: A Guide to Accessing Computer Information Services* (Englewood Cliffs, N.J.: Prentice-Hall, 1984).

PERSONAL INTERVIEWS
Personal interviews are a good source of information on subjects too new to have been discussed in the professional literature or inappropriate for widespread publication (such as local political questions). Most students are inexperienced at interviewing and hence reluctant to conduct interviews. Interviewing, like any other communications skill, requires practice. The following discussion explains how to make interviewing straightforward and productive.

CHOOSING A RESPONDENT Start by defining on paper what you want to find out. Only then can you begin to search for a person who can provide the necessary information.

The ideal respondent can be defined as an expert willing to talk. Many times the nature of your subject will dictate whom you should ask. If you are writing about research being conducted at your university, for instance, the logical choice would be a person involved in the project. Sometimes, however, you might be interested in a topic about which a number of people could speak knowledgeably, such as the reliability of a particular kind of office equipment or the reasons behind the growth in the number of adult students. Use directories, such as local industrial guides, to locate the names and addresses of potential respondents.

Once you have located an expert, find out if he or she is willing to be interviewed. On the phone or in a letter, state what you want to ask about; the person might not be able to help you but might be willing to refer you to someone who can. And be sure to tell the potential respondent why you have decided to ask him or her: a well-chosen compliment will be more effective than an admission that the person you really wanted to interview is out of town. Don't forget to mention what you plan to do with the information: write a report, give a talk, etc. Then, if the person is willing to be interviewed, set up an appointment at his or her convenience.

PREPARING FOR THE INTERVIEW Never give the impression that the reason you are conducting the interview is to avoid library research. If you ask questions that are already answered in the professional literature, the respondent might become annoyed and uncooperative. Make sure you are thoroughly prepared for the interview: research the subject carefully in the library.

Write out your questions in advance, even if you think you know them by heart. Frame your questions so that the respondent won't simply answer yes or no. Instead of "Are adult students embarrassed about being in class with students much younger than themselves?" ask, "How do the adult students react to being in class with students much younger than themselves?" You don't want to give the impression that all you want is a simple confirmation of what you already know.

CONDUCTING THE INTERVIEW Arrive on time for your appointment. Thank the respondent for taking the time to talk with you. Repeat the subject and purpose of the interview and what you plan to do with the information.

If you wish to tape-record the interview, ask permission ahead of time; taping makes some people uncomfortable. Have paper and pens ready. Even if you are taping the interview, you will want to take brief notes as you go along.

Start by asking your first prepared question. Listen carefully to the respondent's answer. Be ready to ask a follow-up question or request a clarification. Have your other prepared questions ready, but be willing to deviate from them. In a good interview, the respondent probably will lead you in directions you had not anticipated. Gently return the respondent to the point if he or she begins straying unproductively, but don't interrupt rudely or show annoyance.

After all your questions have been answered (or you have run out of time), thank the respondent again. If a second meeting would be useful—and you think the person would be willing to talk with you further—this would be an appropriate time to ask. If you might want to quote the respondent by name, ask permission now.

AFTER THE INTERVIEW After you have left, take the time to write up the important information while the interview is still fresh in your mind. (This step is, of course, unnecessary if you have recorded the interview.)

It's a good idea to write a brief thank-you note and send it off within a day or two of the interview. Show the respondent that you appreciate the courtesy extended to you and that the information you learned will be of great value to you. Confirm any previous offers you made, such as to send a copy of the report.

LETTERS OF INQUIRY A letter of inquiry is often a useful alternative to a personal interview. If you are lucky, the person who responds to your inquiry letter will provide detailed and helpful answers. Keep in mind, however, that the person might not understand what information you are seeking or might not want to take the trouble to help you. Also, you can't ask follow-up questions in a letter, as you can in an interview. Although the strategy of the inquiry letter is essentially that of a personal interview—persuading the reader to cooperate and phrasing the questions carefully—inquiry letters in general are less successful, because, unlike respondents in an interview, the readers have not already agreed to provide information.

For a full discussion of inquiry letters, see Chapter 18.

QUESTION-NAIRES Questionnaires enable you to solicit information from a large group of people. Although they provide a useful and practical alternative to interviewing dozens of people spread out over a large geographical area, questionnaires rarely yield completely satisfactory results. For one thing, some of the questions, no matter how carefully constructed, are not going to work: the respondents will misinterpret them or supply useless answers. In addition, you probably will not receive nearly as many responses as you had hoped. The response rate will almost never exceed 50 percent; in most cases, it will be closer to 10 or 20 percent. And you can never be sure how representative the responses will be; in general, people who feel strongly about an issue are much more likely to respond than are those who do not. For this reason, be careful in drawing conclusions based on a small number of responses to a questionnaire.

When you send a questionnaire, you are asking the recipient to do you a favor. If the questionnaire requires only two or three minutes to complete, of course you are more likely to receive a response than if it requires an hour. Your goal, then, should be to construct questions that will elicit the information you need as simply and efficiently as possible.

CREATING EFFECTIVE QUESTIONS   Effective questions are unbiased and clearly phrased. Avoid charged language and slanted questions. Don't ask, "Should we protect ourselves from unfair foreign competition?" Instead, ask, "Are you in favor of imposing tariffs?" Make sure the questions are worded as specifically as possible, so that the reader understands exactly what information you are seeking. If you ask, "Do you favor improving the safety of automobiles?" only an eccentric would answer, "No." However, if you ask, "Do you favor requiring automobile manufacturers to equip new cars with air bags, at a cost to the consumer of $300 per car?" you are more likely to get a useful response.

As you make up the questions, keep in mind that there are several formats from which to choose:

MULTIPLE CHOICE

Would you consider joining a company-sponsored sports team?

    Yes _____   No _____

How do you get to work? (Check as many as apply.)

    my own car _____

    car/van pool _____

    bus _____

    train _____

    walk _____

    other _____ (please specify)

The flextime program has been a success in its first year.

_____	_____	_____	_____
Agree strongly	Agree more than disagree	Disagree more than agree	Disagree strongly

RANKING

Please rank the following work schedules in order of preference. Put a
"1" next to the schedule you would most like to have, a "2" next to
your second choice, etc.

    8:00-4:30 _____

    8:30-5:00 _____

    9:00-5:30 _____

    flexible _____

SHORT ANSWER

What do you feel are the major advantages of the new parts-requisitioning policy?

    1. _____

    2. _____

    3. _____

SHORT ESSAY

The new parts-requisitioning policy has been in effect for a year. How
well do you think it is working?

_____

_____

_____

_____

_____

Remember that you will receive fewer responses if you ask for essay answers; moreover, essays, unlike multiple-choice answers,
cannot be quantified. A simple statement with a specific number
in it—"Seventy-five percent of the respondents own two or more
PCs"—helps you make a clear and convincing case. However
thoughtful and persuasive an essay answer may be, it is subject to
the interpretations of different readers.

After you have created the questions, write a letter or memo to
accompany the questionnaire. A letter to someone outside your organization is basically an inquiry letter (sometimes with the questions themselves on a separate sheet); therefore, it must clearly in-

dicate who you are, why you are writing, what you plan to do with the information, and when you will need it. (See Chapter 18 for a discussion of inquiry letters.) For people within your organization, a memo accompanying a questionnaire should answer the same questions.

Figure 4-4 shows a sample questionnaire.

SENDING THE QUESTIONNAIRE Drafting the questions is only part of the task. The next step is to administer the questionnaire. Determining whom to send it to can be simple or difficult. If you want

FIGURE 4-4

Questionnaire

September 6, 19--

To: All employees

From: William Bonoff, Vice-President of Operations

Subject: Evaluation of the Lunches Unlimited food service

As you may know, every two years we evaluate the quality and cost of the food service that caters our lunchroom. We would like you to help by sharing your opinions about the food service. Your anonymous responses will help us in our evaluation. Please drop the completed questionnaires in the marked boxes near the main entrance to the lunchroom.

1. Approximately how many days per week do you eat lunch in the

    lunchroom?

    0____ 1____ 2____ 3____ 4____ 5____

2. At approximately what time do you eat in the lunchroom?

    11:30-12:30____ 12:00-1:00____ 12:30-1:30____

    varies____

3. Do you have trouble finding a clean table?

    often____ sometimes____ rarely____

4. Are the Lunches Unlimited personnel polite and helpful?

    always____ usually____ sometimes____ rarely____

5. Please comment on the quality of the different kinds of food

    you have had in the lunchroom.

FIGURE 4-4

Questionnaire
(Continued)

a. Hot meals (daily specials)

excellent____   good____   satisfactory____   poor____

b. Hot dogs and hamburgers

excellent____   good____   satisfactory____   poor____

c. Sandwiches

excellent____   good____   satisfactory____   poor____

d. Salads

excellent____   good____   satisfactory____   poor____

e. Desserts

excellent____   good____   satisfactory____   poor____

6. What foods would you like to see served that are not served now?

_____

_____

_____

7. What beverages would you like to see served that are not served now?

_____

_____

_____

to know what the residents of a particular street think about a proposed construction project, your job is easy. But if you want to know what mechanical engineering students in colleges across the country think about their curricula, you will need background in sampling techniques in order to isolate a representative sample.

Before you send *any* questionnaire, show it and the accompanying letter or memo to a few people whose backgrounds are similar to those of your real readers. In this way you can sample your questionnaire's effectiveness before "going public."

Be sure also to include a self-addressed, stamped envelope with questionnaires sent to people outside your organization.

FIGURE 4-4

*Questionnaire
(Continued)*

8. Please comment on the prices of the foods and beverages served.

    a. Hot meals (daily specials)

        too high_____   fair_____   a bargain_____

    b. Hot dogs and hamburgers

        too high_____   fair_____   a bargain_____

    c. Sandwiches

        too high_____   fair_____   a bargain_____

    d. Desserts

        too high_____   fair_____   a bargain_____

    e. Beverages

        too high_____   fair_____   a bargain_____

9. Would you be willing to spend more money for a better-quality lunch, if you thought the price was reasonable?

    yes, often_____   sometimes_____   not likely_____

10. Please provide whatever comments you think will help us evaluate the catering service.

    _____

    _____

    _____

    _____

Thank you for your cooperation.

USING THE INFORMATION

Once you have gathered your books and articles, conducted your interviews, and received responses from your questionnaires, you should start planning a strategy for using the information. The first step in working with any group of materials is to evaluate each source.

EVALUATING THE SOURCES Authority is sometimes difficult to determine. Your respondents are likely to be authoritative (you would not deliberately have chosen suspect sources), but you must still judge the quality of the information they have given. Your questionnaires, once you have discarded eccentric or clearly extreme responses, should contain authoritative material. With books and articles, however, the question is not so clear-cut. Check to see if the author and the publisher or journal are respected.

Many books and journals include biographical sketches. Does the author appear to have solid credentials in the subject he or she is writing about? Look for academic credentials, other books or articles written, membership in professional associations, awards, and so forth. If no biographical information is provided, consult a "Who's Who" of the field.

Evaluate the publisher, too. A book should be published by a reputable trade, academic, or scholarly house. A journal should be sponsored by a university or professional association. Read the list of editorial board members; they should be well-known names in the field. If you have any doubts about the authority of a book or journal, ask the reference librarian or a professor in the appropriate field to comment on the reputation.

Finally, check the date of publication. In high-technology fields, in particular, a five-year-old article or book is likely to be of little value (except, of course, for historical studies).

SKIMMING   Inexperienced writers often make the mistake of trying to read every potential source. The result, all too commonly, is that they get halfway through one of their several books when they realize they have to start writing immediately in order to submit their reports on time. Knowing how to skim is invaluable.

To skim a book, read the prefatory materials—the preface and introduction. This will give you a basic idea of the writer's approach and methods. The acknowledgments section will tell you about any assistance the author received from other experts in the field, or about his or her use of important primary research or other resources. Read the table of contents carefully to get an idea of the scope and organization of the book. A glance at the notes at the ends of chapters or the end of the book will help you understand the nature of the author's research. Check the index for clues about the coverage that the information you need receives.

To skim an article, focus on the abstract. Check the notes and references as you would for a book. Read the headings throughout the article. For both books and articles, sample a few paragraphs from different portions of the text to gauge the quality and relevance of the information.

Skimming will not assure you that a book or article *is* going to be useful. A careful reading of a work that looks useful might prove disappointing. However, skimming can tell you if the work is *not* going to be useful: because it doesn't cover your subject, for example, or because its treatment is too superficial or too advanced. Eliminating the sources that you don't need will give you more time to spend on the ones that you do.

TAKING NOTES   Note taking is often the first step in the actual writing of the report. Your notes will provide the vital link between what your sources have said and what you are going to say. You will refer to them over and over. For this reason, it is smart to take notes logically and systematically.

If you have access to a word processor, start using it at this stage. You will be able to transfer your notes directly to your document, thus saving time and reducing the chances of introducing errors.

If you do not have access to a word processor, buy two packs of note cards: one 4 inches by 6 inches or 5 inches by 8 inches, the other 3 inches by 5 inches. (There is nothing sacred about ready-made note cards; you can make up your own out of scrap paper.) The major advantage of using cards is that they are easy to rearrange later, when you want to start outlining the report.

On the smaller cards, record the bibliographic information for each source from which you take notes. For a book, record the following information:

author
title
publisher
place of publication
year of publication
call number

Figure 4-5 shows a sample bibliography card for a book.

For an article, record the following information:

author
title of the article
periodical
volume
number
date
pages on which the article is included
call number

FIGURE 4-5

*Bibliography Card
for a Book*

KA

31.6   Honeywell, Alfred R.
.H306   The Meaning of Saturn II
        N.Y.: The Intersteller Society,
                                1983

Figure 4-6 shows a sample bibliography card for an article.

On the larger cards, record the notes. To simplify matters, write on one side of a card only and limit each card to a narrow subject or discrete concept so that you can easily reorder the information to suit the needs of your document.

Although many writers have devised their own systems of note taking, most notation involves two different kinds of activities: paraphrasing and quoting.

PARAPHRASING A paraphrase is a restatement, in your own words, of someone else's words. "In your own words" is crucial: if

FIGURE 4-6

*Bibliography Card
for an Article*

HZ
102.3

Hastings, W.
"The Skylab debate"
The Modern Inquirer 19, 2 (Fall, 1983)
                        106-113

you simply copy someone else's words—even a mere two or three in a row—you must use quotation marks.

What kind of material should be paraphrased? Any information that you think *might* be useful in writing the report: background data, descriptions of mechanisms or processes, test results, and so forth.

To paraphrase accurately, you have to study the original and understand it thoroughly. Once you are sure you follow the writer's train of thought, rewrite the relevant portions of the original. If you find it easiest to rewrite in complete sentences, fine. If you want to use fragments or merely list information, make sure you haven't compressed the material so much that you'll have trouble understanding it later. Remember, you might not be looking at the card again for a few weeks. Put a title on the card so that you'll be able to identify its subject at a glance. The title should include the general subject the writer describes—such as "Open-sea pollution-control devices"—and the author's attitude or approach to that subject—such as "Criticism of open-sea pollution-control devices." To facilitate the later documentation process, also include the author's last name, a short title of the article or book, and the page number of the original.

Figure 4-7 shows a paraphrased note card based on the following discussion of computer applications in nursing (Parks et al. 1986: 105). The student has paraphrased each paragraph on a separate note card.

Walker recognized the need for developing educational objectives for nursing computer education. To help prioritize computer learning needs, she surveyed 193 Registered Nurses from some 29 states who were recognized as experts in nursing computerization. A list of 11 categories of computer learning needs were rank ordered by these experts as follows: 1) fundamentals of data processing, 2) importance of nursing involvement, 3) overview of health care applications, 4) developing a systematized nursing data base, 5) systems analysis, 6) affective impact of computerization, 7) confidentiality/legal issues, 8) potential problems in computerized health care systems, 9) basic understanding of statistics and research methods, 10) change theory, and 11) computer programming/programming languages.

Ronald surveyed a national sample of 159 nursing educators on their current and desired levels of knowledge in 16 computer areas (e.g., use of computers in statistical analysis, and how to use a terminal). Although all respondents had at least a master's degree, they evaluated themselves as relatively low in current knowledge across the 16 computer areas ($M = 1.04$ on a 0 to 4 scale). However, these educators aspired to relatively high levels of knowledge for the 16 areas ($M = 3.06$ on the 0 to 4

Figure 4-7

Paraphrased Notes

What nurses should be learning about computers

Walker's survey of experts in computers in nursing yielded a list
of 11 priorities for nursing computer education. They range from
data-processing basics and the role of computers in nursing to
programming skills.

"Faculty and Student Perceptions . . ." p. 105.

Motivation of nursing educators to learn about computers

Ronald's survey of (at least) master's level nursing educators
showed that although most were weak in computers, most wanted to
learn much more.

"Faculty and Student Perceptions . . ." p. 105.

What do program administrators know about computers?

A Southern Council on Collegiate Education for Nursing survey of
nursing-program administrators showed that, although they knew
little or nothing about computers, they used them for administrative work.

"Faculty and Student Perceptions . . ." p. 105.

scale). These findings suggest the existence of high levels of learning needs
for faculty, and presumably of sufficient motivation to engage in com-
puter-related learning experiences.

A more recent survey was conducted by the Southern Council on Collegiate Education for Nursing (SCCEN); the survey involved nursing administrative heads of associate degree and baccalaureate nursing pro-

grams in 14 Southern states. The SCCEN received a total of 257 replies (75 per cent) regarding the administrators' personal interest in computer technology; these administrators also reported their opinions about their faculty's knowledge, experience, and interests concerning computer education. A large majority (83 per cent) of the administrators reported little or limited personal knowledge about computers. However, the majority of the respondents reported using the computer, usually for administrative purposes.

Notice how a heading provides a focus for each card. The student has omitted the information he does not need, but he has recorded the necessary bibliographic information so that he can document his source easily or return to it if he wants to reread it. There is no one way to paraphrase: you have to decide what to paraphrase—and how to do it—on the basis of your analysis of the audience and the purpose of your report.

QUOTING On occasion you will want to quote a source, either to preserve the author's particularly well-expressed or emphatic phrasing or to lend authority to your discussion. In general, do not quote passages more than two or three sentences long, or your report will look like a mere compilation. Your job is to integrate an author's words into your own work, not merely to introduce a series of quotations.

The simplest form of quotation is an author's exact statement:

As Jones states, "Solar energy won't make much of a difference in this century."

To add an explanatory word or phrase to a quotation, use brackets:

As Nelson states, "It [the oil glut] will disappear before we understand it."

Use ellipses (three spaced dots) to show that you are omitting part of an author's statement:

ORIGINAL STATEMENT:

"The generator, which we purchased in May, has turned out to be one of our wisest investments."

ELLIPTICAL QUOTATION:

"The generator . . . has turned out to be one of our wisest investments."

For more details on the mechanics of quoting, see the entries under "Quotation Marks," "Brackets," and "Ellipses" in Appendix A.

For a discussion of the different styles of documentation, see Appendix B.

EXERCISES

1. Choose a topic on which to write a report for this course. Make sure the topic is sufficiently focused that you will be able to cover it in some detail.

   a. Using Sheehy's *Guide to Reference Books*, plan a strategy for researching this topic.
      (1) Which guides, handbooks, dictionaries, and encyclopedias contain the background information you should read first?
      (2) Which basic reference books discuss your topic?
      (3) Which major indexes and abstract journals cover your topic?
   b. Write down the call numbers of the three indexes and abstract journals most relevant to your topic.
   c. Compare two abstract journals in your field on the basis of sponsoring organization, scope, clarity of listings, and timeliness. Also determine who writes the abstracts contained in each.
   d. Make up a preliminary bibliography of two books and five articles that relate to your topic.
   e. Using the bibliography from Exercise 1d, write down the call numbers of the books and journals (those that your library receives).
   f. Find one of the works listed and write a brief assessment of its value.
   g. Using a local industrial guide, make up a list of five persons who might have first-hand knowledge of your topic.
   h. Make up two sets of questions: one for an interview with one of the persons on your list from Exercise 1g, and one for a questionnaire to be sent to a large group of people.
   i. Photocopy the first two pages from an article listed in your bibliography from Exercise 1d. Paraphrase any three paragraphs, each on a separate note card. Also note at least two quotations, each on a separate card: one should be a complete sentence, and one an excerpt from a sentence.

REFERENCE　Parks, P. L. et al. 1986. Faculty and student perceptions of computer applications in nursing. *Journal of Professional Nursing* 2 (March-April): 105.

CHAPTER FIVE

TECHNICAL WRITING STYLE

Perhaps you have heard it said that the best technical writing style is no style at all. This means simply that the readers should not be aware of your presence as a writer. They should not notice that you have a wonderful vocabulary or that your sentences flow beautifully—even if those things are true. In the best technical writing, the readers "notice" nothing. They are aware only of the information being conveyed. The writer fades into the background.

This is as it should be. Few people read technical writing for pleasure. Most readers either *must* read it as part of their work or want to keep abreast of new developments in the field. People read technical writing to gather information, not to appreciate the writer's flair. For this reason, experienced writers do not try to be fancy. The old saying has never been more appropriate: *Write to express, not to impress.*

The word *style*, as it is used in this chapter, encompasses word choice, sentence construction, and paragraph structure. Technical writing, like any other kind of writing, requires conscious decisions—about which words or phrases to use, what kinds of sentences to create, and how to turn those sentences into clear and coherent paragraphs.

DETERMINING THE APPROPRIATE STYLISTIC GUIDELINES

Most successful writers agree that the key to effective writing is revision: coming back to a draft and adding, deleting, and changing. Time permitting, an important document might go through four or five drafts before the writer finally has to stop. In revising, the writer will make many stylistic changes in an attempt to get closer and closer to the exact meaning he or she wishes to convey.

FIGURE 5-1

From the U.S. Government Printing Office Style Manual

12. NUMERALS

(See also Tabular Work; Leaderwork)

12.1. Most rules for the use of numerals are based on the general principle that the reader comprehends numerals more readily than numerical word expressions, particularly in technical, scientific, or statistical matter. However, for special reasons numbers are spelled out in indicated instances.

12.2. The following rules cover the most common conditions that require a choice between the use of numerals and words. Some of them, however, are based on typographic appearance rather than on the general principle stated above.

12.3. Arabic numerals are generally preferable to Roman numerals.

NUMBERS EXPRESSED IN FIGURES

12.4. A figure is used for a single number of *10* or more with the exception of the first word of the sentence. (See also rules 12.9, 12.23.)

50 ballots	24 horses	about 40 men
10 guns	nearly 10 miles	10 times as large

Numbers and numbers in series

☆ **12.5.** Figures are used in a group of 2 or more numbers, or for related numbers, any one of which is *10* or more. The sentence will be regarded as a unit for the use of figures.

Some stylistic matters, however, can be determined before you start to write. Learning the "house style" that your organization follows will cut down the time needed for revision.

An organization's stylistic preferences may be defined explicitly in a company style guide that describes everything from how to write numbers to how to write the complimentary close at the end of a letter. In some organizations, an outside style manual, such as the *U.S. Government Printing Office Style Manual*, is the "rule book." (Figure 5-1 shows a page from this manual.) In many

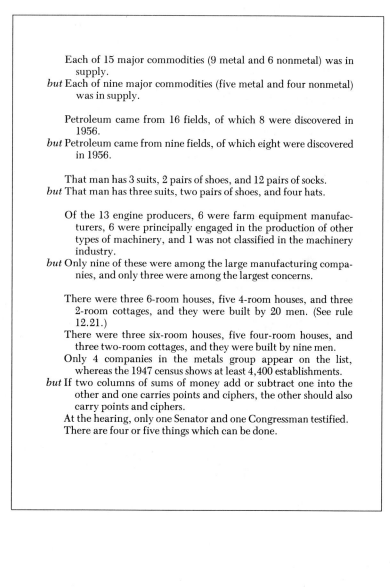

Each of 15 major commodities (9 metal and 6 nonmetal) was in supply.
but Each of nine major commodities (five metal and four nonmetal) was in supply.

Petroleum came from 16 fields, of which 8 were discovered in 1956.
but Petroleum came from nine fields, of which eight were discovered in 1956.

That man has 3 suits, 2 pairs of shoes, and 12 pairs of socks.
but That man has three suits, two pairs of shoes, and four hats.

Of the 13 engine producers, 6 were farm equipment manufacturers, 6 were principally engaged in the production of other types of machinery, and 1 was not classified in the machinery industry.
but Only nine of these were among the large manufacturing companies, and only three were among the largest concerns.

There were three 6-room houses, five 4-room houses, and three 2-room cottages, and they were built by 20 men. (See rule 12.21.)
There were three six-room houses, five four-room houses, and three two-room cottages, and they were built by nine men.
Only 4 companies in the metals group appear on the list, whereas the 1947 census shows at least 4,400 establishments.
but If two columns of sums of money add or subtract one into the other and one carries points and ciphers, the other should also carry points and ciphers.
At the hearing, only one Senator and one Congressman testified.
There are four or five things which can be done.

organizations, however, the stylistic preferences are implicit; no style manual exists, but over the course of years a set of unwritten guidelines has evolved. The best way to learn the unwritten house style is to study some letters, memos, and reports in the files and to ask more-experienced coworkers for explanations. Secretaries, in particular, often are valuable sources of information.

The following discussion covers three important stylistic matters about which organizations commonly have clear preferences: voice, person, and sexist language.

As was discussed in Chapter 4, a word processor is a valuable tool in making stylistic revisions. A number of style programs exist that help isolate many of the topics described in the following discussion. Even without specialized programs, however, you can perform some of the same techniques with the search function of any word-processing program.

ACTIVE AND PASSIVE VOICE

There are two voices: active and passive. In the active voice, the subject of the sentence performs the action expressed by the verb. In the passive voice, the subject receives the action. (In the following examples, the subjects are italicized.)

ACTIVE

Brushaw drove the launch vehicle.

PASSIVE

The launch *vehicle* was driven by Brushaw.

ACTIVE

Many *physicists* support the big-bang theory.

PASSIVE

The big-bang *theory* is supported by many physicists.

In most cases, the active voice is preferable to the passive voice. The active-voice sentence more clearly emphasizes the actor. In addition, the active voice sentence is shorter, because it does not require a form of the *to be* verb and the past participle, as the passive-voice sentence does. In the second example, for instance, the verb is "support," rather than "is supported," and "by" is unnecessary.

The passive voice, however, is generally more appropriate in four cases:

1. The actor is clear from the context.

Students are required to take both writing courses.

The context makes it clear that the college requires that students take both writing courses.

2. The actor is unknown.

The comet was first referred to in an ancient Egyptian text.

We don't know *who* referred to the comet.

3. The actor is less important than the action.

The documents were hand-delivered this morning.

It doesn't matter *who* the messenger was.

4. A reference to the actor is embarrassing, dangerous, or in some other way inappropriate.

Incorrect data were recorded for the flow rate.

It might be inappropriate to say *who* recorded the incorrect data.

Many people who otherwise take little interest in grammar have strong feelings about the relative merits of active and passive. A generation ago, students were taught that the active voice is inappropriate because it emphasizes the person who does the work rather than the work itself and thus robs the writing of objectivity. In many cases, this idea is valid. Why write, "I analyzed the sample for traces of iodine," when you can say, "The sample was analyzed for traces of iodine"? If there is no ambiguity about who did the analysis, or if it is not necessary to identify who did the analysis, a focus on the action being performed is appropriate.

Supporters of the active voice argue that the passive voice creates a double ambiguity. When you write, "The sample was analyzed for traces of iodine," your reader is not quite sure who did the analysis (you or someone else) or when it was done (as part of the project being described or some time previously). Even though a passive-voice sentence can contain all the information found in its active-voice counterpart, often the writer omits the actor.

The best approach to the active-passive problem is to recognize how the two voices differ and use them appropriately. In the following examples, the writer mixes active and passive voice for no good reason.

AWKWARD

He lifted the cage door, and a hungry mouse was seen.

BETTER

He lifted the cage door and saw a hungry mouse.

AWKWARD

The new catalyst produced good-quality foam, and a flatter mold was caused by the new chute-opening size.

BETTER

The new catalyst produced good-quality foam, and the new chute-opening size resulted in a flatter mold.

A number of style programs can help you find the passive voice in your writing. With any word-processing program, however, you can search for *is* and *was*, the forms of the verb *to be* that are most commonly used in passive-voice expressions. In addition, searching for *-ed* and *-en* will isolate the past participles, which also appear in most passive-voice expressions.

FIRST, SECOND, AND THIRD PERSON

Closely related to the question of voice is that of person. The term *person* refers to the different forms of the personal pronoun:

FIRST PERSON

I worked . . . , we worked . . .

SECOND PERSON

You worked . . .

THIRD PERSON

He worked . . . , she worked . . . , it worked . . . , the machine worked . . . , they worked . . .

Organizations that prefer the active voice generally encourage the use of the first-person pronouns: "We analyzed the rate of flow." Organizations that prefer the passive voice often *prohibit* the use of the first-person pronouns: "The rate of flow was analyzed." (Use the search function on the word processor to find *I* and *we*.)

Another question of person that often arises is whether to use the second or the third person in instructions. In some organizations, instructional material—step-by-step procedures—is written in the second person: "You begin by locating the ON/OFF switch." The second person is concise and easy to understand. Other organizations prefer the more formal third person: "The operator begins by locating the ON/OFF switch." Perhaps the most popular version is the second person in the imperative: "Begin by locating the ON/OFF switch." In the imperative, the *you* is implicit. Regardless of the preferred style, however, be consistent in your use of the personal pronoun.

SEXIST
LANGUAGE

Sexist language favors one sex at the expense of the other. Although sexist language can shortchange males—as in some writing about female-dominated professions such as nursing—in most cases the female is victimized. Common examples include nouns such as *workman* and *chairman* and pronouns as used in the sentence "Each worker is responsible for his work area."

In some organizations, the problem of sexist language is still considered trivial; in internal memos and reports, sexist language is used freely. In most organizations, however, sexist language is a serious matter. Unfortunately, it is not easy to eliminate all gender bias from writing.

A number of male-gender words have no standard nongender substitutes, and there is simply no graceful way to get around the pronoun *he*. Over the years different organizations have created synthetic pronouns—such as *thon, tey,* and *hir*—but these pronouns have never caught on. Some writers use *he/she* or *s/he*, although other writers consider these constructions awkward.

However, many organizations have formulated guidelines in an attempt to reduce sexist language.

The relatively simple first step is to eliminate the male-gender words. *Chairman*, for instance, is being replaced by *chairperson* or *chair*. *Firemen* are *firefighters, policemen* are *police officers.*

Rewording a sentence to eliminate masculine pronouns is also effective.

SEXIST

The operator should make sure he logs in.

NONSEXIST

The operator should make sure to log in.

In this revision, an infinitive replaces the *he* clause.

NONSEXIST

Operators should make sure they log in.

In this revision, the masculine pronoun is eliminated through a switch from singular to plural.

Notice that sometimes the plural can be unclear:

UNCLEAR

Operators are responsible for their operating manuals.

Does each operator have one operating manual or more than one?

CLEAR

Each operator is responsible for his or her operating manual.

In this revision, "his or her" clarifies the meaning. *He or she* and *his or her* are awkward, especially if overused, but they are at least clear.

If you use a word processor, search for *he*, *man*, and *men*, words and parts of words most commonly associated with sexist writing. Some style programs search out the most common sexist terms and suggest nonsexist alternatives.

In some quarters, the problem of sexist language is side-stepped. The writer simply claims innocence: "The use of the pronoun *he* does not in any way suggest a male bias." Many readers find this kind of approach equivalent to that of a man who enters a crowded elevator, announces that he knows it is rude to smoke, and then proceeds to light up a cheap cigar. Sexism in language is not a trivial matter, although most people would agree that eliminating sexual discrimination regarding salaries, benefits, and promotions deserves a higher priority. Still, sexist language offends many readers—both men and women.

For a full discussion of nonsexist writing, see *The Handbook of Nonsexist Writing* (Miller and Swift 1980).

CHOOSING THE RIGHT WORDS AND PHRASES

Choosing the right words and phrases is, of course, as important as choosing the appropriate voice or person or avoiding sexist language. Choosing the best word or phrase to convey what you want to say is a different kind of problem, however; it involves constant, sustained concentration. Your organization's style guide can't help you. You're on your own. Most writers don't worry about word choice during the drafting phase, but they are always on the lookout for the better word while revising.

The following discussion includes seven basic guidelines for choosing the right word:

1. Be specific.
2. Avoid unnecessary jargon.
3. Avoid wordy phrases.
4. Avoid clichés.
5. Avoid pompous words.
6. Focus on the "real" subject.
7. Focus on the "real" verb.

BE SPECIFIC   Being specific involves using precise words, providing adequate detail, and avoiding ambiguity.

Wherever possible, use the most precise word you can. A Ford Mustang is an automobile, but it is also a vehicle, a machine, and a thing. In describing the Ford Mustang, the word *automobile* is better than *vehicle*, because the less specific word also refers to trains, hot-air balloons, and other means of transport. As the words become more abstract—from *machine* to *thing*, for instance—the chances for misunderstanding increase.

In addition to using the most precise words you can, be sure to provide enough detail. Remember that the reader probably knows less than you do. What might be perfectly clear to you might be too vague for the reader.

VAGUE

An engine on the plane experienced some difficulties.

What engine? What plane? What difficulties?

CLEAR

The left engine on the Jetson 411 unaccountably lost power during flight.

Avoid ambiguity. That is, don't let the reader wonder which of two meanings you are trying to convey.

AMBIGUOUS

After stirring by hand for ten seconds, add three drops of the iodine mixture to the solution.

Stir the iodine mixture or the solution?

CLEAR

Stir the iodine mixture by hand for ten seconds. Then add three drops to the solution.

CLEAR

Stir the solution by hand for ten seconds. Then add three drops of the iodine mixture.

What should you do if you don't have the specific data? You have two options: to approximate—and clearly tell the reader you are doing so—or to explain why the specific data are unavailable and indicate when they will become available.

VAGUE

The leakage in the fuel system is much greater than we had anticipated.

CLEAR

The leakage in the fuel system is much greater than we had anticipated; we estimate it to be at least five gallons per minute, rather than two.

Several style programs isolate common vague terms and suggest more precise alternatives.

AVOID
UNNECESSARY
JARGON

Jargon is shoptalk. To a banker, *CD* means certificate of deposit; to an audiophile it is a compact disc. The term *CAFE* (Corporate Average Fuel Economy) is a meaningful acronym among auto executives. Although jargon is often held up to ridicule, it is a useful and natural kind of communication in its proper sphere. Two baseball pitchers would find it hard to talk to each other about their craft if they couldn't use terms such as *slider* and *curve*.

In one sense, the abuse of jargon is simply a needless corruption of the language. The best current example is the degree to which computer-science terminology has crept into everyday English. In offices, employees are frequently asked to provide "feedback" or told that the coffee machine is "down." To many people, such words seem dehumanizing; to others, they seem silly. What is a natural and clear expression to one person is often strange or confusing to another. Communication breaks down.

An additional danger of using jargon outside a very limited professional circle is that it sounds condescending to many people, as if the writer is showing off—displaying a level of expertise that excludes most readers. While the readers are concentrating on how much they dislike the writer, they are missing the message.

If some readers are offended by unnecessary jargon, others are intimidated. They feel somehow inadequate or stupid because they do not know what the writer is talking about. When the writer casually tosses in jargon, many readers really *can't* understand what is being said.

If you are addressing a technically knowledgeable audience, feel free to use appropriate jargon. However, an audience that includes managers or the general public will probably have trouble with specialized vocabulary. If your document has separate sections for different audiences—as in the case of a technical report with an executive summary—use jargon accordingly. A glossary (list of definitions) is useful if you suspect that the technical sections will be read by the managers.

AVOID WORDY PHRASES

Wordy phrases weaken technical writing by making it unnecessarily long. Sometimes writers deliberately choose phrases such as "demonstrates a tendency to" rather than "tends to." The long phrase rolls off the tongue easily and appears to carry the weight of scientific truth. But the humble "tends to" says the same thing—and says it better for having done so concisely. The sentence "We can do it" is the concise version of "We possess the capability to achieve it."

Some wordy phrases just pop into writers' minds. We are all so used to hearing *take into consideration* that we don't realize that *consider* gets us there faster. Replacing wordy phrases with concise ones is therefore more difficult than it might seem. Avoiding the temptation to write the long phrase is only half of the solution. The other half is to try to root out the long phrase that has infiltrated the prose unnoticed.

Following are a wordy sentence and a concise translation.

WORDY

I am of the opinion that, in regard to profit achievement, the statistics pertaining to this month will appear to indicate an upward tendency.

CONCISE

I think this month's statistics will show an increase in profits.

A special kind of wordiness to watch out for is unnecessary redundancy, as in *end result, any and all, each and every, completely eliminate,* and *very unique.* Be content to say something once. Use "The liquid is green," not "The liquid is green in color."

REDUNDANT

We initially began our investigative analysis with a sample that was spherical in shape and heavy in weight.

BETTER

We began our analysis with a heavy, spherical sample.

Following is a list of some of the most commonly used wordy phrases and their concise equivalents.

WORDY PHRASE	CONCISE PHRASE
a majority of	most
a number of	some, many
at an early date	soon
at the conclusion of	after, following
at the present time	now
at this point in time	now
based on the fact that	because
despite the fact that	although
due to the fact that	because
during the course of	during
during the time that	during, while
have the capability to	can
in connection with	about, concerning
in order to	to
in regard to	regarding, about
in the event that	if
in view of the fact that	because
it is often the case that	often
it is our opinion that	we think that
it is our understanding that	we understand that
it is our recommendation that	we recommend that
make reference to	refer to
of the opinion that	think that
on a daily basis	daily
on the grounds that	because
prior to	before
relative to	regarding, about
so as to	to
subsequent to	after
take into consideration	consider
until such time as	until

AVOID CLICHÉS   The English writer George Orwell once offered some good advice about writing: "Never use a metaphor, simile or other figure of speech which you are used to seeing in print." Rather than writing, "It's a whole new ball game," write, "The situation has changed completely." Don't write, "I am sure the new manager can cut the mustard"; write, "I am sure the new manager can do his job effectively." If someone suggests that you "go for it," don't. Why would you want to sound like Rocky?

Sometimes, writers further embarrass themselves by getting their clichés wrong: expressions become so timeworn that users forget what the words mean. The phrase "a new bag of worms" has found its way into print; the producer Sam Goldwyn was famous for such statements as "An oral agreement isn't worth the paper it's written on." And the phrase "I could care less" often is used when the writer means just the opposite. The best solution to this problem is, of course, not to use clichés.

Following are a cliché-filled sentence and a translation into plain English.

TRITE

Afraid that we were between a rock and a hard place, we decided to throw caution to the winds with a grandstand play that would catch our competition with its pants down.

PLAIN

Afraid that we were in a grave situation, we decided on a risky and aggressive move that would surprise our competition.

AVOID POMPOUS WORDS   Writers sometimes try to impress their readers by using pompous words, such as *initiate* for *begin*, *perform* for *do*, and *prioritize* for *rank*. When asked why they use big words where small ones will do, writers say that they want to make sure their readers know they have a strong vocabulary, that they are well educated.

Undoubtedly, pompous words have a role in some kinds of communication. Sports commentator Howard Cosell and conservative columnist William F. Buckley, Jr., both understand the comic potential of pomposity. But in technical writing, plain talk is best. If you know what you're talking about, be direct and simple. Even if you're not so sure of what you're talking about, say it plainly; big words won't fool anyone for more than a few seconds.

Following are a few pompous sentences translated into plain English.

POMPOUS

The purchase of a minicomputer will enhance our record maintenance capabilities.

PLAIN

Buying a minicomputer will help us maintain our records.

POMPOUS

It is the belief of the Accounting Department that the predicament was precipitated by a computational inaccuracy.

PLAIN

The Accounting Department thinks a math error caused the problem.

Following is a list of some of the most commonly used fancy words and their plain equivalents:

FANCY WORD	PLAIN WORD
advise	tell
ascertain	learn, find out
attempt (verb)	try
commence	start, begin
demonstrate	show
employ	use
endeavor (verb)	try
eventuate (verb)	happen
evidence (verb)	show
finalize	end, settle, agree
furnish	provide, give
impact (verb)	affect
initiate	begin
manifest (verb)	show
parameters	variables, conditions
perform	do
prioritize	rank
procure	get, buy
quantify	measure
terminate	end, stop
utilize	use

Several style programs isolate fancy words and expressions. Of course, you can use any word-processing program to search for terms that you tend to use inappropriately.

In the long run, your readers will be impressed with your clarity and accuracy. Don't waste your time thinking up fancy words.

FOCUS ON THE "REAL" SUBJECT

The conceptual or "real" subject of the sentence should also be the sentence's grammatical subject, and it should appear prominently in technical writing. Don't bury the real subject in a prepositional phrase following a useless or "limp" grammatical subject. In the following examples, notice how the limp subjects disguise the real subjects. (The grammatical subjects are italicized.)

WEAK

The *use* of this method would eliminate the problem of motor damage.

STRONG

This *method* would eliminate the problem of motor damage.

WEAK

The *presence* of a six-membered lactone ring was detected.

STRONG

A six-membered lactone *ring* was detected.

Another way to make the subject of the sentence prominent is to eliminate grammatical expletives. You can almost always remove expletive constructions—*it is* . . . , *there is* . . . , and *there are* . . .—without changing the meaning of the sentence.

WEAK

There are many problems that must be worked out.

STRONG

Many problems must be worked out.

WEAK

It is with great pleasure that I welcome you to our annual development seminar.

STRONG

With great pleasure I welcome you to our annual development seminar.

STRONG

I am pleased to welcome you to our annual development seminar.

Using the search function of any word-processing program, you can find most weak subjects: usually they're right before the word *of*. Expletives are also easy to find.

FOCUS ON THE "REAL" VERB

A "real" verb, like a "real" subject, should be prominent in every sentence. Few stylistic problems weaken a sentence more than nominalizing verbs. To nominalize the real verb, you convert it into a noun; you must then supply another, usually weaker, verb to convey your meaning. "To install" becomes "to effect an installation," "to analyze" becomes "to conduct an analysis." Notice how nominalizing the real verbs makes the following sentences both awkward and unnecessarily long. (The nominalized verbs are italicized.)

WEAK

Each *preparation* of the solution is done twice.

STRONG

Each solution is prepared twice.

WEAK

An *investigation* of all possible alternatives was undertaken.

STRONG

All possible alternatives were investigated.

WEAK

Consideration should be given to an acquisition of the properties.

STRONG

We should consider acquiring the properties.

Some software programs search for the most common nominalizations. With any word-processing program you can catch most of the nominalizations if you search for character strings such as *tion*, *ment*, and *ance*. Many nominalized verbs are used with the preposition *of*.

SENTENCE STRUCTURE AND LENGTH

In addition to recognizing word and phrase problems that can weaken a sentence, you should also understand how to use different sentence structures and lengths to make your writing more effective.

TYPES OF SENTENCES

There are four basic types of sentences:

1. simple (one independent clause)

The technicians soon discovered the problem.

2. compound (two independent clauses, linked by a semicolon or by a comma and one of the seven common coordinating conjunctions: *and, or, nor, for, so, yet,* and *but*)

The technicians soon discovered the problem, but they found it difficult to solve.

3. complex (one independent clause and at least one dependent clause, linked by a subordinating conjunction)

Although the technicians soon discovered the problem, they found it difficult to solve.

4. compound-complex (at least two independent clauses and at least one dependent clause)

Although the technicians soon discovered the problem, they found it difficult to solve, and they finally determined that they needed some additional parts.

Most commonly used in technical writing is the simple sentence, because it is clear and direct. However, a series of three or four simple sentences can bore and distract the reader. The main shortcoming of the simple sentence is that it can communicate only one basic idea, because it is made up of only one independent clause.

Compound and complex sentences communicate more sophisticated ideas. The compound sentence works on the principle of coordination: the two halves of the sentence are roughly equivalent in importance. The complex sentence uses subordination: one half of the sentence is less important than the other.

Compound-complex sentences are useful in communicating very complicated ideas, but their length and difficulty make them unsuitable for most readers.

One common construction deserves special mention: the compound sentence linked by the coordinating conjunction *and.* In many cases, this construction is a lazy way of linking two thoughts that could be linked more securely. For instance, the sentence "Enrollment is up, and the school is planning new course offerings" represents a weak use of the *and* conjunction, because it doesn't show the cause-effect relationship between the two clauses. A stronger link would be *so:* "Enrollment is up, so the school is planning new course offerings." A complex sentence would sharpen the relationship even more: "Because enrollment is up, the school is planning new course offerings." Another example of the weak *and* conjunction is the sentence "The wires were inspected, and none was found to be damaged." In this case, a simple sentence would be more effective: "The inspection of the wires revealed that none was damaged" or "None of the wires inspected was damaged."

A number of software programs classify sentences according to type, so that during revision you can see the kinds of sentences you have written. On the basis of your analysis of audience and purpose, you can then decide whether your mixture of sentence types is appropriate.

SENTENCE LENGTH   Although it would be artificial and distracting to keep a count of the number of words in your sentences, sometimes sentence length can work against effective communication. In revising a draft, you might want to compute an average sentence length for a page of writing. (Many software programs compute sentence length.)

There are no firm guidelines covering appropriate sentence length. In general, 15 to 20 words is effective for technical writing. A succession of 10-word sentences would be abrupt and choppy; a series of 35-word sentences would probably be too demanding.

The best approach to determining an effective sentence length is to consider the audience and purpose.

AUDIENCE

The more the readers know about the subject, the more easily they will be able to handle longer sentences.

PURPOSE

If you are writing a set of instructions or some other kind of information that your reader will be working from directly, short sentences are more effective. In addition, short sentences emphasize a particularly important point.

The following four examples show how the same information can be conveyed in sentences of varying type and length.

The ruling on dumping was expected to have widespread implications for the chemical industry. However, problems in interpreting the ruling have led to protracted legal battles. As a result, the government has not yet won a single conviction.

This example contains three simple sentences, of 14, 12, and 12 words. This version would be suitable for most general audiences.

The ruling on dumping was expected to have widespread implications for the chemical industry, but problems in interpreting the ruling have led to protracted legal battles. As a result, the government has not yet won a single conviction.

In this version, the first two simple sentences have been combined into a compound sentence—two independent clauses linked by a coordinating conjunction—of 26 words. Although the material has remained the same, this version is more difficult than the previous one because it asks the readers to remember more information as they read the longer first sentence.

The ruling on dumping was expected to have widespread implications for the chemical industry. Because problems in interpreting the ruling have led to protracted legal battles, however, the government has not yet won a single conviction.

Here, the second and third sentences of the first version have been combined in a complex sentence. The original second sentence is now a dependent clause; the original third sentence is now an independent clause. This version, of 14 and 22 words, is about as difficult as the second version.

Although the ruling on dumping was expected to have widespread implications for the chemical industry, problems in interpreting the ruling have led to protracted legal battles, and the government has not yet won a single conviction.

In this version, the three original sentences have been combined into a 36-word compound-complex sentence: a dependent clause and two independent clauses. This is the most sophisticated and most demanding version of the information.

A NOTE ON READABILITY FORMULAS
Readability formulas are mathematical techniques used to determine how "readable" a piece of writing is; that is, how difficult it is for someone to read and understand it. More than 100 different readability formulas exist, and they are being used increasingly by government agencies and private businesses in an attempt to improve writing. For this reason, you should become familiar with the concept behind them.

Most readability formulas are based on the idea that short words and sentences are easier to understand than long ones. One of the more popular formulas, Robert Gunning's "Fog Index," works like this:

1. Find the average number of words per sentence, using a 100-word passage.

2. Find the number of "difficult words" in that same passage. "Difficult words" are words of three or more syllables, except for proper names, combinations of simple words (such as *manpower*), and verbs whose third syllable is *-es* or *-ed* (such as *contracted*).
3. Add the average sentence length and the number of "difficult words."
4. Multiply this sum by 0.4.

The figure obtained from this procedure represents the approximate grade level that the reader must have reached in order to understand the writing.

Average number of words per sentence:	13.9
Number of difficult words:	16

$$29.9 \times 0.4 = 12$$

The average twelfth-grade student could understand the writing.

Readability formulas are easy to use, and it is appealing to think you can be objective in assessing your writing, but unfortunately they have not been proven to work. They simply have not been shown to reflect accurately how difficult it is to read a piece of writing (Selzer 1981; Battison and Goswami 1981). The problem with readability formulas is that they attempt to evaluate words on paper—without considering the reader. It is possible to measure how well a specific person understands a specific writing sample. But words on paper don't communicate until somebody reads them, and everyone is different. A "difficult word" for a lawyer might not be difficult for a biologist, and vice versa. And someone who is interested in the subject being discussed will understand more than the reluctant reader will.

Readability formulas offer a false sense of security by suggesting that short words and sentences will make writing easy to read. Things are not so simple. Good writing has to be well thought out and carefully structured. The sentences have to be clear, and the vocabulary has to be appropriate for the readers. *Then* the writing will be readable. Nonetheless, many organizations use readability formulas, and you should be familiar with them.

Many software programs compute readability formulas.

USING HEADINGS AND LISTS

Much of the preceding discussion of style applies to any kind of nonfiction writing. Headings and lists, although not unique to technical writing, are a major stylistic feature of reports, memos, and letters. Although headings—and to a lesser extent, lists—might at first appear to be mechanical elements that need be considered only during revision, they are in fact fundamental, and they determine content as well as form.

HEADINGS   The main purpose of a heading is to announce the subject of the discussion that follows it: the word HEADINGS, for instance, tells you that this discussion will cover the subject of headings. For the writer, headings eliminate the need to announce the subject in a sentence such as "Let us now turn to the subject of headings."

For the reader, headings clarify the hierarchical relationships within the document. This text, for example, has four main parts, each of which is subdivided into chapters. Each chapter, in turn, is subdivided into major units (such as USING HEADINGS AND LISTS). Most of the major units are subdivided again into smaller units (such as HEADINGS). Understanding the level of importance of the various components of the discussion helps the reader concentrate on the discussion itself.

To signify different hierarchical levels in headings in typewritten manuscript, use capitalization, underlining, and indention. Capital (uppercase) letters are more emphatic than small (lowercase) letters; therefore, an all-capitals heading (PROCEDURE) signals a more important category of information than an initial-capital heading (Procedure). Similarly, underlined headings (PROCEDURE) are more emphatic than nonunderlined headings (PROCEDURE). Finally, a heading centered horizontally on the page is more emphatic than one that begins at the left margin. And a heading beginning at the left margin and placed on a line by itself is more emphatic than one beginning at the left margin and followed by text on the same line.

Many word processors offer additional formatting options, including boldface, italics, and different sizes of type. Boldface, of course, is more emphatic than regular type, just as large letters are more emphatic than small letters.

Figure 5-2 shows how capitalization, underlining, and indention can be used to create headings that show different levels of hierarchy.

To further reinforce the hierarchical levels, you can add a numbering scheme, such as the traditional outline system or the decimal system:

```
I.                    1.
   A.                    1.1
      1.                    1.1.1
      2.                    1.1.2
```

```
B.                    1.2
   1.                    1.2.1
   2.                    1.2.2
      a.                    1.2.2.1
      b.                    1.2.2.2
        (1)                    1.2.2.2.1
        (2)                    1.2.2.2.2
```

Notice that the decimal system can easily accommodate a large number of hierarchical levels—an advantage when you are writing complicated technical documents.

When you make up the headings, keep in mind that you are not restricted to simple phrases. Your assessment of the writing situation might indicate that a heading such as "What Is Wrong with the Present System?" would be more effective than one such as "Problem." Although general terms such as *problem, results,* and *conclusions* are well known by engineers and scientists, general readers and upper-level managers might have an easier time with phrases or questions that are more informative.

FIGURE 5-2

A Four-Level Heading System

FIRST LEVEL   <u>REPAIRING ASPHALT STREETS</u>

SECOND LEVEL   INTRODUCTION

    This manual is intended ....

....

SECOND LEVEL   TYPES OF DEFECTS

    Three basic types of defects in asphalt streets ....

....

THIRD LEVEL   <u>Cracks</u>

    A crack is defined as ....

....

FOURTH LEVEL	<u>Identifying Cracks</u>. To identify a crack, ...
FOURTH LEVEL	<u>Repairing Cracks</u>. The first step ....
THIRD LEVEL	<u>Grade Depressions</u> A grade depression is defined as ....
FOURTH LEVEL	<u>Identifying Grade Depressions</u>. To identify a grade depression, ...
FOURTH LEVEL	<u>Repairing Grade Depressions</u>. The first step ....
THIRD LEVEL	<u>Potholes</u> A pothole is defined as ....
FOURTH LEVEL	<u>Identifying Potholes</u>. To identify a pothole, ...
FOURTH LEVEL	<u>Repairing Potholes</u>. The first step ....

LISTS   Like headings, lists let you manipulate the placement of words on the page to improve the effectiveness of the communication.

Many sentences in technical writing are long and complicated:

We recommend that more work on heat-exchanger performance be done with a larger variety of different fuels at the same temperature, with similar fuels at different temperatures, and with special fuels such as diesel fuel and shale-oil-derived fuels.

Here readers cannot concentrate on the information because they must worry about remembering all the *with* phrases following "done." If they could "see" how many phrases they had to remember, their job would be easier.

Revised as a list, the sentence is easier to follow:

We recommend that more work on heat-exchanger performance be done:
1. with a larger variety of different fuels at the same temperature
2. with similar fuels at different temperatures
3. with special fuels such as diesel fuels and shale-oil-derived fuels

In this version, the placement of the words on the page reinforces the meaning. The readers can easily see that the sentence contains three items in a series. And the fact that each item begins at the same left margin helps, too.

Make sure the items in the list are presented in a parallel structure. (See Appendix A for a discussion of parallelism.)

NONPARALLEL

Here is the schedule we plan to follow:

1. construction of the preliminary proposal
2. do library research
3. interview with the Bemco vice-president
4. first draft
5. revision of the first draft
6. after we get your approval, typing of the final draft

PARALLEL

Here is the schedule we plan to follow:

1. write the preliminary proposal
2. do library research
3. interview the Bemco vice-president
4. write the first draft
5. revise the first draft
6. type the final draft, after we receive your approval

In this example, the original version of the list is sloppy, a mixture of noun phrases (items 1, 3, 4, and 5), a verb phrase (item 2), and a participial phrase preceded by a dependent clause (item 6). The revision uses parallel verb phrases and deemphasizes the dependent clause in item 6 by placing it after the verb phrase.

Note that reports, memos, and letters do not have to look "formal," with traditional sentences and paragraphs covering the whole page. Headings and lists make writing easier to read and understand.

PARAGRAPH STRUCTURE AND LENGTH

A paragraph can be defined as a group of sentences (or sometimes a single sentence) that is complete and self-sufficient but that also contributes to a larger discussion. The challenge of creating an effective paragraph in technical writing is to make sure, first, that all the sentences clearly and directly substantiate one main point, and second, that the whole paragraph follows logically from the material that precedes it. Readers tend to pause between paragraphs (not between sentences) to digest the information given in one paragraph and link it with that given in the previous paragraphs. For this reason, the paragraph is the key unit of composition. Readers might forgive or at least overlook a slightly fuzzy sentence. But if they can't figure out what a paragraph says or why it appears where it does, communication is likely to break down.

PARAGRAPH STRUCTURE Too often in technical writing, paragraphs seem to be written for the writer, not the reader. They start off with a number of details: about who worked on the problem before and what equipment or procedure they used; about the ups and downs of the project, the successes and setbacks; about specifications, dimensions, and computations. The paragraph winds its way down the page until, finally, the writer concludes: "No problems were found."

This structure—moving from the particular details to the general statement—accurately reflects the way the writer carried out the activity he or she is describing, but it makes the paragraph difficult to follow. As you put a paragraph together, focus on your readers' needs. Do they want to "experience" your writing, to regret your disappointments and celebrate your successes? Probably not. They just want to find out what you have to say.

Help your readers. Put the point—the topic sentence—up front. Technical writing should be clear and easy to read, not full of suspense. If a paragraph describes a test you performed on a piece of equipment, include the result in your first sentence: "The point-to-point continuity test on Cabinet 3 revealed no problems." Then go on to explain the details. If the paragraph describes a complicated idea, start with an overview: "Mitosis occurs in five stages: (1) interphase, (2) prophase, (3) metaphase, (4) anaphase, and (5) telophase." Then describe each stage. In other words, put the "bottom line" on top.

Notice, for instance, how difficult the following paragraph is, because the writer structured the discussion in the same order she performed her calculations:

Our estimates are based on our generating power during eight months of the year and purchasing it the other four. Based on the 1985 purchased power rate of $0.034/KW (January through April cost data) inflating at 8 percent annually, and a constant coal cost of $45–$50, the projected 1988 savings resulting from a conversion to coal would be $225,000.

Putting the bottom line on top makes the paragraph much easier to read. Notice how the writer adds a numbered list after the topic sentence.

The projected 1988 savings resulting from a conversion to coal are $225,000. This estimate is based on three assumptions: (1) that we will be generating power during eight months of the year and purchasing it the other four, (2) that power rates inflate at 8 percent from the 1985 figure of $0.034/KW (January through April cost data), and (3) that coal costs remain constant at $45–$50.

The topic sentence in technical writing functions just as it does in any other kind of writing: it summarizes or forecasts the main point of the paragraph.

Why don't writers automatically begin with the topic sentence if that helps the readers understand the paragraph? One reason is that it is easier to present the events in their natural order, without first having to decide what the single most important point is. Perhaps a more common reason is that writers feel uncomfortable "exposing" the topic sentence at the start of a paragraph. Most people who do technical writing were trained in science, engineering, or technology; they were taught that they must not reach a conclusion before obtaining sufficient evidence to back it up. Putting the topic sentence up top *looks* like stating a conclusion without "proving" it, even though the proof follows the topic sentence directly in the rest of the paragraph. Beginning the paragraph with the sentence "It was concluded that human error caused the overflow" somehow seems risky. It isn't. The real risk is that you might frustrate or bore your readers by making them hunt for the topic sentence.

After the topic sentence comes the support. The purpose of the support is to make the topic sentence clear and convincing. Sometimes a few explanatory details can provide all the support needed. In the paragraph about estimated fuel savings presented earlier, for example, the writer simply fills in the assumptions used in making the calculation: the current energy rates, the inflation rate, and so forth. Sometimes, however, the support must carry a heavier load: it has to clarify a difficult thought or defend a controversial one.

Because every paragraph is unique, it is impossible to define the exact function of the support. In general, however, the support fulfills one of the following roles:

1. to define a key term or idea included in the topic sentence
2. to provide examples or illustrations of the situation described in the topic sentence
3. to identify factors that led to the situation
4. to define implications of the situation
5. to defend the assertion made in the topic sentence

The techniques used in developing the support include those used in most nonfiction writing: definition, comparison and contrast, classification and partition, and causal analysis. These techniques are described further in Part II.

PARAGRAPH LENGTH

How long should a paragraph of technical writing be? In general, a length of 75 to 125 words will provide enough space for a topic sentence and four or five supporting sentences. Long paragraphs are more difficult to read than short paragraphs, for the simple reason that the readers have to concentrate longer. Long paragraphs also intimidate many readers by presenting long, unbroken stretches of type. Some readers actually will skip over long paragraphs.

Don't let an arbitrary guideline about length take precedence over your analysis of the audience and purpose. Often you will need to write very brief paragraphs. You might need only one or two sentences—to introduce a graphic aid, for example. A transitional paragraph—one that links two other paragraphs—also is likely to be quite short. If a brief paragraph fulfills its function, let it be. Do not combine two ideas in one paragraph in order to achieve a minimum word count.

While it is confusing to include more than one basic idea in a paragraph, the concept of unity can be violated in the other direction. Often you will find it necessary to divide one idea into two or more paragraphs. A complex idea that would require 200 or 300 words probably should not be squeezed into one paragraph.

The following example shows how a writer addressing a general audience divided one long paragraph into two:

High-tech companies have been moving their operations to the suburbs for two main reasons: cheaper, more modern space and a better labor pool. A new office complex in the suburbs will charge anywhere from half to two-thirds of the rent charged for the same square footage in the city. And that money goes a lot further, too. The new office complexes are bright and airy, with picture windows looking out on lush landscaping. New office space is already wired for the computers; and exercise clubs, shopping centers, and even libraries are often on-site.

The second major factor attracting high-tech companies to the suburbs is the availability of experienced labor. Office workers and middle managers are abundant; many suburbanites, especially women returning to the labor force after their children start school, are highly trained and willing to make the short trip to the office complex. In addition, the engineers and executives, who tend to live in the suburbs anyway, are happy to forgo the commuting, the city wage taxes, and the noise and stress of city life.

A strict approach to paragraphing would have required one paragraph, not two, because all the information presented supports the topic sentence that opens the first paragraph. Many readers, in fact, could easily understand a one-paragraph version. However, the writer found a logical place to create a second paragraph and thereby increased the effectiveness of his communication.

Another writer might have approached the problem differently, making each "reason for moving to the suburbs" a separate paragraph.

High-tech companies have been moving their operations to the suburbs for two main reasons: cheaper, more modern space and a better labor pool.

Office space is a bargain in the suburbs. A new office complex will charge anywhere from half to two-thirds of the rent charged for the same square footage in the city. And that money goes a lot further, too. The new office complexes are bright and airy, with picture windows looking out on lush landscaping. New office space is already wired for the computers; and exercise clubs, shopping centers, and even libraries are often on-site.

The second major factor attracting high-tech companies to the suburbs is the availability of experienced labor. Office workers and middle managers are abundant; many suburbanites, especially women returning to the labor force after their children start school, are highly trained and willing to make the short trip to the office complex. In addition, the engineers and executives, who tend to live in the suburbs anyway, are happy to forgo the commuting, the city wage taxes, and the noise and stress of city life.

The original topic sentence becomes a transitional paragraph that leads clearly and logically into the two explanatory paragraphs.

MAINTAINING COHERENCE WITHIN AND BETWEEN PARAGRAPHS

After you have blocked out the main structure of the paragraph—the topic sentence and the support—make sure the paragraph is coherent. In a coherent paragraph, thoughts are linked together logically and clearly. Parallel ideas are expressed in parallel grammatical constructions. If the paragraph moves smoothly from sentence to sentence, emphasize the coherence by adding transitional words and phrases, repeating key words, and using demonstratives.

Maintaining coherence *between* paragraphs is the same process as maintaining coherence *within* paragraphs: place the transitional device as close as possible to the beginning of the second element. For example, the link between two sentences within a paragraph should be near the start of the second sentence:

The new embossing machine was found to be defective. *However*, the warranty on the machine will cover replacement costs.

The link between the two paragraphs should be near the start of the second paragraph:

The complete system would be too expensive for us to purchase now...
....
....
In addition, a more advanced system is expected on the market within six months..
....
....

Transitional words and phrases help the reader understand a discussion by pointing out the direction the thoughts are following. Here is a list of the most common logical relationships between two thoughts and some of the common transitions that express those relationships:

RELATIONSHIP	TRANSITIONS
addition	also, and, finally, first (second, etc.), furthermore, in addition, likewise, moreover, similarly
comparison	in the same way, likewise, similarly

contrast	although, but, however, in contrast, nevertheless, on the other hand, yet
illustration	for example, for instance, in other words, to illustrate
cause-effect	as a result, because, consequently, hence, so, therefore, thus
time or space	above, around, earlier, later, next, to the right (left, west, etc.), soon, then
summary or conclusion	at last, finally, in conclusion, to conclude, to summarize

In the following examples, the first versions contain no transitional words and phrases. Notice how much clearer the second versions are.

WEAK

Neurons are not the only kind of cell in the brain. Blood cells supply oxygen and nutrients.

IMPROVED

Neurons are not the only kind of cell in the brain. *For example*, blood cells supply oxygen and nutrients.

WEAK

The project was originally expected to cost $300,000. The final cost was $450,000.

IMPROVED

The project was originally expected to cost $300,000. *However*, the final cost was $450,000.

WEAK

The manatee population of Florida has been stricken by an unknown disease. Marine biologists from across the nation have come to Florida to assist in manatee-disease research.

IMPROVED

The manatee population of Florida has been stricken by an unknown disease. *As a result*, marine biologists from across the nation have come to Florida to assist in manatee-disease research.

Repetition of key words—generally, nouns—helps the reader follow the discussion. Notice in the following example how the first version can be confusing.

UNCLEAR

For months the project leaders carefully planned their research. The cost of the work was estimated to be over $200,000. (*What is the* work, *the planning or the research?*)

CLEAR

For months the project leaders carefully planned their *research*. The cost of the *research* was estimated to be over $200,000.

Out of a misguided desire to be "interesting," some writers keep changing their important terms. *Plankton* becomes *miniature seaweed*, then *the ocean's fast food*. Leave this kind of word game to TV sportscasters; technical writing must be clear.

In addition to transitional words and phrases and repetition of key phrases, demonstratives—*this, that, these,* and *those*—can help the writer maintain the coherence of a discussion by linking ideas securely. Demonstratives should in almost all cases serve as adjectives rather than as pronouns. In the following examples, notice that a demonstrative pronoun by itself can be confusing.

UNCLEAR

New screening techniques are being developed to combat viral infections. These are the subject of a new research effort in California.

What is being studied in California, new screening techniques or viral infections?

CLEAR

New screening techniques are being developed to combat viral infections. *These techniques* are the subject of a new research effort in California.

UNCLEAR

The task force could not complete its study of the mine accident. This was the subject of a scathing editorial in the union newsletter.

What was the subject of the editorial, the mine accident or the task force's inability to complete its study of the accident?

CLEAR

The task force failed to complete its study of the mine accident. *This failure* was the subject of a scathing editorial in the union newsletter.

Even when the context is clear, a demonstrative pronoun used without a noun refers the reader to an earlier idea and therefore interrupts the reader's progress.

INTERRUPTIVE

The law firm advised that the company initiate proceedings. This resulted in the company's search for a second legal opinion.

FLUID

The law firm advised that the company initiate proceedings. *This advice* resulted in the company's search for a second legal opinion.

Transitional words and phrases, repetition, and demonstratives cannot *give* your writing coherence: they can only help the reader to appreciate the coherence that already exists. Your job is, first, to make sure your writing is coherent and, second, to highlight that coherence.

TURNING A "WRITER'S PARAGRAPH" INTO A "READER'S PARAGRAPH"

The best way to demonstrate what has been said in this discussion is to take a weak paragraph and improve it. The following paragraph is an excerpt from a status report written by a branch manager of the utility company mentioned in Chapter 1. The subject of the paragraph is how the writer decided on a method to increase the company's business within his particular branch.

There were two principal alternatives considered for improving Montana Branch. The first alternative was to drill and equip additional sources of supply with sufficient capacity to provide for the present and projected system deficiencies. The second alternative was to provide for said deficiencies through a combination of additional sources of supply and a storage facility. Unfortunately, groundwater studies which were conducted in the Southeast Montana area by the consulting firm of Smith and Jones indicated that although groundwater is available within this general area of our system, it is limited as to quantity, and considerable separation between well sources is necessary in order to avoid interference between said sources. This being the case, it becomes necessary to utilize the sources that are available or that can be developed in the most efficient manner, which means operating them in conjunction with a storage facility. In this way, the sources only have to be capable of providing for the average demand on a maximum day, and the storage facility can be utilized to provide for the peaking requirements plus fire protection. Consequently, the second alternative as mentioned hereinabove was determined to be the more desirable alternative.

First, let's be fair. The paragraph has been taken out of its context—a 17-page report—and was never meant to stand alone on a page. Also, it was not written for the general reader, but for an executive of the water company—someone who, in this case, is technically knowledgeable in the writer's field. Still, an outsider's analysis of an essentially private communication can at least isolate the weaknesses.

The most important element of a paragraph is the topic sentence, so you look for it in the usual place: the beginning. At first glance, the topic sentence looks good. The statement that two principal alternatives were considered appears to function effectively as an introduction to the rest of the paragraph. You assume that the paragraph will define the two alternatives.

The more you think about the topic sentence, however, the less good it looks. The problem is suggested by the word *principal*. Most complex decisions eventually come down to a choice between the two best alternatives; why bother labeling the two best alternatives "principal"? The writer almost sounds as if he is congratulating himself for weeding out the undesirable alternatives. (Further, he begins with the expletive "there are" and uses a passive construction that weakens the sentence.)

Structured as it is, the paragraph focuses on the process of choosing an alternative rather than on the choice itself. Perhaps the writer thinks his readers are curious about how he does his job; perhaps he wants to suggest that he's done his best, so that if it turns out that he made the wrong choice, at least he cannot be criticized for negligence; or perhaps the writer is unintentionally recreating the scientific method by describing how he examined the two alternatives and finally came to a decision.

For whatever reasons, the writer has built a lot of suspense into the paragraph. It proceeds like a horse race, with first one alternative gaining, then the other. Consequently, the readers end up worrying about the outcome and can't concentrate on the specific reasons that one choice prevailed over the other. Quite likely, the only thing you will remember about the paragraph is the final emphatic statement: that the writer decided to go ahead with the second alternative. And given the complexity of the paragraph, it would be easy to forget what that alternative is.

Without too much trouble the paragraph can be rewritten so that it is much easier to understand. A careful topic sentence—one that defines the choice of a technical solution to the problem—is a constructive start. Then the writer should elaborate on the necessary details of the solution. Only after he has finished his discussion of the solution should he define and explain the alternative that was rejected. Using this structure, the writer clearly emphasizes what he *did* decide to do, putting what he decided *not* to do in a subordinate position.

Here is the writer's paragraph translated into a reader's paragraph:

We found that the best way to improve the Montana branch would be to add a storage facility to our existing supply sources. Currently, we can handle the average demand on a maximum day; the storage facility will enable us to meet peaking requirements and fire-protection needs. In conducting our investigation, we considered developing new supply sources with sufficient capacity to meet current and future needs. This alternative was rejected, however, when our consultants (Smith and Jones) did groundwater studies that revealed that insufficient groundwater is available and that the new wells would have to be located too far apart if they were not to interfere with each other.

One clear advantage of the revision is that it is about half as long as the original. The structure of the first version is largely responsible for its length: the writer announced that two alternatives were considered; defined them; explained why the first was impractical and why it led to the second alternative; and, finally, announced the choice of the second alternative. A direct structure eliminates all this zigzagging and clarifies the paragraph.

The only possible objection to the streamlined version is that it is *too* clear, that it leaves the writer vulnerable in case his decision turns out to have been wrong. But if the decision doesn't work out, the writer will be responsible no matter how he described it, and he will end up in more trouble if a supervisor has to investigate who made the decision. Good writing is the best bet under any circumstances.

EXERCISES

1. The following sentences are vague. Revise them by using specific information for the vague elements. Make up any reasonable details.

   a. The results won't be available for a while.
   b. The fire in the laboratory caused extensive damage.
   c. Analysis using a highlighting fluid revealed an abnormality in the tissue culture.

2. Consider the voice (active or passive) in the following sentences. Which aspect of each sentence is emphasized? In what way would changing each sentence to the other voice change the emphasis?

   a. The proposal was submitted to the Planning Commission by Dr. Hendrick.

    b. A study of the three available flood-control systems was approved by the Planning Commission.

    c. On Tuesday, employee John Pawley drove the one-millionth Plymouth Reliant off the assembly line.

3. In the following sentences, the real subjects are buried in prepositional phrases or obscured by expletives. Revise the sentences so that the real subjects appear prominently.

    a. The creation of the Energy Task Force will decrease our overhead costs.

    b. It is on the basis of recent research that I recommend the newer system.

    c. There is the need for new personnel to learn to use the computer.

    d. The completion of the new causeway will enable 40,000 cars to cross the ravine each day.

    e. There has been a decrease in the number of students enrolled in two of our training sessions.

    f. The use of point-of-purchase video presentations has resulted in a dramatic increase in business.

4. In the following sentences, unnecessary nominalization has obscured the real verb. Revise the sentences to focus on the real verb.

    a. Pollution constitutes a threat to the Wilson Wildlife Reserve.

    b. The switch from our current system to the microfilm can be accomplished in about two weeks.

    c. The construction of each unit will be done by three men.

    d. The ability to expand both the working memory and the mass storage must be possible.

5. The following sentences contain wordy phrases. Revise the sentences to make them more direct.

    a. As far as experimentation is concerned, much work has to be done on animals.

    b. The analysis should require a period of three months.

    c. The second forecasting indicator used is that of energy demand.

    d. In Kevin McCarthy's article "Computerized Tax Returns," he argues that within a few years no firms will be computing returns "by pencil."

6. The following sentences contain clichés. Revise the sentences to eliminate the clichés.

    a. I would like to thank each and every one of you.

    b. With our backs to the wall, we decided to drop back and punt.

    c. If we are to survive this difficult period, we are going to have to keep our ears to the ground and our noses to the grindstone.

7. The following sentences contain pompous words. Revise the sentences to eliminate the pomposity.

   a. This state-of-the-art beverage procurement module is to be utilized by the personnel associated with the Marketing Department.

   b. It is indeed a not unsupportable inference that we have been unsuccessful in our attempt to forward the proposal to the proper agency in advance of the mandated date by which such proposals must be in receipt.

   c. This system will facilitate a reduction in employee time in reference to filing materials.

   d. Our aspiration is the expedition of the research timetable.

8. The following sentences contain sexist language. Revise the sentences to eliminate the sexism.

   a. Each doctor is asked to make sure he follows the standard procedure for handling Medicare forms.

   b. Policemen are required to live in the city in which they work.

   c. Two of the university's distinguished professors—Prof. Henry Larson and Ms. Anita Sebastian—have been elected to the editorial board of *Modern Chemistry*.

9. In each of the following compound sentences, the coordinating conjunction *and* links two clauses. Revise the sentences to make the link between the two clauses stronger.

   a. We interviewed George Karney, president of Whelk Industries, and he said that our suggestions seem very interesting.

   b. This new schedule will give the employees more freedom, and they will be more productive.

   c. A section of the upper layer of the tissue is seen in Figure 3W, and it is identical to the original sample.

10. The following sentences might be too long for some readers. Break each one into two or more sentences. If appropriate, add transitional words and phrases or other coherence devices.

   a. In the event that we get the contract, we must be ready by June 1 with the necessary personnel and equipment to get the job done, so with this end in mind a staff meeting, which all group managers are expected to attend, is scheduled for February 12.

   b. Once we get the results of the stress tests on the 125-Z fiberglass mix, we will have a better idea where we stand in terms of our time constraints, because if it isn't suitable we will really have to hurry to find and test a replacement by the Phase I deadline.

   c. Although we had a frank discussion with Becker's legal staff, we were not able to get them to discuss specifics on what they would be looking for in an out-of-court settlement, but they gave us a strong impression that they would rather not take the matter to court.

11. The following examples contain choppy and abrupt sentences. Combine sentences to create a smoother prose style.

   a. I need a figure on the surrender value of a policy. The policy number is A6423146. Can you get me this figure by tomorrow?
   b. There are advantages to having your tax return prepared by a professional. There are also disadvantages. One of the advantages is that it saves you time. One of the disadvantages is that it costs you money.
   c. We didn't get the results we anticipated. The program obviously contains an error. Please ask Paul Davis to go through the program.

12. The information contained in the following sentences could be conveyed better in a list. Rewrite each sentence in the form of a list.

   a. The freezer system used now is inefficient in several ways: the chef cannot buy in bulk or take advantage of special sales, there is a high rate of spoilage because the temperature is not uniform, and the staff wastes time buying provisions every day.
   b. The causes of burnout can be studied from three areas: physiological—the roles of sleep, diet, and physical fatigue; psychological—the roles of guilt, fear, jealousy, and frustration; environmental—the roles of the physical surroundings at home and at work.
   c. There are many problems with the on-line registration system currently used at Dickerson. First, lists of closed sections cannot be updated as often as necessary. Second, students who want to register in a closed section must be assigned to a special terminal. Third, the computer staff is not trained to handle the student problems. Fourth, the Computer Center's own terminals cannot be used on the system; therefore, the university has to rent fifteen extra terminals to handle registration.

13. Provide a topic sentence for each of the following paragraphs.

   a. _____
   _____.

   All service centers that provide gas and electric services in the tri-county area must register with the TUC, which is empowered to carry out unannounced inspections periodically. Additionally, all service centers must adhere to the TUC's Fair Deal Regulations, a set of standards that encompasses every phase of the service-center operations. The Fair Deal Regulations are meant to guarantee that all centers adhere to the same standards of prompt, courteous, and safe work at a fair price.

   b. _____
   _____.

   The reason for this difference is that a larger percentage of engineers working in small firms may be expected to hold high-level positions. In firms with fewer then twenty engineers, for example, the median income was $33,200. In firms of twenty to two hundred engineers, the median income was $30,345. For the largest firms, the median was $27,600.

14. Develop the following topic sentences into full paragraphs.

   a. Job candidates should not automatically choose the company that offers the highest salary.
   b. Every college student should learn at least the fundamentals of computer science.
   c. The one college course I most regret not having taken is _____.
   d. Sometimes two instructors offer contradictory advice about how to solve the same kind of problem.

15. In this exercise, several paragraphs have been grouped together into one long paragraph. Recreate the separate paragraphs by marking where the breaks would appear.

BOOKS ON COMPACT DISCS

Compact discs, which have revolutionized the recorded music industry, are about to do the same for the book publishing industry. A number of reference books—such as trade directories and multivolume encyclopedias—are already available in compact disc format. And now trade publishers are working out the legal issues involved in publishing their books in compact disc format. How is the compact disc technology applied to books? The heart of the system is the same as that used for recorded music. Any kind of information that can be digitized—converted to the numbers 0 and 1—can be transferred to the 4.7 in. diameter compact discs. Words and pictures, of course, are digitized in the common personal computer. Instead of outputting the digitized information exclusively as sound, the new technology hooks up a compact disc player to a computer. The information stored on the disc is then output as words and pictures on the screen and as sound emitted through a speaker. Compact discs offer several important advantages over traditional delivery systems for printed information. First, compact discs can hold a tremendous amount of information. A 100-volume encyclopedia could fit on a single disc. The space storage advantages are considerable. Second, compact discs offer the accessing ease of an on-line system. If the user wants information on subatomic particles, he or she simply types in the phrase and the system finds every reference to the subject. On request, the citations or even the entries themselves can be printed out on paper. And third, information stored on compact discs can be updated much less expensively than paper information. The subscriber or purchaser simply receives an updated disc periodically.

16. In the following paragraphs, transitional words and phrases have been removed. Add an appropriate transition in each blank space. Where necessary, add punctuation.

a. As you know, the current regulation requires the use of conduit for all cable extending more than 18″ from the cable tray to the piece of equipment. _____ conduit is becoming increasingly expensive: up 17% in the last year alone. _____ we would like to determine whether the NRC would grant us any flexibility in its conduit regulations. Could we _____ run cable without conduit for lengths up to 3′ in low-risk situations such as wall-mounted cable or low-traffic areas? We realize _____ that conduit will always remain necessary in high-risk situations. The cable specifications for the Unit Two report to the NRC are due in less than two months; _____ we would appreciate a quick reply to our request, as this matter will seriously affect our materials budget.

b. Several characteristics of personal computers limit the kind of local area network (LAN) we can use. _____ personal computers are portable. Losing the ability to move the personal computers would be an unacceptable loss. This is especially true in light of our plans to reorganize the office next year. _____ the number of personal computers is increasing and is expected to continue to increase at least for another decade. The LAN must be easily expandable. _____ we do not have the time or expertise to maintain or repair a LAN. Any LAN we lease _____ would have to be virtually maintenance-free. _____ the choice of a LAN will require careful analysis.

17. The following paragraphs are poorly organized and developed. Rewrite the paragraphs, making sure to include in each a clear topic sentence, adequate support, and effective transitions.

a. The cask containing the spent nuclear fuel is qualified to sustain a 30-foot fall to an unyielding base. If the cask is allowed to pierce a floor after a 30-foot fall, we can safely assume that the cask is not qualified to maintain its integrity. This could result in spent fuel being released from the cask, resulting in high radioactive exposure to personnel.

b. The use of a single-phase test source results in partial energization of the polarizing and operating circuits of relay units other than the one under test. It is necessary that these "unfaulted" relay units be disabled to assure that only the unit being tested is providing information regarding operation or nonoperation. The disabling is accomplished by disengaging printed circuit cards from the card sockets in the relay logic unit. The following table lists the cards to be disengaged by card address in the logic unit under test with respect to STF position.

c. The results indicate that the plaques are not chemically homogeneous: the percentage of NILOC varied among the various sections. This fact represents a problem with determinations such as were conducted in this study. Since only a portion of each plaque submitted was analyzed, differences in the amount of materials extracted could be solely a function of the portion of the plaque examined. Another problem in these determinations was the suspended materials present in the samples. It is not known if any of the compounds of interest occluded onto or interacted with the precipitate. For these reasons, the results of the tests should be considered as having a ±10 percent error.

REFERENCES   Battison, R., and D. Goswami. 1981. Clear writing today. *Journal of Business Communication* 18, no. 4: 5–16.

Miller, C., and K. Swift. 1980. *The handbook of nonsexist writing.* New York: Lippincott & Crowell.

Selzer, J. 1981. Readability is a four-letter word. *Journal of Business Communication* 18, no. 4: 23–34.

PART TWO

TECHNIQUES OF TECHNICAL WRITING

CHAPTER
SIX

DEFINITIONS

No technique is more basic to technical writing than that of definition. Because technical writing requires clear, objective communication of factual information, you will often have to define objects, processes, and ideas.

The world of business and industry depends on clear and effective definitions. Without written definitions, the working world would be chaotic. For example, suppose that you learn at a job interview that the potential employer pays tuition and expenses for its employees' job-related education. That's good news, of course, if you are planning to continue your education. But until you study the employee-benefits manual, you will not know with any certainty just what the company will pay for. Who, for instance, is an *employee*? You would think an employee would be anybody who works for and is paid by the company. But you might find that for the purposes of the manual, an employee is someone who has worked for the company in a full-time capacity (35 hours per week) for at least six uninterrupted months. A number of other terms would have to be defined in the description of tuition benefits. What, for example, is *tuition*? Does the company's definition include incidental laboratory or student fees? What is *job-related education*? Does a course on methods of dealing with stress qualify under the company's definition? What, in fact, constitutes *education*? All these terms and many others must be defined for the employees to understand their rights and responsibilities.

Definitions play a major role, therefore, in communicating policies and standards "for the record." When a company wants to purchase air-conditioning equipment, it might require that the supplier provide equipment certified by a professional organization of air-conditioner manufacturers. The organization's definitions of acceptable standards of safety, reliability, and efficiency will provide some assurance that the equipment is of high quality.

Definitions, of course, have many uses outside legal or contractual contexts. Two such uses occur very frequently. Definitions can help the writer clarify a description of a new technology or a new development in a technical field. When a new animal species is discovered, for instance, it is named and defined. When a new laboratory procedure is devised, it is defined and then described in an article printed in a technical journal. Definitions can also help a specialist communicate with a less knowledgeable audience. A manual that explains how to tune up a car will include definitions of parts and tools. A researcher at a manufacturing company will use definitions in describing a new product to the sales staff.

Definitions, then, are crucial in all kinds of technical writing, from brief letters and memos to technical reports, manuals, and journal articles. All kinds of readers, from the layperson to the expert, need effective definitions to carry out their jobs every day.

Definitions, like every other technique in technical writing, require thought and planning. Before you can write a definition to include in a document, you must carry out three steps:

1. Analyze the writing situation.
2. Determine the kind of definition that will be appropriate.
3. Decide where to place the definition.

THE WRITING SITUATION

The first step in writing effective definitions is to analyze the writing situation: the audience and purpose of your document.

Unless you know whom you are addressing and how much that audience already knows about the subject, you cannot know which terms need to be defined or the kind of definition to write. A group of physicists wouldn't need a definition of the word *entropy*, but a group of lawyers might. Builders know what a Molly bolt is, but some insurance agents might not. If you are aware of your audience's background and knowledge, you can easily devise effective informal definitions. For example, if you are describing a cassette deck to a group of readers who understand automobiles, you can use a familiar analogy: "The PAUSE button is the brake pedal of the cassette deck."

Keep in mind, too, your purpose in writing. If you want to give your readers only a basic understanding of a concept—say, time-sharing vacation resorts—a brief, informal definition will usually be sufficient. However, if you want your readers to understand an object, process, or concept thoroughly and be able to carry out tasks based on what you have written, then a more formal and elaborate definition is called for. For example, a definition of a "Class 2 Alert" written for operators at a nuclear power plant will have to be comprehensive, specific, and precise.

TYPES OF DEFINITIONS

The preceding analysis of the writing situation has suggested that definitions can be short or long, informal or formal. Three basic types of definitions can be isolated:

1. parenthetical
2. sentence
3. extended

PAREN-THETICAL DEFINITION   A parenthetical definition is a brief clarification placed unobtrusively within a sentence. Sometimes a parenthetical definition is a mere word or phrase:

The crane is located on the starboard (right) side of the ship.

Summit Books announced its desire to create a new colophon (emblem or trademark).

United Engineering is seeking to purchase the equity stock (common stock) of Minnesota Textiles.

A parenthetical definition can also take the form of a longer explanatory phrase or clause:

Motorboating is permitted in the Jamesport Estuary, the portion of the bay that meets the mouth of the Jamesport River.

The divers soon discovered the kentledge, the pig-iron ballast.

Before the metal is plated, it is immersed in the pickle, the acid bath that removes scales and oxides from the surface.

Parenthetical definitions are not, of course, authoritative. They serve mainly as quick and convenient ways of introducing new terms to the readers. Because parenthetical definition is particularly common in writing addressed to general readers, make sure that the definition itself is clear. You have gained nothing if your readers don't understand your clarification:

Next, check for blight on the epicotyl, the stem portion above the cotyledons.

If your readers are botanists, this parenthetical definition will be clear (although it might be unnecessary, for if they know the meaning of *cotyledons*, they are likely to know *epicotyl*). However, if you are addressing the general reader, this definition will merely be frustrating.

SENTENCE DEFINITION

A sentence definition is a one- or two-sentence clarification. It is more formal than the parenthetical definition. Usually, the sentence definition follows a standard pattern: the item to be defined (the *species*) is placed in a category of similar items (the *genus*) and then distinguished from the other items (by the *differentia*):

SPECIES	=	GENUS	+	DIFFERENTIA
A flip flop	is	a circuit		containing active elements that can assume either one of two stable states at any given time.
An electrophorus	is	an instrument		used to generate static electricity.
Hypnoanalysis	is	a psychoanalytical technique		in which hypnosis is used to elicit unconscious information from a patient.
A Bunsen burner	is	a small laboratory burner		consisting of a vertical metal tube connected to a gas source.
An electron microscope	is	a microscope		that uses electrons rather than visible light to produce magnified images.

Sentence definitions are useful when your readers require a more formal or more informative clarification than parenthetical definition can provide. Sentence definitions are often used to establish a working definition for a particular document: "In this report, the term *electron microscope* will be used to refer to any microscope that uses electrons rather than visible light to produce magnified images."

In sentence definitions, keep several points in mind.

First, be as specific as you can in writing the *genus* and the *differentia*. Remember, your purpose is not merely to provide some information about the item you are defining; you are trying to distinguish it from all other similar items. If you write, "A Bunsen burner is a burner that consists of a vertical metal tube connected to a gas source," the imprecise *genus*—"a burner"—defeats the purpose of your definition: there are many types of large-scale burners that use vertical metal tubes connected to gas sources. If you write, "Hypnoanalysis is a psychoanalytical technique used to elicit unconscious information from a patient," the imprecise *differentia*—"used to elicit . . ."—ruins the definition: there are many psychoanalytical techniques used to elicit a patient's unconscious information. If more than one *species* is described by your definition, you have to sharpen either the *genus* or the *differentia*, or both.

Second, avoid writing circular definitions: that is, definitions that merely repeat the key words of the *species* (the ones being defined) in the *genus* or *differentia*. In "A required course is a course that is required," what does "required" mean? Required of whom, by whom? The word is never defined.

Similarly, "A balloon mortgage is a mortgage that balloons" is useless. However, you can use *some* kinds of words in the *species* as well as in the *genus* and *differentia*; in the definition of *electron microscope* given earlier, for example, the word *microscope* is repeated. Here, *microscope* is not the "difficult" part of the species; readers know what a microscope is. The purpose of defining *electron microscope* is to clarify the *electron* part of the term.

Third, be sure the *genus* contains a noun or a noun phrase rather than a phrase beginning with *when, what,* or *where.*

INCORRECT

A brazier is what is used to . . .

CORRECT

A brazier is a metal pan used to . . .

INCORRECT

An electron microscope is when a microscope . . .

CORRECT

An electron microscope is a microscope that . . .

INCORRECT

Hypnoanalysis is where hypnosis is used . . .

CORRECT

Hypnoanalysis is a psychoanalytical technique in which . . .

EXTENDED DEFINITION An extended definition is a long (one- or several-paragraph), detailed clarification of an object, process, or idea. Often an extended definition begins with a sentence definition, which is then elaborated. For instance, the sentence definition "An electrophorus is an instrument used to generate static electricity" tells you the basic function of the device, but it leaves many questions unanswered: How does it work? What does it look like? An extended definition would answer these and other questions.

Extended definitions are useful, naturally, when you want to give your readers a reasonably complete understanding of the item. And the more complicated or more abstract the item being defined, the greater the need for an extended definition.

There is no one way to "extend" a definition. Your analysis of the audience and purpose of the communication will help to indicate which method to use. In fact, an extended definition will sometimes employ several different methods. Often, however, one of the following techniques will work effectively:

1. exemplification
2. analysis
3. principle of operation
4. comparison and contrast
5. negation
6. etymology

Exemplification, using an example (or examples) to clarify an object or idea, is particularly useful in making an abstract term easy to understand. The following paragraph is an extended definition of the psychological defense mechanism called *conversion* (Wilson 1964: 84).

A third mechanism of psychological defense, "conversion," is found in hysteria. Here the conflict is converted into the symptom of a physical illness. In a case of conversion made famous by Freud, a young woman went out for a long walk with her brother-in-law, with whom she had fallen in love. Later, on learning that her sister lay gravely ill, she hurried to her bedside. She arrived too late and her sister was dead. The young woman's grief was accompanied by sharp pain in her legs. The pain kept recurring without any apparent physical cause. Freud's explanation was that she felt guilty because she desired the husband for herself, and unconsciously converted her repressed feelings into an imaginary physical ailment. The pain struck her in the legs because she unconsciously connected her feelings for the husband with the walk they had taken together. The ailment symbolically represented both the unconscious wish and a penance for the feelings of guilt which it engendered.

Notice that the first two sentences in this paragraph are essentially a sentence definition that might be paraphrased as follows: "Conversion is a mechanism of psychological defense by which the conflict is converted into the symptoms of a physical illness."

This extended definition is effective because the writer has chosen a clear and interesting (and therefore memorable) example of the subject he is describing. No other examples are necessary in this case. If conversion were a more difficult concept to describe, an additional example might be useful.

Analysis is the process of dividing a thing or idea into smaller parts so that the reader can more easily understand it.

A load-distributing hitch is a trailer hitch that is designed to distribute the hitch load to all axles of the tow vehicle and the trailer. The crucial component of the load-distributing hitch is the set of spring bars that attaches to the trailer. For a complete understanding of the load-distributing hitch, however, the following other components should be explained first:

1. the shank
2. the ball mount
3. the sway control
4. the frame bracket

The shank is the metal bar that is attached to the frame of the tow vehicle . . .

This extended definition of a load-distributing hitch uses analysis as its method of development. The hitch is divided into its major components, each of which is then defined and described.

The *principle of operation* is an effective way to develop an extended definition, especially that of an object or process. The following extended definition of a thermal jet engine is based on the mechanism's principle of operation.

A thermal jet engine is a jet-propulsion device that uses air, along with the combustion of a fuel, to produce the propulsion. In operation, the thermal jet engine draws in air, increases its pressure, heats it by combustion of the fuel, and finally ejects the heated air (and the combustion gases). The increased velocity of the ejected mixture determines the thrust: the greater the difference in velocity between air entering and leaving the unit, the greater the thrust.

Note that this extended definition begins with a sentence definition.

Comparison and contrast is another useful technique for developing an extended definition. With this technique, the writer discusses similarities or differences between the item being defined and an item with which the readers are more familiar. The following definition of a bit brace begins by comparing and contrasting it to the more common power drill.

A bit brace is a manual tool used to drill holes. Cranked by hand, it can theoretically turn a bit to bore a hole in any material that a power drill can bore. Like a power drill, a bit brace can accept any number of different sizes and shapes of bits. The principal differences between a bit brace and a power drill are:

1. A bit brace drills much more slowly.
2. A bit brace is a manual tool, and so it can be used where no electricity is available.
3. A bit brace makes almost no noise in use.

The bit brace consists of the following parts. . . .

A special kind of contrast is sometimes called *negation* or *negative statement*. Negation is the technique of clarifying a term by distinguishing it from a different term with which the reader might have confused it.

An ambulatory patient is *not* a patient who must be moved by ambulance. On the contrary, an ambulatory patient is one who can walk without assistance from another person. . . .

Negation is rarely the only technique used in an extended definition. In fact, negation is used most often in a sentence or two at the start of a definition: for after you state what the item is *not*, you still have to define what it *is*.

Etymology, the history of a word, is often a useful and interesting way to develop a definition.

No-fault auto insurance, as the name suggests, is a type of insurance that ignores who or what caused the accident. When the accident damage is less than a specific amount set by the state, the people involved in the accident may not use legal means to determine who was at fault. What this means to the driver is that . . .

The word *mortgage* was originally a compound of *mort* (dead) and *gage* (pledge). The meaning of the word has not changed substantially since its origin in Old French. A mortgage is still a pledge that is "dead" upon either the payment of the loan or the forfeiture of the collateral and payment from the proceeds of its sale.

Etymology, like negation, is rarely used alone in technical writing, but it is an effective way to introduce an extended definition.

The following extended definition of *nowcasting*, a weatherforecasting technique, shows some of the common ways to elaborate a sentence definition. The definition is directed to a general audience.

Nowcasting is a new word used to describe a short-term weather forecasting technique: describing the current weather and forecasting over the next 2 to 12 hours for a limited geographical area. Nowcasting relies on the most modern data-processing technology: computer systems that let the user manipulate and interpret tremendous amounts of data almost instantaneously.

Traditional weather forecasting is more accurate for periods of 12 to 36 hours in the future than it is for the 2-to-12-hour period. The reason for this is that traditional forecasting relies on data in the atmosphere. Temperature, humidity, and wind are measured by satellites and by radiosondes carried by balloons. These data are then fed into mathematical models that measure motion, mass, and thermodynamics. The resulting forecasts are reasonably accurate, but they cannot focus on a limited geographic area.

By contrast, nowcasting supplements satellite data with numerous ground-based readings of clouds, temperature, and humidity. The advantage of using ground-based readings is that the nowcaster can make predictions on effects caused by the peculiarities of the local terrain. For example, the nowcaster can predict squall lines, fronts, mountain-valley wind patterns, and land-sea breezes that are too small and local to be measured by the satellites or upper-atmosphere radiosondes. The resulting data are highly accurate for a limited geographic area—but only for a few hours.

The first paragraph contains what is essentially a sentence definition: "Nowcasting is a weather forecasting technique that describes the current weather and forecasts the weather for the next 2 to 12 hours for a limited geographical area." Also notice that the first sentence uses etymology in referring to the origin of the term *nowcasting*.

The two paragraphs that follow compare and contrast forecasting and nowcasting. Each paragraph describes the principle of operation of the technique, and then comments on its effectiveness.

PLACEMENT OF DEFINITIONS

In many cases, the writer does not need to decide where to place a definition. In writing your first draft, for instance, you may realize that most of your readers will not be familiar with a term you want to use. If you can easily provide a parenthetical definition that will satisfy your readers' needs, simply do so.

Often, however, in assessing the writing situation before beginning the draft, you will conclude that one or more complicated terms will have to be introduced, and that your readers will need more detailed and comprehensive clarifications, perhaps sentence definitions and extended definitions. In these cases, you should plan—at least tentatively—where you are going to place them.

Definitions can be placed in four different locations:

1. in the text
2. in footnotes
3. in a glossary
4. in an appendix

The text itself can accommodate any of the three kinds of definitions. Parenthetical definitions, because they are brief and unobtrusive, are almost always included in the text; even if a reader already knows what the term means, the slight interruption will not be annoying. Sentence definitions are often placed within the text. If you want all your readers to see your definition or you suspect that many of them need the clarification, the text is the appropriate location. Keep in mind, however, that unnecessary sentence definitions can easily become bothersome to your readers. Extended definitions are rarely placed within the text, because of their length. The obvious exception, of course, is the extended definition of a term that is central to the discussion; a discussion of recent changes in workmen's compensation insurance will likely begin with an extended definition of that kind of insurance.

Footnotes are a logical location for an occasional sentence definition or extended definition. The reader who needs the definition can find it easily at the bottom of the page; the reader who doesn't need it is not interrupted. Footnotes are difficult to type, however, and they make the page look choppy. (Some word-processing software can set up footnotes automatically.) If you are going to need more than one footnote for every two or three pages, consider creating a glossary.

A glossary, which is simply an alphabetized list of definitions, can handle sentence definitions and extended definitions of fewer than three or four paragraphs. A glossary can be placed at the start of a document (such as after the executive summary in a report) or at the end, preceding the appendixes. The principal advantage of a glossary is that it provides a convenient collection of definitions that otherwise might clutter up the document. For more information on how to set up a glossary, see Chapter 11.

An appendix is an appropriate place to put an extended definition, one of a page or longer. A definition of this length would be cumbersome in a glossary or in a footnote, and—unless it explains a crucial term—too distracting in the text. Because the definition is an appendix in the document, it will be listed in the table of contents. In addition, it can be referred to—with the appropriate page number—in the text.

WRITER'S CHECKLIST

This checklist covers parenthetical, sentence, and extended definitions.

1. Are all terms that need definitions defined?

2. Are the parenthetical definitions
    a. Appropriate for the audience?
    b. Clear?

3. Does each sentence definition
    a. Contain a sufficiently specific *genus* and *differentia*?
    b. Avoid circular definition?
    c. Contain a noun in the genus?

4. Are the extended definitions developed logically and clearly?

5. Are the definitions placed in the most useful location for the readers?

EXERCISES

1. Add parenthetical definitions of the italicized terms in the following sentences.
    a. Reluctantly, he decided to *drop* the physics course.
    b. Last week the computer was *down*.
    c. The culture was studied *in vitro*.

    d. The tire plant's managers hope they do not have to *lay off* any more employees.

    e. The word processor comes complete with a *printer*.

2. Write a sentence definition for each of the following terms:

    a. a catalyst
    b. a digital watch
    c. a job interview
    d. a cassette deck
    e. flextime

3. Write an extended definition for each of the following terms:

    a. flextime
    b. binding arbitration
    c. energy
    d. an academic major (don't focus on any particular major; define what a major is)
    e. quality control

4. Revise any of the following sentence definitions that need revision.

    a. Dropping a course is when you leave the class.
    b. A thermometer measures temperature.
    c. The spark plugs are the things that ignite the air-gas mixture in a cylinder.
    d. Double parking is where you park next to another car.
    e. A strike is when the employees stop working.

5. Identify the techniques used to create the following extended definition:

    Holography, from the Greek *holos* (entire) and *gram* (message), is a method of photography that produces images that appear to be three dimensional. A holographic image seems to change as the viewer moves in relation to it. For example, as the viewer moves, one object on the image appears to move in front of another object. In addition, the distances between objects on the image also seem to change.

    Holographs are produced by coherent light, that is, light of the same wavelength, with the waves in phase and of the same amplitude. This light is produced by laser. Stereoscopic images, on the other hand, are created by incoherent light—random wavelengths and amplitudes, out of phase. The incoherent light, which is natural light, is focused by a lens and records the pattern of brightness and color differences of the object being imaged.

    How are holographic images created? The laser-produced light is divided as it passes through a beam splitter. One portion of the light, called the reference beam, is directed to the emulsion—the "film." The other portion, the object beam, is directed to the subject and then reflected back to the emulsion. The reference beam is coherent light, whereas the object beam becomes incoherent because it is reflected off the irregular surface of the subject. The resulting dissonance between the reference beam and the object beam is encoded; it records not only the brightness of the different parts of the subject but also the different distances from the laser. This encoding creates the three-dimensional effect of holography.

REFERENCE   Wilson, J. R. 1964. *The mind*. New York: Time, Inc.

CHAPTER SEVEN

ANALYSIS

Different writing situations demand different strategies. Often you will have to describe for your readers two pieces of equipment, both of which are designed to satisfy a need within your organization. Comparing and contrasting them would be a logical way to structure your discussion. Or you might want to create categories, to help your readers make sense of a large number of items. In describing photocopy machines, for example, you might want to classify them by size (desk-top models and floor models) or by features (those that collate and those that do not). Often in technical writing you will have to describe causality—what caused some occurrence, or what will be the effect of some occurrence. For instance, you might have to predict how your company's business will be affected by the opening of a competing retail store nearby. Or you might be asked to investigate what caused a recent decline in sales of one of your company's products.

The term *analysis*, as used in this chapter, refers to these three basic techniques of writing:

1. comparison and contrast
2. classification and partition
3. cause and effect

COMPARISON AND CONTRAST

Much technical writing is structured according to the technique of comparison and contrast—the same technique by which most of us learned as children. What is a zebra? A zebra looks like a horse (comparison), but it has stripes (contrast). In technical writing, comparison-and-contrast discussions are developed much more fully, of course, but the technique is essentially the same.

Comparison and contrast is a common writing technique in the working world because organizations constantly make decisions. Should we purchase the model A computer, or the B or the C? Should we hire person X or person Y? Should we carry out a comprehensive analysis of this potential site for a new plant, or would a more limited analysis be preferable? Probably the most common kind of report in technical writing is a feasibility report, in which the writer investigates a course of action and explains whether it is possible and practical. Almost always, this kind of report requires a substantial comparison-and-contrast section.

The first step in comparing and contrasting two or more items is to establish a basis (or several bases) on which to evaluate the items. If you have only one basis, comparing and contrasting the items is a simple task. Next semester you might need to take a course that is offered at ten o'clock on Mondays, Wednesdays, and Fridays. Your one basis of comparison and contrast is the time at which it is offered—nothing else matters.

Almost always, however, you will have several bases (or criteria) by which to carry out your comparison and contrast. For example, you might need a three-credit science course that meets at ten o'clock on Mondays, Wednesdays, and Fridays. In this case, you have three criteria: number of credits, general subject area, and schedule. When you are comparing and contrasting items for a study on the job, you are likely to have six, eight, or even more criteria. For a recommendation report on which kind of computer your company should buy, for example, your comparison and contrast would probably include all the following criteria:

1. availability of software
2. amount and type of storage capacity
3. ease of operation
4. reliability
5. availability of peripherals
6. accessibility of maintenance and service personnel
7. initial costs and maintenance costs

The more criteria you use in your comparison and contrast, the more precise your analysis will be. If you want to buy a used car but your only criterion is price, you don't have to ponder whether to buy the $1,000 car or the $2,000 car. However, when you add more and more criteria—such as age, reliability, and features— you will probably conclude that the more expensive car is "better"—according to the analysis.

Choose sufficient and appropriate criteria when you compare and contrast items. Don't overlook an important criterion. A beautifully written comparison and contrast of office computers is of little value if you have neglected price. If your company can spend no more than $35,000 and you carry out a detailed comparison and contrast of models that cost more than $50,000, you will have wasted everyone's time. Be careful, too, that the criteria you choose are sensible. If, for instance, your company has a large, empty storeroom whose temperature is just right for the computer, don't use size as one of your criteria: there is no reason to spend more money for a compact unit if a larger model would be perfectly acceptable.

Once you have chosen your criteria and carried out your research, choose a structure for writing the comparison and contrast. The two structures from which to choose are called *whole-by-whole* and *part-by-part*.

In the whole-by-whole structure, the first item is discussed, then the second, and so on:

Item 1
   Aspect A
   Aspect B
   Aspect C
Item 2
   Aspect A
   Aspect B
   Aspect C
Item 3
   Aspect A
   Aspect B
   Aspect C

The whole-by-whole structure is easy to read and understand because it focuses on each item as a separate entity. Therefore, the whole-by-whole structure is effective if you want to focus on each item as a complete unit rather than on individual aspects of different items. For example, you are writing a feasibility report on purchasing an expensive piece of laboratory equipment. You have narrowed your choice to three different models, each of which fulfills your basic criteria. Each model has its advantages and disadvantages. Because your organization's decision on which model to buy will depend on an overall assessment rather than on any single aspect of the three models, you choose the whole-by-whole structure. This pattern gives your readers a good overview of the different models.

If you are using the more-important-to-less-important pattern (see Chapter 3), you will discuss the recommended model first. If you are using the problem-methods-solution pattern, you will discuss the recommended model last.

In the part-by-part structure, each aspect is discussed separately:

Aspect A
   Item 1
   Item 2
   Item 3
Aspect B
   Item 1
   Item 2
   Item 3
Aspect C
   Item 1
   Item 2
   Item 3

The part-by-part structure lets you focus on an individual aspect of the different items. Detailed comparisons and contrasts are more effective in the part-by-part structure. For example, you are comparing and contrasting the three pieces of laboratory equipment. The one factor that distinguishes the three models is reliability: one model has a much better reliability record than the other two. The part-by-part structure lets you create a section on reliability to highlight this aspect. Comparing and contrasting the three models in one place in the document rather than in three places makes your point more emphatic. You sacrifice a coherent overview of the different items for a clear, forceful comparison and contrast of an aspect of the three models.

As with the whole-by-whole pattern, your overall pattern of development—such as more-important-to-less-important or problem-methods-solution—will determine which aspect you treat first.

The following comparison of passive and active solar-energy systems, addressed to a general audience, employs the whole-by-whole structure.

A passive solar-energy system is one that uses the greenhouse effect: the sun shines through the windows of the house, and the heat is retained in the house. A new house designed for a passive system is generally situated so that its long dimension is on the east-west plane. Most of its glass is on the southern exposure; the north has few windows. Often, a greenhouse is added to the southern exposure.

An active solar-energy system consists of solar collectors mounted on a wall or the roof and a delivery system to carry the heat to the different rooms of the house or to a storage system. An active system can be incorporated easily into most new or existing houses, because almost every house has an appropriate site for a collector.

Notice in this example that each paragraph has a two-part structure. First, each type of solar energy system is described. Next, the siting requirements are discussed. Of course, the whole-by-whole pattern can be expanded beyond a few paragraphs. A detailed discussion of active and passive solar-energy systems, for instance, might require several chapters of a book.

The following two paragraphs, taken from a longer discussion of the two beetles that transmit Dutch elm disease in the United States (Strel and Lanier 1981), exemplify a comparison-and-contrast passage that uses the part-by-part structure:

European and native bark beetles are similar in size (between two and 3.5 millimeters in length) but dissimilar in appearance. The European species is shiny and distinctly two-toned, the wing covers are dark reddish brown and the thorax is black. The native beetle is a drab grayish brown and has a rough texture. The immature stages of the two kinds of beetle can easily be distinguished by the gallery systems in which they are found. The female European beetle bores its egg gallery along the grain of the wood, and the tunnels made by the larvae radiate from the egg gallery, resulting in an elliptical pattern. The egg gallery of the native beetle crosses the grain, and the larvae tend to bore along the grain, creating a butterfly-shaped pattern.

The native elm bark beetle inhabits the entire natural range of elms in North America, but it is apparently absent in the Rocky Mountain states and westward, where elms were introduced in comparatively recent times. The European beetle is now found wherever elms grow in North America except for some northern areas where its over-wintering larvae are destroyed by temperatures lower than about minus 25 degrees Fahrenheit.

In the first paragraph, the two beetles are compared and contrasted according to the criteria of size and appearance and then of gallery systems. In the second, the beetles are compared and contrasted according to the criterion of location of habitation.

CLASSIFICATION AND PARTITION

CLASSIFICA-
TION

Classification (sometimes called classification and division) is the process of establishing categories, of grouping items that share certain characteristics. Classification also works in reverse: sometimes you will wish to break categories down into smaller units. Consider, for example, a computer science major whose courses for the semester include operating systems, file structures, discrete structures, American literature since 1860, and physical education. These courses all belong to the general class *college courses* and to the more specific class *courses that this student is taking this semester*. However, you might wish to divide this more specific class into two smaller classes, such as *major requirements* and *electives*. The first subclass would include operating systems, file structures, and discrete structures; the second would include American literature since 1860 and physical education. Accordingly, classification can cover a range of divisions. The process involves both the creation of larger categories so that similar items can be understood as a group and the division of larger groups into individual items. Although these exercises would appear to be opposites, they are in fact the same, as they explore the same relations between items and groups of items.

You classify when you define; you also classify when you outline. When you select the information from your brainstorming list that is relevant to your topic, you are creating a class—relevant information—to which the selected items belong. Looking at the process from another angle, you are dividing the bulk of your information into two classes, relevant and irrelevant information. You then subclassify the relevant information. In this way you can organize an understandable discussion, regardless of whether you are producing a report, a memo, a letter, or any other kind of technical document.

PARTITION

Partition, the process of separating a unit into its components, shares some ground with classification and follows many of the same rules. However, partition is best thought of as an independent technique that is used frequently in technical writing. For example, if you wished to discuss a stereo system, you might partition it into the following components: amplifier, tuner, turntable, and speakers. Each component is separate; the only relation among them is that together they form a stereo system. Similarly, you cannot have a complete stereo system that lacks any of these components. Each component can of course be partitioned further. Partition underlies descriptions of objects, mechanisms, and processes (see Chapter 8).

SOME GUIDE-
LINES FOR
CLASSIFICA-
TION AND
PARTITION

The most important concept to keep in mind when you are classifying or partitioning is that you are establishing a variety of levels. As you know from outlining, you must avoid confusing major and minor categories (faulty coordination). An item that is out of place can destroy any system that you employ. Consider the following attempt to classify keyboard instruments:

pianos

organs

harpsichords

synthesizers

concert grand pianos

Clearly, "concert grand pianos" is a subcategory of pianos and does not belong in this system of classification. The following partition of a small sailboat is similarly faulty:

hull

sails

rudder

centerboard

mast

mainsail

jib

"Mainsail" and "jib" are subcategories of sails and represent a level of partition beyond that defined by the rest of this list.

The following guidelines should help you prevent this error:

1. Use only one basis at a time.
2. Choose a basis consistent with your audience and purpose.
3. Avoid overlap.
4. Be inclusive.
5. Arrange the categories in a logical sequence.

1. *Use only one basis of classification or partition at a time.* If you are classifying fans according to movement, do not include another basis of classification:

oscillating

stationary

inexpensive

"Inexpensive" is inappropriate here because it has a different basis of classification—price—from the other categories. Also, do not mix classification and partition. If you are partitioning a fan into its major components, do not include a stray basis of classification:

motor

blade

casing

metal parts

"Metal parts" is inappropriate here because motor, blade, and casing are partitions based on function; "metal parts," as opposed to "plastic parts," might be a useful classification, but it does not belong in this partition.

2. *Choose a basis of classification or partition that is consistent with your audience and purpose.* Never lose sight of your audience—their backgrounds, skills, and reasons for reading—and your purpose. If you were writing a brochure for hikers in a particular state park, you would probably classify the local snakes according to whether they are poisonous or nonpoisonous, not according to size, color, or any other basis. If you were writing a manual describing do-it-yourself maintenance for a particular car, you would most likely partition the car into its component parts so that the owner could locate specific discussions.

3. *Avoid overlap.* Make sure that no single item could logically be placed in more than one category of your classification or, in partition, that no listed component includes another listed component. Overlapping generally results from changing the basis of classification or the level at which you are partitioning a unit. In the following classification of bicycles, for instance, the writer introduces a new basis of classification that results in overlapping categories:

racing bikes

touring bikes

ten-speed bikes

The first two categories here have use as their basis; the third category is based on number of speeds. The third category is illogical, because a particular ten-speed bike could be a touring bike or a racing bike. In the following partition of a guitar, the writer changes the level of his focus:

body

bridge

neck

frets

The first three items here are basic parts of a guitar. "Frets," as part of the neck, represents a further level of partition.

4. *Be inclusive.* When classifying or partitioning, be sure to include all the categories. For example, a classification of music according to type would be incomplete if it included popular and classical music but not jazz. A partition of an automobile by major systems would be incomplete if it included the electrical, fuel, and drive systems but not the cooling system.

If your purpose or audience requires that you omit a category, tell your readers what you are doing. If, for instance, you are writing in a classification passage about recent sales statistics of General Motors, Ford, and Chrysler, don't use "American car manufacturers" as a classifying tag for the three companies. Although all three belong to the larger class "American car manufacturers," there are a number of other American car manufacturers. Rather, classify General Motors, Ford, and Chrysler as "the big three" or refer to them by name.

5. *Arrange the categories in a logical sequence.* After establishing your categories and subcategories of classification or partition, arrange them according to some reasonable plan: time (first to last), space (top to bottom), importance (most to least), and so on. For example, a classification of record turntables might begin with the simplest kind and proceed to the most sophisticated. See Chapter 3 for a discussion of patterns of development.

Following are several examples of classification and partition.

The first example is taken from a description of background-sound systems used in industry.

The various background-sound systems available for industrial use can be classified according to whether the client has any control over the content of the programs. Piped-in phone line systems and FM radio systems are examples of those services that are not controlled by the client. In the piped-in phone line system, a special phone hook-up brings in the music, which is programmed entirely by the vendor. The FM radio system uses a special receiver to edit out news and other nonmusic broadcasts. Sound systems that can be controlled by the client include on-the-premise systems and customized external systems. An on-the-premise system is any arrangement in which a person or persons at the job site selects and broadcasts recorded music. A customized external system lets the client choose from a number of different programs or, for an extra fee, stipulate what the vendor is to broadcast.

This classification is based on who controls the music being broadcast. The various music systems could easily be classified according to other bases: for example, according to the source of the music (FM radio and nonradio); or according to cost (expensive and inexpensive systems); or according to where the music physically originates (on the job site or off).

The next example, a classification of nondestructive testing techniques, shows how the writer uses classification and subclassification effectively. She is introducing nondestructive testing to a technical audience.

Nondestructive testing of structures permits early detection of stresses that can cause fatigue and ultimately structural damage. The least sensitive tests isolate macrocracks. More sensitive tests identify microcracks. The most sensitive tests identify slight stresses. All sensitivities of testing are useful because some structures can tolerate large amounts of stress— or even cracks—before their structural integrity is threatened.

Currently there are four techniques for nondestructive testing:

1. body-wave reflection
2. surface-wave reflection
3. ultrasonic attenuation
4. acoustic emission

BODY-WAVE REFLECTION   In this technique, a transducer sends an ultrasonic pulse through the test material. When the pulse strikes a crack, part of the pulse's energy is reflected back to the transducer. Body-wave reflection cannot isolate stresses; the pulse is sensitive only to relatively large cracks.

SURFACE-WAVE REFLECTION   The transducer generates an ultrasonic pulse that travels along the surface of the test material. Cracks reflect a portion of the pulse's energy back to the transducer. Like body-wave reflection, surface-wave reflection picks up only macrocracks. Because cracks often begin on interior surfaces of materials, surface-wave reflection is a poor predictor of serious failures.

ULTRASONIC ATTENUATION   The transducer generates an ultrasonic pulse either through or along the surface of the test material. When the pulse strikes cracks or the slight plastic deformations associated with stress, part of the pulse's energy is scattered. Thus, the amount of the pulse's energy decreases. Ultrasonic attenuation is a highly sensitive method of nondestructive acoustic testing.

Two methods of ultrasonic attenuation exist. One technique reflects the pulse back to the transducer. In the other technique, a second transducer can be set up to receive the pulses sent through or along the surface of the material.

ACOUSTIC EMISSION   When a test specimen is subjected to a great amount of stress, it begins to emit waves, some of which are in the ultrasonic range. A transducer attached to the surface of the test specimen records these waves. Current technologies make it possible to interpret these waves accurately for impending fatigue and cracks.

Notice that the writer has classified nondestructive testing by technology. (One method, ultrasonic attenuation, is subclassified.) Another basis for classification would be the sensitivity of the techniques: very sensitive and less sensitive.

Following is an example of partition.

Preventing computer abuse requires a three-pronged attack. First, issue a policy statement that defines specifically what can and cannot be done with the computer. This in itself won't prevent most computer abuse, which is deliberate and planned, but it will at least establish guidelines under which abusers who are caught can be prosecuted. Second, investigate potential employees carefully. A middle manager requires a more substantial investigation than a keypunch clerk, of course. A number of firms specialize in investigating departmental personnel; their fees range from $25 to $800, depending on the detail required. Third, form an information security department to prevent, detect, and deal with computer abuse. Again, a number of consulting companies specialize in instructing clients how to set up this kind of department.

This paragraph partitions the process of guarding against computer abuse into three phases, thus making the process easier to understand and remember.

For more examples of partition, see Chapter 8, which covers descriptions of objects, mechanisms, and processes.

CAUSE AND EFFECT

Technical writing often involves cause-and-effect reasoning. Sometimes you will reason forward: cause to effect. If we raise the price of a particular product we manufacture (cause), what will happen to our sales (effect)? The government has a regulation that we may not use a particular chemical in our production process (cause); what will we have to do to keep the production process running smoothly (effect)? Sometimes you will reason backward: from effect to cause. Productivity went down by 6 percent in the last quarter (effect); what factors led to this decrease (causes)? The federal government has decided that used-car dealers are not required to tell potential customers about the cars' defects (effect); why did the federal government reach this decision (causes)?

Causality, the use of cause-and-effect reasoning, is therefore the way to answer the following two questions:

What will be the effect(s) of X?

What caused X?

Causality requires an understanding of the two basic forms of reasoning: inductive and deductive.

In inductive reasoning, the writer moves from the specific to the general by adding up a series of facts and coming to a conclusion. If a laboratory rat dies after being injected with a chemical, you might suspect that the chemical caused the death, but you couldn't be sure. But if a significant number of rats—say, 20—all reacted in the same way to exactly the same treatment, you could be reasonably sure that the chemical was responsible. Inductive reasoning is the bedrock of the modern scientific method, the basis of trial-and-error experimentation. However, it can be challenged, and often is.

Generally, challenges to inductive reasoning are aimed not at the validity of the reasoning itself but at its usefulness. For example, defenders of the chemical saccharin assert that in experiments mice were given a quantity of the chemical several hundred times greater than a human could possibly ingest and, therefore, that the experiments did not demonstrate that saccharin is harmful to humans. Sometimes, the very nature of inductive reasoning is challenged. Tobacco lobbyists assert that the scientific community has not "proven" that cigarette smoking is harmful to the smoker. The lobbyists do not dispute the data that link cigarette smoking with cancer, heart disease, and a number of other disorders; they just don't accept the surgeon general's inductive conclusion: that smoking has been found to be hazardous to your health. Inductive reasoning does produce only circumstantial evidence, but arguments that it doesn't "prove" anything are usually considered self-serving.

In deductive reasoning, the writer moves from the general to the specific. The common illustration of deductive reasoning is the syllogism, in which a minor premise is added to a major premise to yield a conclusion:

MAJOR PREMISE

All men are mortal.

MINOR PREMISE

Dworski is a man.

CONCLUSION

Dworski is mortal.

Notice that syllogisms cannot be restructured arbitrarily:

Dworski is mortal.
All men are mortal.
Dworski is a man.

Dworski could be a fish or a carrot. Similarly:

Rain makes the streets wet.
The streets are wet.
It rained.

Maybe a water main burst.

Like inductive reasoning, deductive reasoning can be challenged. In most cases, the challenge is based on a dispute over the validity of the major premise. Those who believe the Bible is a scientifically valid document might create a syllogism such as this:

The Bible is scientifically accurate.
The Bible says that God created all the animals as fixed forms.
God created all the animals as fixed forms.

Obviously, they would find it hard to agree with an evolutionist, who would not accept the Bible as a scientifically accurate document.

The relationship between inductive and deductive reasoning should by now be clear: inductive reasoning creates the major premises used in deductive reasoning. Mischievous people will infer from this relationship that it is impossible ever to know anything—that when we toss a ball into the air, we can't count on its coming down (the results of all those other ball tosses were just co-incidence). If gravity is just a rumor, all deductions based on the existence of gravity are nonsense.

These games can be fun, but they hide the fact that establishing causal links is a serious task that requires great care. When you use inductive logic, you must be sure to abide by the principles of research in your field: if the specifics you collect are inaccurate or insufficient, the conclusion you derive from them might be invalid. And when using deductive logic, you must make sure your major premise is clear and indisputable (at least within your professional community) and your syllogistic reasoning is correct.

When you try to establish causality, be careful to avoid three basic logical errors:

1. inadequate sampling
2. *post hoc, ergo propter hoc*
3. begging the question

1. *Inadequate sampling.* Don't draw conclusions about causality unless you have an adequate sample of evidence. It would be illogical to draw the following conclusion:

The new Gull is an unreliable car. Two of my friends own Gulls, and both have had reliability problems.

The writer would have to study a much larger sample and compare the findings with those for other cars in the Gull's class before he or she could reach any valid conclusions. However, sometimes incomplete data are all you have to work with. It is proper in these cases to qualify your conclusions:

The Martin Company's Collision Avoidance System has performed according to specification in extensive laboratory tests. However, the system cannot be considered effective until it has performed satisfactorily in the field. At this time, Martin's system must be considered very promising.

2. *Post hoc, ergo propter hoc.* This phrase means "After this, therefore, because of this." The fact that A precedes B does not mean that A caused B. The following statement is illogical:

There must be something wrong with the new circuit breaker in the office. Ever since we had it installed, the air conditioners haven't worked right.

The air conditioners' malfunctioning *might* be caused by a problem in the circuit breaker; on the other hand, it might be caused by inadequate wiring or by problems in the air conditioners themselves.

3. *Begging the question.* To beg the question is to assume the truth of what you are trying to prove. The following statement is illogical:

The Matrix 411 printer is the best on the market because it is better than the other printers.

The writer here states that the Matrix 411 is the best because it is the best; he does not show why he thinks it is the best.

The following passage, from a discussion of why mammals survived but dinosaurs did not (Jastrow 1977: 104–105), illustrates a very effective use of cause-and-effect reasoning.

Effect
Cause no. 1

Why were the ancestral mammals brainier than the dinosaurs? Probably because they were the underdogs during the rule of the reptiles, and the pressures under which they lived put a high value on intelligence in the struggle for survival. These little animals must have lived in a state of constant anxiety—keeping out of sight during the day, searching for food at night under difficult conditions, and always outnumbered by their enemies. They were Lilliputians in the land of Brobdingnag. Small and physically vulnerable, they had to live by their wits.

Cause no. 2

The nocturnal habits of the early mammal may have contributed to his relatively large brain size in another way. The ruling reptiles, active during the day, depended mainly on a keen sense of sight; but the mammal, who moved about in the dark much of the time, must have depended as much on the sense of smell, and on hearing as well. Probably the noses and ears of the early mammals were very sensitive, as they are in modern mammals such as the dog. Dogs live in a world of smells and sounds, and, accordingly, a dog's brain has large brain centers devoted solely to the interpretation of these signals. In the early mammals, the parts of the brain concerned with the interpretation of strange smells and sounds also must have been quite large in comparison to their size in the brain of the dinosaur.

Effect
Cause

Intelligence is a more complex trait than muscular strength, or speed, or other purely physical qualities. How does a trait as subtle as this evolve in a group of animals? Probably the increased intelligence of the early mammals evolved in the same way as their coats of fur and other bodily changes. In each generation, the mammals slightly more intelligent than the rest were more likely to survive, while those less intelligent were likely to become the victims of the rapacious dinosaurs. From generation to generation, these circumstances increased the number of the more intelligent and decreased the number of the less intelligent, so that the average intelligence of the entire population steadily improved, and their brains grew in size. Again the changes were imperceptible from one generation to the next, but over the course of many millions of generations the pressures of a hostile world created an alert and relatively large-brained animal.

Notice that in this discussion, which is addressed to a general audience, the author has sequenced the two causal relationships just as he has structured the cause and effect within each relationship. Each relationship begins with a question that states the effect and reasons backward to the cause. Similarly, the first causal relationship—the fact that ancestral mammals were more intelligent than the dinosaurs—is the "effect" of the second causal relationship—that natural selection "caused" the greater intelligence among the mammals.

WRITER'S CHECKLIST
The following checklist covers comparison and contrast, classification and partition, and cause and effect.

COMPARISON AND CONTRAST

1. Does the comparison-and-contrast passage
   a. Include the necessary criteria?
   b. Employ an appropriate structure?

CLASSIFICATION AND PARTITION

1. Does the classification or partition passage
   a. Use only one basis at a time?
   b. Use a basis that is consistent with the audience and purpose?
   c. Avoid overlap?
   d. Include all of the appropriate categories?
   e. Arrange the categories in a logical sequence?

CAUSE AND EFFECT

1. Does the cause-and-effect passage
   a. Arrange the items in a logical sequence?
   b. Avoid inadequate sampling?
   c. Avoid *post hoc, ergo propter hoc* reasoning?
   d. Avoid begging the question?

EXERCISES

1. Write a comparison-and-contrast paper on one of the following topics. Be sure to indicate the audience and purpose for your paper.

    a. lecture classes and recitation classes
    b. the record reviews in *Stereo Review* and *Rolling Stone*
    c. manual transmission and automatic transmission automobiles
    d. black-and-white and color photography

2. Rewrite the following comparison-and-contrast paragraph so that it contains a clear topic sentence.

Both the "Gorilla" and the "Hunter" welding shields can be worn without a hard hat and allow the welder to move freely in a congested area. Neither provides head protection. The "Gorilla" is a leather hood. It is uncomfortable because it fogs up easily and absorbs unpleasant odors. Most welders prefer the "Hunter" because it does not fog up and because, being made of plastic, it does not absorb odors. Inspectors have found that welders wear it even when not required to do so. The result has been that sometimes it is difficult to find them. No problem is found in locating the "Gorilla" shields.

3. Write a classification paper on one of the following topics. Be sure to indicate your audience and purpose.

    a. foreign cars
    b. smoke alarms
    c. college courses
    d. personal computers
    e. insulation used in buildings
    f. cameras

4. Write a partition paper about a piece of equipment or machinery you are familiar with, or about one of the following topics. Be sure to indicate your audience and purpose.

    a. a student organization on your campus
    b. a tape cassette
    c. a portable radio
    d. a bicycle
    e. an electric clock

5. Rewrite the following classification paragraph to improve its structure, development, and writing style.

There are three basic types of tape decks that are available. The first type is the reel to reel. The reel to reel is the best type of tape deck on the market today, simply because they have been designed to be more versatile than the other types. The next type of tape deck is the eight-track tape player. This type of unit uses a cartridge to make tape loading easier. However, this type of unit sacrifices the quality of sound for ease of operation. The next type of tape deck is the cassette. This type of unit is perhaps the least likely choice for hi-fi since it uses a slower tape speed, which in turn sacrifices quality. In both eight-track and cassette tape decks, quality has been sacrificed for ease of operation. However, the amount of sound quality sacrificed will not be able to be detected by the human ear, because of its lack of sensitivity. There-fore, these types of units are competitive on today's market.

6. Write a causality paper—reasoning either forward or backward—on one of the following topics. Be sure to indicate your audience and pur-pose.

   a. the number of women serving in the military
   b. the price of gasoline
   c. the emphasis on achieving high grades in college
   d. the prospects for employment in your field

7. In one paragraph each, describe the logical flaws in the following items.

   a. The election couldn't have been fair—I don't know anyone who voted for the winner.
   b. 1. Endangered animals should be protected.
      2. Alligators are plentiful.
      3. Alligators should not be protected.
   c. Environmentalists ought to be ignored because we don't need any more cranks in this world.
   d. 1. Fish swim.
      2. Bill swims.
      3. Bill is a fish.
   e. Since the introduction of cola drinks at the start of this century, cancer has become the second greatest killer in the United States. Cola drinks should be outlawed.

REFERENCES   Jastrow, R. 1977. *Until the sun dies.* New York: Norton.

Strel, G. A., and G. N. Lanier. 1981. Dutch elm disease. *Scientific Amer-ican* 245, no. 2 (August): 60.

CHAPTER
EIGHT

DESCRIPTIONS

Technical writing is filled with descriptions of objects, mechanisms, and processes. For our purpose here, a *description* is a verbal and visual representation.

What is an object? The range is enormous: from volcanoes and other kinds of natural phenomena to synthetic artifacts such as common hand tools. A tomato plant is an object, as is an automobile tire or a book.

A mechanism could be thought of as a synthetic object consisting of a number of identifiable parts that work together as a system. A compact disc player is a mechanism, as are a voltmeter, a lawn mower, a submarine, and a steel mill. A complex mechanism, such as a submarine, could be thought of as a mechanism because, despite its immense complexity, it can be described as a synthetic object consisting of several different "parts." Each of these parts—such as the heating, ventilating, and air conditioning systems—is of course made up of dozens of different, smaller mechanisms.

A process is an activity that takes place over time: how the Earth was formed, how steel is made, how animals evolve, how plants perform photosynthesis. Descriptions of processes should be distinguished from instructions: process descriptions explain how something happens; instructions explain how to do something.

The readers of a process description want to understand the process, *but they don't want to perform the process.* A set of instructions, on the other hand, is a step-by-step guide intended to enable the readers to perform the process. A process description answers the question "How is wine made?" A set of instructions answers the question "How do I go about making wine?"

Instructions are discussed in Chapter 9.

THE ROLE OF DESCRIPTION IN TECHNICAL WRITING

Descriptions of objects, mechanisms, and processes appear in virtually every kind of technical writing. A company doing a feasibility study on whether to renovate an old factory or build a new one starts by describing the existing facility. That old factory is a complex mechanism. A rational decision can be made only if the company understands it thoroughly.

A writer wants to persuade his readers to authorize the purchase of some equipment; part of his proposal will be a mechanism description. An engineer is trying to describe to her research-and-development department the features she would like incorporated in a piece of equipment that is now being designed; she will include a mechanism description in her report. An engineer is trying to describe to the sales staff how a product works so that they can advertise it effectively; he will include a mechanism description. Mechanism descriptions are used frequently by companies communicating with their clients and customers; operating instructions, for instance, often include mechanism descriptions.

Process descriptions are common in technical writing because often readers who are not directly involved in performing a process nevertheless need to understand it. If, for example, a new law is enacted that places limits on the amount of heated water that a nuclear power plant may discharge into a river, the plant managers have to understand how water is discharged. If the plant is in violation of the law, engineers have to devise alternative solutions—ways to reduce the quantity or temperature of the discharge—and describe them to their supervisors so that a decision can be reached on which alternative to implement. Or consider a staff accountant who performs an audit of one of her company's branches and then writes up a report describing the process and its results. Her readers will review her report to determine whether she performed the audit properly and to learn her findings.

Notice that a description rarely appears as a separate document. Almost always, it is part of a larger report. For example, a maintenance manual for a boiler system might begin with a mechanism description, to help the reader understand how the system operates. However, the ability to write an effective description is so important in technical writing that it is discussed here as a separate technique.

THE WRITING SITUATION

Before you begin to write a description as part of a longer document, consider carefully how the audience and purpose of the document will affect the way you describe the object, mechanism, or process.

What does the audience already know about the general subject? If, for example, you are to describe an electron microscope, you first have to know whether your readers understand what a microscope is. If you want to describe how the use of industrial robots will affect the process of manufacturing cars, you first have to know whether your readers already understand the current process. In addition, you need to know whether they understand the basics of robotics.

The audience will determine your use of technical vocabulary as well as your sentence and paragraph length. Another audience-related factor is the use of graphic aids. Less knowledgeable readers rely on simple illustrations, but they might have trouble understanding sophisticated schematics or decision charts.

What are you trying to accomplish in the description? If you want your readers to understand how a basic computer works, you will want to write a general description: that is, a description that applies to several different varieties of computer. If, however, you want to enable your readers to understand the intricacies of a particular model of computer, you will want to write a particular description. A general description of a typewriter will use classification to point out that there are manual, electric, and electronic typewriters and then go on to a fundamental partition of typical models of each, describing the basic parts. A particular description of a Smith Corona Model 2100 will describe only that one typewriter and most likely include a highly detailed partition of that model. An understanding of your purpose will determine every aspect of the description, including length, amount of detail, and number and type of graphic aids included.

THE STRUCTURE OF OBJECT
AND MECHANISM DESCRIPTIONS

Object and mechanism descriptions have the same basic structure. In the following discussion, the word *item* will refer to both objects and mechanisms.

Most descriptions of items have a three-part structure:

1. a general introduction that tells the reader what the item is and what it does (definition/classification)
2. a part-by-part description of the item (partition)
3. a conclusion that summarizes the description and tells how the parts work together

THE GENERAL   The general introduction provides the basic information that your
INTRODUCTION   readers will need in order to understand the detailed description that follows. In writing the introduction, answer these five questions about the item:

1. What is it?
2. What is its function?
3. What does it look like?
4. How does it work?
5. What are its principal parts?

1. *What is the item?* Generally, the best way to answer this question is to provide a sentence definition (see Chapter 6): "An electron microscope is a microscope that uses electrons rather than visible light to produce magnified images." Or "The Aleutian Islands are a chain of islands curving southwest from the tip of Alaska toward the tip of the Kormandorski Islands of the Soviet Union." Then elaborate if necessary.

2. *What is the function of the item?* State clearly what the item does: "Electron microscopes are used to magnify objects that are smaller than the wavelengths of visible light." Often, the function of a mechanism is implied in its sentence definition: "A hydrometer is a sealed, graduated tube, weighted at one end, that sinks in a fluid to a depth that indicates the specific gravity of the fluid." Of course, some objects have no "function." The Aleutian Islands, as valuable as they might be, have no function in the sense that microscopes do.

3. *What does the item look like?* Include a photograph or drawing if possible (see Chapter 10). If not, use an analogy or a

comparison to an item that would be familiar to your readers: "The cassette that encloses the tape is a plastic shell, about the size of a deck of cards." Mention the material, texture, color, and the like, if relevant.

Sometimes, an object is best pictured with both graphics and words. For example, a map showing the size and location of the Aleutian Islands would be useful. But a verbal picture would be useful, too: "The Aleutian Islands, a rugged string of numerous small islands, show signs of volcanic action. The islands are tree-less and fog enshrouded. The United States military installations on the Aleutians are the main source of economic activity. . . ."

4. *How does the item work?* In a few sentences, define the operating principle of the item:

A tire pressure gauge is essentially a calibrated rod fitted snugly within an open-ended metal cylinder. When the cylinder end is placed on the tire nozzle, the pressure of the air escaping from the tire into the cylinder pushes against the rod. The greater the pressure, the farther the rod is pushed.

Sometimes, objects do not "work"; they merely exist.

5. *What are the principal parts of the item?* Few objects or mechanisms described in technical writing are so simple that you could discuss all the parts. And, in fact, you rarely will need to describe them all. Some parts will be too complicated or too unimportant to be mentioned; others will already be understood by your readers. A description of a bicycle would not mention the dozens of nuts and bolts that hold the mechanism together, nor would it mention the bicycle seat. The focus of the description would be the chain, the pedals, the wheels, and the frame.

Similarly, a description of the Aleutian Islands would mention the four main groups of islands but would not mention all the individual islands, some of which are very small.

At this point in the description, the principal parts should merely be mentioned; the detailed partition comes in the next section. The parts can be named in a sentence or (if there are many parts) in a list. Make sure that you have named the parts in the order in which they will be described in the body of the description.

The introduction is an excellent place to put a graphic aid. Your goal is to orient the readers so that they will be able to follow the part-by-part description easily. Therefore, try to create a graphic aid that complements your listing of major parts. If you partition the Aleutian Islands into four main island groups, your map should identify those same four groups. See Chapter 10 for a discussion of graphic aids.

The information provided in the introduction generally follows this five-part pattern. However, don't feel you have to answer each question in order, or in separate sentences. If your readers' needs or the nature of the item suggests a different sequence—or different questions—adjust the introduction accordingly.

THE PART-
BY-PART
DESCRIPTION

The body of the description—the part-by-part description—is essentially like the introduction in that each major part is treated as if it were a separate item. That is, in writing the body you define what each part is, then, if applicable, describe its function, operating principle, and appearance. The discussion of the appearance should include shape, dimension, material, and physical details such as texture and color (if essential). For some descriptions, other qualities, such as weight or hardness, might also be appropriate. If a part has any important subparts, describe them in the same way. A description of an item therefore resembles a map with a series of detailed insets. A description of a stereo set would include the turntable as one of its parts. The description of the turntable, in turn, would include the tonearm as one of *its* parts. And finally, the description of the tonearm would include the cartridge as one of *its* parts. This process of ever-increasing specificity would continue as required by the complexity of the item and the needs of the readers.

In planning to partition the item into its parts, don't forget to discuss them in a logical sequence. The most common structure reflects the way the item works or is used. In a stereo set, for instance, the "sound" begins at the turntable, travels into the amplifier, and then flows out through the speakers. Another common sequence is based on the physical structure of the item: from top to bottom, outside to inside, and so forth. A third sequence is to move from more-important to less-important parts. (Be careful when you use this sequence: Will your readers understand and agree with your assessment of "importance"?) Most descriptions could be organized in a number of different ways. Just make sure you consciously choose a pattern; otherwise you might puzzle and frustrate your readers.

Graphic aids should be used liberally throughout the part-by-part description. In general, try to create a graphic aid for each major part. Use photographs to show external surfaces, line drawings to emphasize particular items on the surface, and cutaways and exploded diagrams to show details beneath the surface. Other kinds of graphic aids, such as graphs and charts, are often useful supplements to your discussion. If, for instance, you are describing the Aleutian Islands, you might want to create a table that shows total land area and habitable land area of the different island groups.

THE CONCLUSION   Descriptions of items generally do not require elaborate or long conclusions. A brief conclusion is necessary, however, if only to summarize the description and prevent the readers from placing excessive emphasis on the part discussed last in the part-by-part description.

A common technique for accomplishing these goals in descriptions of mechanisms and of some objects is to describe briefly how the different parts interact as the item performs its function. The conclusion of a description of a telephone, for example, might include a paragraph such as the following:

> When the phone is taken off the hook, a current flows through the carbon granules. The intensity of the speaker's voice causes a greater or lesser movement of the diaphragm and thus a greater or lesser intensity in the current flowing through the carbon granules. The phone receiving the call converts the electrical waves back to sound waves by means of an electromagnet and a diaphragm. The varying intensity of the current transmitted by the phone line alters the strength of the current in the electromagnet, which in turn changes the position of the diaphragm. The movement of the diaphragm reproduces the speaker's sound waves.

The following general description of a record turntable is addressed to the lay reader (Isganitis 1977). Marginal comments have been added to the description.

Because the turntable is well known, the writer uses an informal parenthetical definition rather than a formal definition. The definition concentrates on the function rather than the nature of the mechanism.

GENERAL DESCRIPTION OF A RECORD TURNTABLE

INTRODUCTION

The music that comes out of the speakers of a record player is encoded in the grooves of the phonograph record. The first step in getting the music out of the record and into the air is performed by the record turntable, the component that "reads" the grooves of the record and passes the encoded signal on to be amplified and finally transmitted.

Figure 1 shows the four basic components of the record turntable.

The writer describes the operating principle of the mechanism.

The operating principle of the turntable is simple. The motor and the drive system rotate the platter, on which the record rests. As the record rotates, the tonearm picks up the recorded signals and transmits them to the amplifier.

FIGURE 1

Basic Components of a Turntable

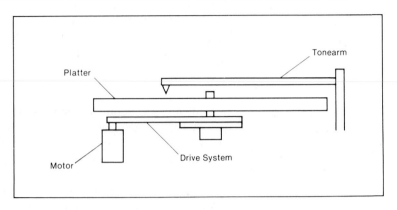

THE COMPONENTS

The writer describes the role of the motor in the operation of the whole mechanism. The mechanism description is general, not particular.

THE MOTOR   The function of the motor is to provide a source of power to turn the platter at an accurate and consistent speed. Speed accuracy is important in preventing "wow and flutter," the short-term fluctuations that give the recorded music a harsh and muddy sound.

Although several different kinds of motors are used in turntables, the most common kind is the AC synchronous motor, which operates at a fixed speed determined by the frequency of the AC voltage from the power line. A good turntable will have a speed accuracy better than ± 0.10 percent.

The description of the drive system focuses on its function rather than its nature. Because the writer is going to describe each of the three types of drive systems, he introduces them in a list.

THE DRIVE SYSTEM   The drive system transmits the power from the motor to the platter. A good drive system will not introduce any mechanical vibrations that cause "rumble," a humming sound.

Three basic drive systems are used in turntables:

1. belt-drive
2. idler-rim drive
3. direct-drive

The writer uses sketches to clarify his descriptions of each of the three types of drive systems.

In the belt-drive system (see Figure 2), a pulley is mounted on the end of the motor shaft. This pulley drives a flexible rubber belt coupled to the underside of the platter. The advantage of the belt-drive system is that the belt acts as a mechanical damper, reducing the rumble and the wow and flutter. However, the belt eventually deteriorates, varying the speed at which the platter turns.

After describing each drive system, the writer briefly points out its advantages and (if applicable) its disadvantages.

In the idler-rim system (see Figure 3), the motor shaft is linked to the platter by a hard rubber wheel. Idler-rim drive is a sturdy system, but because the platter is directly linked to the motor, rumble is relatively high.

In the direct-drive system (see Figure 4), the platter rests directly on the motor shaft. The direct-drive is powered by a DC servocontrolled motor, which can be operated at the exact platter speed required: 33⅓ or 45 RPM. Direct-drive requires no linkage, and thus it produces the least rumble and wow and flutter. Direct-drive is a sturdy and reliable system, but it is expensive.

FIGURE 2
Belt-Drive System

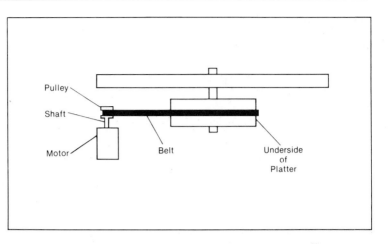

The platter requires only a brief discussion because it is a relatively simple component.

The sentence definition of the tonearm mentions its function.

THE PLATTER   The platter is the machined metallic disc (sometimes covered with a soft material such as felt) on which the record rests while it is being played. There is little variation in the quality of different platters.

THE TONEARM   The tonearm houses the cartridge, the removable case that contains the stylus (the "needle") and the electronic circuitry, which converts the mechanical vibrations of the record grooves into electrical signals. The cartridge is the crucial component of the turntable because it actually touches the grooves of the record, thus affecting not only the sound quality but also the record wear.

The design of every kind of tonearm is unique. However, the operating characteristics of a tonearm can be measured. Four factors are generally considered in describing the performance of a tonearm:

1. tracking force
2. pivot friction

FIGURE 3
Idler-Rim Drive System

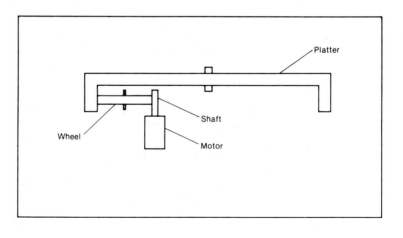

FIGURE 4
*Direct-Drive
System*

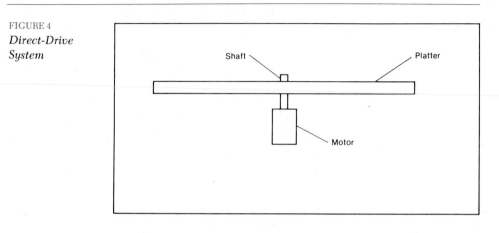

3. tracking error

4. tonearm resonance

"Tracking force" is
defined effectively
in a brief extended
definition.

Tracking force is the force that the stylus and tonearm exert upon the record being played. The amount of tracking force required by a particular cartridge depends on the quality of its stylus construction. High-quality cartridges require low tracking forces from ½ to 1½ grams. A tonearm should therefore be capable of operating at low tracking forces so that high-quality cartridges may be used.

"Pivot friction" is
also defined effec-
tively.

Pivot friction is horizontal and vertical friction in the pivot of the tonearm. Excessive pivot friction requires higher tracking forces to prevent the stylus from jumping the record groove. This in turn increases record and stylus wear.

"Tracking error" is
not defined clearly.
The reader does not
know what tracking
error is until almost
the end of the para-
graph.

Tracking error (or tangent error) results when the cartridge is not held exactly tangent to the record circumference, as shown in Figure 5. The tracking error figure for a tonearm is a measure of the error in the angle of the cartridge with respect to the exact tangent. Depending on the arm geometry, tracking error will vary from point to point along the record radius. When tracking error occurs, the stylus does not ride properly in the record groove, distorting the sound and damaging the record grooves. The maximum value of tracking error for a high-quality tonearm is approximately 5 degrees at the outer edge of a 12-inch record.

"Tonearm reso-
nance" could also be
defined more
clearly.

Tonearm resonance occurs when the tonearm encounters a signal (or vibration) of a frequency equal to its own resonant frequency. At this resonant frequency, the arm undergoes abnormally large vibrations, resulting in mistracking of the record groove by the stylus. Resonance is a common problem in many mechanical devices and is not easily avoided. However, proper design can limit the range of frequencies over which resonance will occur. For tonearms, the ideal range for

FIGURE 5
*Illustration of
Tracking Error
(Top View)*

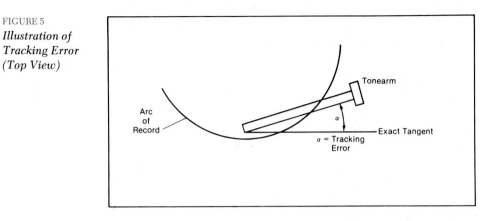

resonant frequencies is between 10 and 15 Hz, because it is below the audible frequency range (20 to 20,000 Hz).

CONCLUSION

Here the writer summarizes his description by showing how the various parts of the turntable work together.

Turntables can cost from about thirty dollars to more than ten times that amount. They range from single-play models, which require that the operator lift the tonearm and place it on the record, to sophisticated systems that the operator can program to search out particular songs without playing the entire side of the record. However, the basic principle of all turntables is the same: a motor produces power that rotates the platter. The cartridge picks up and deciphers the signals embedded in the grooves of the record that is resting on the platter. The signal is then sent along to the other components—the preamplifier, amplifier, and speakers—to produce audible sound.

Following is a description of a modern long-distance running shoe (based on Kyle 1985). Marginal notes have been added.

GENERAL DESCRIPTION OF A LONG-DISTANCE RUNNING SHOE

The writer provides a brief background.

When track and field events became sanctioned sports in the modern world some hundred and fifty years ago, the running shoe was much like any other: a heavy, high-topped leather shoe with a leather or rubber sole. In the last decade, however, advances in technology have combined with increased competition among manufacturers to create long-distance running shoes that fulfill the two goals of all runners: decreased injuries and increased speed.

INTRODUCTION

The writer states the subject and kind of description.

This paper is a generalized description of a modern, high-tech shoe for long-distance running.

The modern distance running shoe has five major components:

The major components are listed.

1. the outsole
2. the heel wedge
3. the midsole
4. the insole
5. the shell

The writer refers to the graphic aid.

Figure 1 is an exploded diagram of the shoe.

The writer explains the organization of the description.

THE COMPONENTS

The five principal components of the shoe will be discussed from bottom to top.

The writer's approach is to provide a brief physical description of the component followed by its function.

THE OUTSOLE   The outsole is made of a lightweight, rubberlike synthetic material. Its principal function is to absorb the runner's energy safely as the foot lands on the surface.

Here the writer sketches the complete shoe, then shows its components in an exploded diagram. (See Chapter 10 for a discussion of graphic aids.)

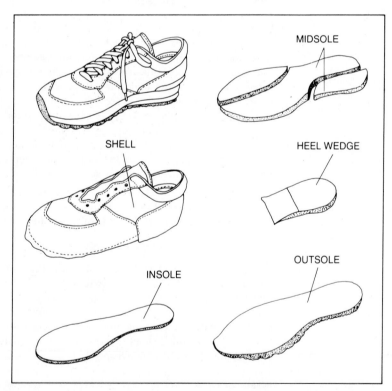

FIGURE 1
Exploded Diagram of a Long-Distance Running Shoe

The writer provides brief parenthetical definitions.

As the runner's foot approaches the surface, it supinates—rolls outward. As the foot lands, it pronates—rolls inward. Through tread design and increased stiffness on the innerside, the outsole helps reduce inward rolling.

Inward rolling is a major cause of foot, knee, and tendon injuries because of the magnitude of the force generated during running. The force on the foot as it touches the running surface can be up to three times the runner's weight. And the acceleration transmitted to the leg can be 10 times the force of gravity.

THE HEEL WEDGE   The heel wedge is a flexible platform that absorbs shock. Its purpose is to prevent injury to the Achilles tendon. Like the outsole, it is constructed of increasingly stiff materials on the inner side to reduce foot rolling.

THE MIDSOLE   The midsole is made of expanded foam. Like the outsole and the heel wedge, it reduces foot rolling. But it also is the most important component in absorbing the shock.

From the runner's point of view, running efficiency and shock absorption are at odds. The safest shoe would have a midsole of thick padding that would crush uniformly as the foot hit the running surface. A constant rate of deceleration would ensure the best shock absorption.

However, absorbing all the shock would mean absorbing all the energy. As a result, the runner's next stride would require more energy. The most efficient shoe would have a foam insole that is perfectly elastic. It would return all the energy back to the foot, so that the next stride required less energy. Currently, distance shoes have midsoles designed to return 40 percent of the runner's energy back to the foot.

THE INSOLE   The insole, on which the runner's foot rests, is another layer of shock-absorbing material. Its principal function, however, is to provide an arch support, a relatively new feature in running shoes.

THE SHELL   The shell is made of leather and synthetic materials such as nylon. It holds the soles on the runner's foot and provides ventilation. The shell accounts for about one-third of the nine ounces a modern shoe weighs.

CONCLUSION

The conclusion summarizes the major points of the description.

Today, scientific research on the way people run has led to great improvements in the design and manufacture of different kinds of running shoes. With the vast numbers of runners, more and more manufacturers have entered the market. The results are a lightweight, shock-absorbing running shoe that balances the needs of safety and increased speed.

THE STRUCTURE OF THE PROCESS DESCRIPTION

The structure of the process description is essentially the same as that of the object or mechanism description. The only real difference is that "steps" replace "parts": the process is partitioned into a reasonable (usually chronological) sequence of steps. Most process descriptions contain the following three components:

1. a general introduction that tells the reader what the process is and what it is used for
2. a step-by-step description of the process
3. a conclusion that summarizes the description and tells how the steps work together

THE GENERAL INTRODUCTION   The general introduction gives your readers the basic information they will need to understand the detailed description that follows. In writing the introduction, answer these six questions about the process:

1. What is the process?
2. What is its function?
3. Where and when does it take place?
4. Who or what performs it?
5. How does it work?
6. What are its principal steps?

1. What is the process? Generally, the best way to answer this question is to provide a sentence definition (see Chapter 6): "Debugging is the process of identifying and eliminating any errors within the program." Then elaborate if necessary.

2. What is the function of the process? State clearly the function of the process: "The central purpose of performing a census is to obtain up-to-date population figures on which to base legislative redistricting and revenue-sharing revisions." Make sure the function is clear to your readers; if you are unsure whether they already know the function, state it anyway. Few things are more frustrating for your readers than not knowing why the process should be performed.

3. Where and when does the process take place? State clearly the location and occasion for the process: "The stream is stocked at the hatchery in the first week of March each year." These details can generally be added simply and easily. Again, omit reference to these facts only if you are certain your readers already know them.

4. *Who or what performs the process?* Most processes are performed by people, by natural forces, by machinery, or by some combination of the three. In most cases, you do not need to state explicitly that, for example, the young trout are released into the stream by a person; the context makes that point clear. In fact, much of the description usually is written in the passive voice: "The water temperature is then measured." Do not assume, however, that your readers already know what agent performs the process, or even that they understand you when you have identified the agent. Someone who is not knowledgeable about computers, for instance, might not know whether a compiler is a person or a thing. The term *word processor* often refers ambiguously to a piece of equipment and to the person who operates it. Confusion at this early stage of the process description can ruin the effectiveness of the writing.

5. *How does the process work?* In a few sentences, define the principle or theory of operation of the process:

The four-treatment lawn-spray plan is based on the theory that the most effective way to promote a healthy lawn is to apply different treatments at crucial times during the growing season. The first two treatments—in spring and early summer—consist of quick-acting organic fertilizers and weed- and insect-control chemicals. The late summer treatment contains postemergence weed-control and insect-control chemicals. The last treatment—in the fall—uses long-feeding fertilizers to encourage root growth over the winter.

6. *What are the principal steps of the process?* Name the principal steps of the process, in one or several sentences or in a list. Make sure you have named the steps in the order they will be described in the body of the description. The principal steps in changing an automobile tire, for instance, include jacking up the car, replacing the old tire with the new one, and lowering the car back to the ground. The process of changing a tire also includes some secondary steps—such as placing blocks behind the tires to prevent the car from moving once it is jacked up, and assembling the jack—that you should explain or refer to at the appropriate points in the description.

The information given in the introduction to a process description generally follows this pattern. However, don't feel you must answer each question in order, or in separate sentences. If your readers' needs or the nature of the process suggests a different sequence—or different questions—adjust the introduction.

As is the case with object and mechanism descriptions, process descriptions benefit from clear graphic aids in the introduction. Flow charts that identify the major steps of the process are particularly common.

THE STEP-BY-STEP DESCRIPTION

The body of the process description—the step-by-step description—is essentially like the introduction, in that it treats each major step as if it were a process. Of course, you do not repeat your answer to the question about who or what performs the action unless a new agent performs the action at a particular step of the process. The other principal questions—what the step is, what its function is, and when, where, and how it occurs—should be answered. In addition, if the step has any important substeps that the reader will need to know in order to understand the process, they should be explained clearly.

The structure of the step-by-step description should be chronological: discuss the initial step first and then discuss each succeeding step in the order in which it occurs in the process. If the process is a closed system and hence has no "first" step—such as the cycle of evaporation and condensation—explain to your readers that the process is cyclical. Then simply begin with any principal step.

Although the structure of the step-by-step description should be chronological, don't present the steps as if they were individual occurrences that have nothing to do with one another. In many cases, one step leads to another causally (see Chapter 7 for a discussion of cause and effect). In the operation of a four-cycle gasoline engine, for instance, each step sets up the conditions under which the next step can occur. In the compression cycle, the piston travels upward in the cylinder, compressing the mixture of fuel and air. This compressed mixture is ignited by the spark from the spark plug in the next step, the power cycle. Your readers will find it easier to understand and remember your description if you clearly explain the causality in addition to the chronology.

A word about tense. The steps should be discussed in the present tense, unless, of course, you are writing about a process that occurred in the historical past. For example, a description of how the Earth was formed would be written in the past tense: "The molten material condensed. . . ." However, a description of how steel is made would be written in the present tense: "The molten material is then poured into. . . ."

Whenever possible, use graphic aids within the step-by-step description to clarify each point. Additional flow charts are useful, but you will often want to create other kinds of graphic aids, such as photographs, drawings, and graphs. For instance, in the description of a four-cycle gasoline engine, each step could be illustrated as well as the position of the valves. The activity occurring during each step could also be illustrated: the explosion during the ignition step could be shown with arrows pushing the piston down within the cylinder.

THE CONCLUSION   Process descriptions usually do not require long conclusions. If the description itself is brief—less than a few pages—a short paragraph summarizing the principal steps is all that is needed. For longer descriptions, a discussion of the implications of the process might be appropriate.

Following is the conclusion from a description of the four-cycle gasoline engine in operation:

In the intake stroke, the piston moves down, drawing the air-fuel mixture into the cylinder from the carburetor. This mixture is compressed as the piston goes up in the compression stroke. In the power stroke, a spark from the spark plug ignites the mixture, which burns rapidly, forcing the piston down. In the exhaust stroke, the piston moves up, expelling the burned gases.

Following is a particular description of the stages involved in the construction of the tunnel under the English Channel. The description is addressed to a general audience. Marginal notes have been added.

INTRODUCTION

The writer provides the background: the purpose of the process and who is doing it

Almost two hundred years ago, the idea of linking England and France by a tunnel under the English Channel was proposed to Napoleon. The plan was not pursued. In the intervening years the idea has been revived several times, and in 1875 a one-mile tunnel was dug at the foot of the White Cliffs of Dover. Fears of invasion from the continent led to repeated protests from the British military. In 1973, Britain and France agreed to a 32-mile rail link, but austerity measures forced Britain to cancel. The British and French have finally agreed to a plan for a tunnel that is expected to be operational by 1993.

The writer lists the stages of the process.

This paper describes the stages that have occurred—and will occur—in building the English Channel tunnel:

1. determining the objectives
2. determining the constraints
3. selecting a plan

4. constructing the initial tunnel

5. improving the initial tunnel

The tunnel will link Dover to Calais, as shown in Figure 1.

The writer provides a map of the area.

THE STAGES OF THE ENGLISH CHANNEL TUNNEL PROJECT

This overview clarifies the sequence of the stages.

Currently, the objectives and restraints have been determined and a plan has been selected. Construction of the original tunnel is expected to begin in mid-1987. The initial tunnel, which will be a rail link, is expected to be supplemented with a roadway sometime early in the next century. These five stages will be discussed in their chronological sequence.

DETERMINING THE OBJECTIVES   The main goal of the tunnel is to reduce the travel time between London and Paris. Figure 2 shows that the tunnel is expected to draw the capitals much closer. This will increase trade and tourism.

A second objective for both nations is to increase employment and thereby improve their governments' political fortunes. The tunnel project is expected to provide 60,000 jobs for some five or six years.

DETERMINING THE CONSTRAINTS   In addition to the obvious technical questions that both parties considered, the main constraint was offered by Britain's conservative Prime Minister, Margaret Thatcher. Her government insisted that the project be financed privately. The approved project calls for a consortium of three British banks, three French ones, and ten construction companies.

The writer discusses the unsuccessful proposals before the successful one.

SELECTING A PLAN   The competition came down to the proposals offered by four finalists. The British favored the Channel Expressway, which consisted of twin road and rail tunnels. However, objections based on its cost and the difficulties involved in ventilation doomed the idea.

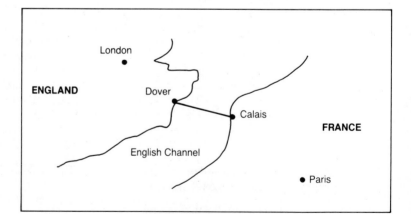

FIGURE 1

The Route of the English Channel Tunnel

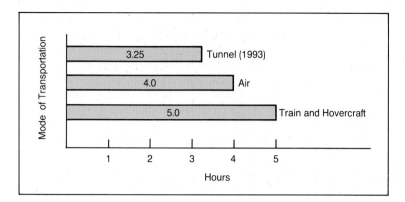

FIGURE 2
Travel Time Between London and Paris (Downtown to Downtown)

The French backed Euroroute, which called for the construction of two artificial islands in the channel. The islands were to be linked by a tunnel and linked to the shores by bridges. Because of concerns over the enormous price—$15 billion—and the islands' vulnerability to terrorist attacks, Euroroute was abandoned.

A third proposal called for the construction of a bridge across the 23-mile waterway. However, questions about the new composite material that was supposed to replace steel in the bridge hurt the plan.

The successful proposal is discussed in more detail than the others.

The proposal that won will cost 6 billion dollars. It calls for constructing a 31-mile twin-tube rail tunnel. The two tubes will be linked by a service corridor. Specially designed rail cars will carry autos, buses, and trucks under the channel in about half an hour. Passengers will be able to stay in their vehicles or in passenger compartments. Trains will leave every three minutes and are expected to carry 4,000 vehicles an hour each way. Estimates are that some 67 million passengers will cross the channel each year—half by the tunnel. Currently, the annual passenger rate is 47 million.

CONSTRUCTING THE INITIAL TUNNEL   The tunnel will be bored by 11 Tunnel-Boring Machines (TBM's), equipped with steel cutting disks that will eat into the chalk that makes up the seabed. The chalk is the perfect material for a tunnel: impervious to water yet easy to cut. Lasers linked to microprocessors will guide the cutters. Once the tunnel is cut, it will be lined with 23-inch-thick prefabricated concrete joined by steel bands.

The service tunnel is expected to be completed by 1990; the rail tunnels, by 1993.

IMPROVING THE INITIAL TUNNEL   Both France and Britain were reluctant to give up the idea of a roadway. Therefore, one clause of the contract calls for the construction of a roadway by the year 2000. If the consortium fails to complete the roadway by that date, the governments may open up bidding to other companies.

CONCLUSION

The English Channel Tunnel project is the most ambitious engineering undertaking of the twentieth century. Since the Second World War, the value of the English Channel as a protection of the British from the Continent has all but disappeared. The tunnel is expected to strengthen and symbolize the economic interdependence of the British and the Europeans.

WRITER'S CHECKLIST

The following questions cover descriptions of objects and mechanisms:

1. Does the introduction to the object or mechanism description
   a. Define the item?
   b. Identify its function (where appropriate)?
   c. Describe its appearance?
   d. Describe its principle of operation (where appropriate)?
   e. List its principal parts?
   f. Include a graphic aid identifying all the principal parts?

2. Does the part-by-part description
   a. Answer, for each of the major parts, the questions listed in item 1?
   b. Describe each part in the sequence in which it was listed in the introduction?
   c. Include graphic aids for each of the major parts of the mechanism?

3. Does the conclusion
   a. Summarize the major points made in the part-by-part description?
   b. Include (where appropriate) a description of the item performing its function?

The following questions cover process descriptions:

1. Does the introduction to the process description
   a. Define the process?
   b. Identify its function?
   c. Identify where and when the process takes place?
   d. Identify who or what performs it?
   e. Describe how the process works?
   f. List its principal steps?
   g. Include a graphic aid identifying all the principal steps?

2. Does the step-by-step description
   a. Answer, for each of the major steps, the questions listed in item 1?
   b. Discuss the steps in chronological order or in some other logical sequence?
   c. Make clear the causal relationships among the steps?
   d. Include graphic aids for each of the principal steps?

3. Does the conclusion

    a. Summarize the major points made in the step-by-step description?

    b. Discuss, if appropriate, the importance or implications of the process?

EXERCISES

1. Write a description of one of the following items or of a piece of equipment used in your field. Be sure to specify your audience and indicate the type of description (general or particular) you are writing. Include appropriate graphic aids.

    a. a carburetor
    b. a locking bicycle rack
    c. a deadbolt lock
    d. a folding card table
    e. a lawn mower
    f. a photocopy machine
    g. a cooling tower
    h. a jet engine
    i. a telescope
    j. an ammeter
    k. a television set
    l. an automobile jack
    m. a stereo speaker
    n. a refrigerator
    o. a computer
    p. a cigarette lighter

2. In an essay, evaluate the effectiveness of one of the following descriptions.

DESCRIPTION OF A CAMERA

A camera is really just a box that does not admit light. On one end is the lens, on the other is the film. The lens focuses an image of what it sees on the film. As seen in Figure 1, this image is upside down.

How does the image become reversed again? The film is made of silver halide, which undergoes a chemical change when it is exposed to light—the more light, the greater the change. During the process of developing the film, the change is emphasized further. When a positive is made from this negative, the image comes out correct.

SPARK PLUGS

A spark plug is a stationary pair of points inside the combustion chamber. One of the electrodes is connected to ground; the other is insulated. When high voltage flows from the coil to the insulated electrode, it will jump across the gap to the grounded electrode. This is the

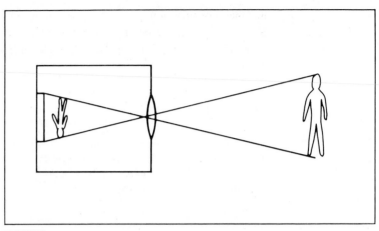

FIGURE 1

spark. The center electrode is insulated by the porcelain that sur-
rounds it. It is sensitive to oil deposits. If enough oil collects on the porcelain, the current will flow down the outside of the porcelain without
jumping the gap. This is called a dead spark plug, and if enough of
them are dead the engine won't start. For this reason, it is necessary to
clean the plugs periodically.

3. Rewrite one of the descriptions in Exercise 2.

4. Write a description of one of the following processes or a similar process with which you are familiar. Be sure to indicate your audience
and the type of description (general or particular) you are writing. Include appropriate graphic aids.

   a. how steel is made
   b. how an audit is conducted
   c. how a nuclear power plant works
   d. how a bill becomes a law
   e. how a suspension bridge is constructed
   f. how a microscope operates
   g. how we hear
   h. how a dry battery operates
   i. how a baseball player becomes a free agent
   j. how cells reproduce

5. In an essay, evaluate the effectiveness of one of the following process
descriptions.

THE FOUR-STROKE POWER CYCLE

The power to drive an engine comes from the four-stroke cycle.
   Intake Stroke: Inside the cylinder, the piston moves down, creating a vacuum that draws in the air/fuel mixture through the intake
valve.

Compression Stroke: The intake valve shuts and the cylinder moves back up, compressing the air/fuel mixture in what is called the combustion chamber. The purpose of compression is to increase the power of the explosion that will occur in the next stroke.

Power Stroke: The spark across the electrodes of the spark plug ignites the air/fuel mixture, pushing the piston down. This in turn moves the wheels of the car.

Exhaust Stroke: The burned gases escape through the exhaust valve as the piston moves up again.

THE PROCESS OF XEROGRAPHY

Xerography—"dry writing"—is the process used in most photocopy machines. Photocopy machines enable us to produce high-quality, inexpensive copies of printed matter.

Xerography depends on electricity and photoconductivity. A photoconductive material is one that maintains an electric charge in the darkness but loses it when exposed to light.

The first step in today's photocopy machine is to give an electric charge to a selenium roll.

Next, a bright light is shined on the page to be copied. Using lenses, it is projected on the selenium roll. Where the image is bright, the electric charge disappears. Where it is dark, the charge is maintained. Thus, the electric charge on the selenium roll corresponds to the page.

Then, the selenium roll receives a coating of a special dust that sticks only where there is an electric charge. This is then transferred to a piece of paper, which is the photocopy.

6. Rewrite one of the process descriptions in Exercise 5.

REFERENCES   Isganitis, E. J. 1977. Unpublished report.

Kyle, C. R. 1986. Athletic clothing. *Scientific American* 254, no. 3: 104–110.

CHAPTER NINE

INSTRUCTIONS
AND MANUALS

A set of instructions is a process description written to help the reader perform a specific task. A process description discusses how a tire is changed. A set of instructions explains how to change the tire. Whereas a process description states that the car is jacked up so that the tire can be removed, a set of instructions goes into more detail: it explains how to assemble and use the jack effectively and safely.

A manual is a document consisting primarily of instructions. Often it is printed and bound, like a book. Manuals can be classified according to function. One common type is a procedures manual. Like a set of instructions, its function is to instruct. For example, it explains how to use a software program, maintain inventory for a particular store, or operate a piece of machinery. There are installation manuals, which tell technicians how to install a device or set of devices. There are maintenance manuals and repair manuals.

Effective instructions and manuals look easy to write, but they aren't. You have to make sure not only that your readers will be able to understand and follow your directions easily, but also that in performing the tasks they won't damage any equipment or, more important, injure themselves or other people.

THE ROLE OF INSTRUCTIONS
AND MANUALS IN TECHNICAL WRITING

Instructions and manuals are central to technical writing, and you will probably be asked to write them or contribute to them often in your career. Many factors have tremendously increased the number and complexity of instructions and manuals produced every day.

Consumer products are much more sophisticated than they used to be. A radio sold 30 years ago needed a simple sheet listing repair shops and perhaps a simple guarantee. Today a modular stereo system—complete with dual cassette, programmable turntable, compact disc player, 5-band graphic equalizer, and ready-to-assemble walnut veneer rack—requires a lot of documentation. Personal computers and video cassette recorders are now commonplace.

At the workplace, procedures are also much more sophisticated. Buying a screwdriver at Sears requires that the clerk enter a string of numbers and letters so that the electronic cash register can help maintain an accurate inventory. Employee salaries and raises are determined by a complicated formula intended to make the process as fair as possible. Learning how to use the formula requires a manual. In manufacturing, workers routinely operate and monitor computers and robots.

Just a decade or two ago, little attention was paid to the quality of documentation—the instructions and manuals. The technical people who created the product or system were the experts. If the product did its job, no more was asked. The users had to cope as well as they could. Often companies did not provide any documentation at all, or they provided it six months or a year after the product was introduced. The main sources of information on how to use the product were a telephone number to call and the service representative—the person who went out to the job site to help the user.

Today the goal is to make systems "user friendly." The competition for our dollars is so great that we can require that the product be easy to set up and use. When computers are compared these days, the quality of the manuals is one of the prime considerations. Major advertising campaigns for all kinds of products stress clear, simple, easy-to-use manuals.

For this reason, documentation is no longer an afterthought but rather an integral part of the planning process. The documentation writers are part of the research-and-development team. And they take questions of audience and purpose seriously.

THE WRITING SITUATION

The purpose of a set of instructions or of a manual is usually self-evident. Assembly instructions for a backyard swing are intended to help the purchaser set up the swing. A maintenance manual for a computerized conveyor belt is intended to explain how and when to perform maintenance on it.

The question of audience, however, is difficult and subtle. In fact, if instructions and manuals are ineffective, chances are that the writer has inaccurately assessed the audience. Performing a function—such as assembling the backyard swing or maintaining the conveyor belt—is easy for the expert but not so easy for the layperson reading the documentation. If the writer has incorrectly assumed that the reader understands what a self-locking washer is or what the calibrations on a timing control mean, the reader might not be able to complete the process.

Before you start to write a set of instructions or a manual, think carefully about your audience's background and skill level. If you are writing to people who are experienced in the field, you can use the technical vocabulary and make any reasonable assumptions about their knowledge. But if you are addressing people who are unfamiliar with the field, define your terminology and explain your directions in more detail. Don't be content to write, "Make sure the tires are rotated properly." Define proper rotation and describe how to achieve it.

The best way to make sure you have assessed your audience effectively is to find someone whose background is similar to that of your intended readers. Give this person a draft of your document and watch as he or she tries to perform the tasks. This process will give you valuable information about how clear your writing is.

One other aspect of audience analysis should be mentioned. Inexperienced writers assume that their readers will read through the document carefully before beginning the first step. They won't. Most people will merely glance at the whole set of instructions or the manual to see if anything catches their eye before they get started. For this reason, you should organize the document in a strict chronological sequence. This is particularly important when you want to call your readers' attention to important points. Use "notes" to inform them, "cautions" to prevent damage to equipment or failure of the process, and "warning" or "danger" to prevent injury.

Note: Two different size screws are provided. Be sure to use the 3/8″ long screws here.

CAUTION: Do not attempt to use non-rechargeable batteries in this charging unit; the charging unit could be damaged.

WARNING: TO PREVENT SERIOUS EYE INJURY, WEAR SAFETY GOGGLES AT ALL TIMES!

If a note, caution, or warning applies to the whole process, state it emphatically at the start of the instructions. If it applies only to a particular step, however, insert it before that step.

INSTRUCTIONS

Instructions can be brief—a small sheet of paper—or extensive—20 pages or more. Brief instructions could be produced by two people: a writer and an artist. Sometimes a technical expert is added to the team. For more extensive instructions, other people—marketing and legal personnel, for example—could be added. The team could consist of 10 or 20 professionals working with a budget of many thousands of dollars.

THE STRUCTURE OF THE INSTRUCTIONS   Regardless of the size of the project, most instructions are structured like process descriptions (see Chapter 8). The main difference is that the conclusion of a set of instructions is less a summary than an explanation of how to make sure the reader has followed the instructions correctly. Most sets of instructions contain the following three components:

1. a general introduction that prepares the reader for performing the instructions
2. the step-by-step instructions
3. a conclusion

THE GENERAL INTRODUCTION The general introduction gives the readers the preliminary information they will need to follow the instructions easily and safely. In writing the introduction, answer these three questions about the process:

1. Why should this task be carried out?
2. What safety measures should the reader take, and what background information is necessary before he or she begins the process?
3. What tools and materials will be needed?

Why should this task be carried out? Often, the answer to this question is obvious and should not be stated. For example, the purchaser of a backyard barbeque grill does not need an explanation of why it should be assembled. Sometimes, however, the readers need to be told why they should carry out the task. Many preventive maintenance chores—such as changing radiator antifreeze every two years—fall into this category.

If appropriate, answer two more questions. One, "When should the task be carried out?" Some tasks, such as rotating tires or planting crops, need to be performed at particular times. Two, "By whom should the task be carried out?" Sometimes it is necessary to describe the person or persons who are to carry out a task. Some kinds of aircraft maintenance, for example, may be carried out only by persons certified to do that maintenance.

What safety measures or other concerns should the reader understand? In addition to the safety measures that apply to the whole task, state any tips that will make your reader's job easier. For example:

NOTE: For ease of assembly, leave all nuts loose.
   Give only 3 or 4 complete turns on bolt threads.

What tools and materials will be needed? The list of necessary tools and equipment is usually included in the introduction so that the readers do not have to interrupt their work to hunt for another tool.

Following is a list of tools and materials from a set of instructions on replacing broken window glass:

You will need the following *tools* and *materials:*

TOOLS	MATERIALS
glass cutter	putty
putty knife	glass of proper size
window scraper	paint
metal shield	hand cleaner
chisel	work gloves
electric soldering iron	linseed oil
razor blade	glazier's points
pliers	
paint brush	

THE STEP-BY-STEP INSTRUCTIONS The step-by-step instructions are essentially like the body of the process description. There are, however, two differences.

First, the instructions should always be numbered. Each step should define a single task that the reader can carry out easily, without having to refer back to the instructions. Don't overload the step:

1. Mix one part of the cement with one part water, using the trowel. When the mixture is a thick consistency without any lumps bigger than a marble, place a strip of about 1″ high and 1″ wide along the face of the brick.

On the other hand, don't make the step so simple that the reader will be annoyed:

1. Pick up the trowel.

Second, the instructions should always be stated in the imperative mood: "Attach the red wire. . . ." The imperative is more direct and economical than the indicative mood ("You should attach the red wire . . ." or "The operator should attach the red wire . . ."). Make sure your sentences are grammatically parallel. Avoid the passive voice ("The red wire is attached . . ."), because it can be ambiguous: Is the red wire already attached?

Keep the instructions simple and direct. However, do not omit the articles (*a, an, the*) to save space. Omitting the articles makes the instructions hard to read and, sometimes, unclear. In the sentence "Locate midpoint and draw line," for example, the reader cannot tell if "draw line" is a noun ("the draw line") or a verb and its object ("draw the line").

Be sure to include graphic aids in your step-by-step instructions. In appropriate cases, each step should be accompanied by a photograph or diagram that shows what the reader is supposed to do. Some kinds of activities—such as adding two drops of a reagent to a mixture—do not need illustration. However, steps that require manipulating physical objects—such as adjusting a chain to a specified tension—can be clarified by graphic aids.

Figure 9-1 shows the extent to which a set of instructions can integrate words and graphics. This excerpt is from the operating-instructions booklet for a video cassette recorder. The booklet has already described the basic controls and functions of the device. At this point the writers use numbered steps clearly indicating where to find the different controls on the VCR. See Chapter 10 for a discussion of graphic aids.

THE CONCLUSION Instructions generally do not require conclusions. Sometimes, however, the instructions conclude with main-

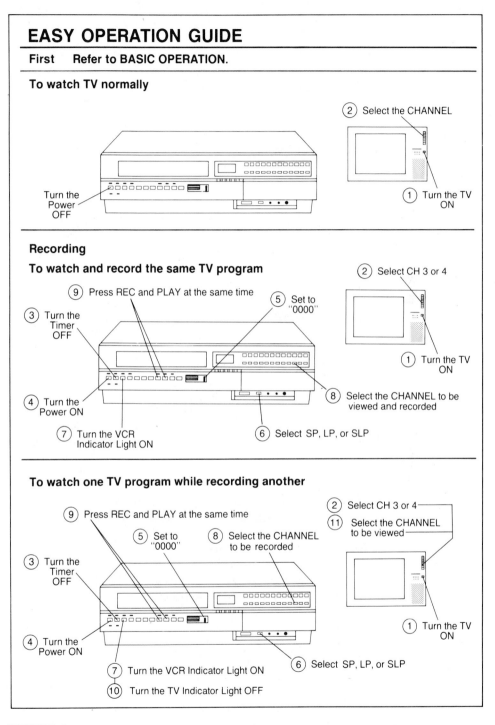

EASY OPERATION GUIDE

First   Refer to BASIC OPERATION.

To watch TV normally

(2) Select the CHANNEL

Turn the Power OFF

(1) Turn the TV ON

Recording

To watch and record the same TV program

(2) Select CH 3 or 4

(9) Press REC and PLAY at the same time

(5) Set to "0000"

(3) Turn the Timer OFF

(1) Turn the TV ON

(4) Turn the Power ON

(8) Select the CHANNEL to be viewed and recorded

(7) Turn the VCR Indicator Light ON

(6) Select SP, LP, or SLP

To watch one TV program while recording another

(9) Press REC and PLAY at the same time

(2) Select CH 3 or 4

(11) Select the CHANNEL to be viewed

(5) Set to "0000"

(8) Select the CHANNEL to be recorded

(3) Turn the Timer OFF

(1) Turn the TV ON

(4) Turn the Power ON

(6) Select SP, LP, or SLP

(7) Turn the VCR Indicator Light ON

(10) Turn the TV Indicator Light OFF

FIGURE 9-1

Instructions That Integrate Words and Graphics

tenance tips. Another popular way to conclude instructions is to present a troubleshooter's checklist. This checklist, usually in the form of a table, identifies and tells how to solve common problems associated with the mechanism or process described in the instructions.

Following is a portion of the troubleshooter's guide included in the operating instructions of a lawn mower.

PROBLEM	CAUSE	CORRECTION
Mower does not start.	1. Out of gas.	1. Fill gas tank.
	2. "Stale" gas.	2. Drain tank and refill with fresh gas.
	3. Spark plug wire disconnected from spark plug.	3. Connect wire to plug.
Mower loses power.	1. Grass too high.	1. Set mower in "higher-cut" position.
	2. Dirty air cleaner.	2. Replace air cleaner.
	3. Buildup of grass, leaves, and trash.	3. Disconnect spark plug wire, attach to retainer post, and clean underside of mower housing.

Some troubleshooter's checklists refer the reader back to the page that discusses the action being described. For example, the Correction column in the lawn-mower troubleshooting checklist might say "1. Fill gas tank. See page 4."

Figure 9-2 shows a set of instructions on how to apply panels to stud walls. The instructions were written by the Home Center Institute of the National Hardware Association.

MANUALS

Most of what has been said about instructions applies to manuals. For example, you have to understand your audience and purpose. You have to explain procedures clearly and complement your descriptions with graphic aids.

However, because manuals are usually much more ambitious projects than instructions, they require much more careful plan-

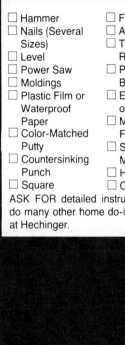

Checklist of tools and materials required

- ☐ Hammer
- ☐ Nails (Several Sizes)
- ☐ Level
- ☐ Power Saw
- ☐ Moldings
- ☐ Plastic Film or Waterproof Paper
- ☐ Color-Matched Putty
- ☐ Countersinking Punch
- ☐ Square
- ☐ Fine-Tooth Saw
- ☐ Adhesive
- ☐ Tape or Folding Rule
- ☐ Proper Saw Blades
- ☐ Electrical Tape or Wire Nuts
- ☐ Magnetic Stud Finder
- ☐ Stain for Wood Molding
- ☐ Hand Cleaner
- ☐ Cleanup Cloths

ASK FOR detailed instructions on how to do many other home do-it-yourself projects at Hechinger.

HOW TO APPLY PANELS TO STUD WALLS

ANOTHER CUSTOMER
SERVICE FROM

HECHINGER
The World's Most Unusual Lumber Yards

ANOTHER CUSTOMER
SERVICE FROM

HECHINGER
The World's Most Unusual Lumber Yards

© 1976 by the Home Center Institute of National Retail Hardware Association. Reproduced by special permission.

FIGURE 9-2

Instructions with Graphics and Checklist

Here are some tips and instructions on installing paneling on stud walls. Take the time to read them thoroughly. Following these instructions can save time, money, and effort. It can also help you to end up with a neater, more satisfactory installation—with less waste.

1. STORING PANELS UNTIL THEY ARE INSTALLED

A.   Store all panels in room temperature comparable to rooms where they are to be applied. This permits each panel to adjust to the temperature and humidity of the area.

B.   Store flat with scraps of wood inserted between panels. Cover the stack. Storing flat reduces warping. Scraps of wood between the panels let each panel breathe and absorb room moisture. A cover over the stack keeps the panels clean and prevents marring.

2. MEASURING TO DETERMINE PANELS NEEDED

A.   Measure each wall carefully to determine the number of panels required. Measure both height and width. Here's a good rule to follow:

B.   Multiply total length of all walls by height. Subtract total area not to be paneled (windows, doors, etc.) from previous total. Add 5% for waste. If ceiling height is 8' or under, the table in Figure 1 can be used.

Figure 1

DETERMINING THE NUMBER OF 4 x 8 PANELS REQUIRED BASED ON PERIMETER OF ROOM

For example, if your room walls measured 16' x 16' x 18' x 18', the perimeter would be 68' requiring 17 4' x 8' panels based on the chart.

Room Perimeter	4' x 8' Panels Required	Room Perimeter	4' x 8' Panels Required
36 ft.	9	60 ft.	15
40 ft.	10	64 ft.	16
44 ft.	11	68 ft.	17
48 ft.	12	72 ft.	18
52 ft.	13	76 ft.	20
56 ft.	14	92 ft.	23

To allow for areas such as doors, windows, etc. use the deductions below and subtract from the total panels required.

For each door deduct $\frac{1}{3}$ panel
For each window deduct $\frac{1}{4}$ panel
For each fireplace deduct $\frac{1}{2}$ panel

Always use the next highest number of panels when the total perimeter is between the ranges shown in the table.

These figures are based on rooms with ceiling heights of 8' or less.

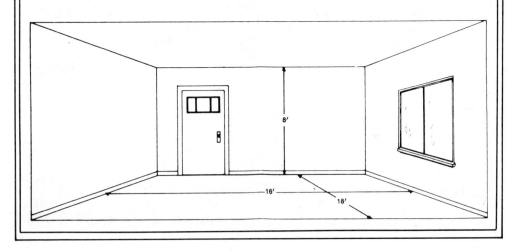

FIGURE 9-2

Instructions with Graphics and Checklist (Continued)

3. PREPARING FOR PANELING ON STUD WALLS

A.   Studs in walls are usually placed 16″ apart. To locate a stud, measure 16″ from wall and test by driving a 6 penny nail into the wall in an unobtrusive location, or use a magnetic stud finder to locate nails in studs (Figure 2).

B.   Locate all studs and mark the location with a chalk line. Use a level on a 2 x 4 to get all markings plumb and straight. Continue lines 2″ or 3″ out on the floor and ceiling (Figure 2).

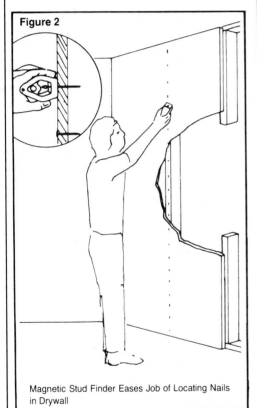

Figure 2

Magnetic Stud Finder Eases Job of Locating Nails in Drywall

C.   Remove all ceiling and floor moldings.

D.   Remove the light fixtures and all wall plates. Tape the exposed wire ends or use wire nuts. Be sure you turn off the electricity at the main control box panel before you begin.

4. PREPARING FOR PANELING ON NEW STUDS

E.   Be sure the studs are plumb and evenly spaced.

F.   Be sure there is a horizontal 2 x 4 at top and bottom (Figure 3).

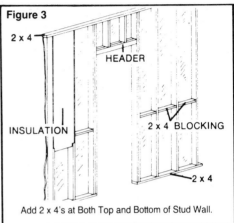

Figure 3

2 x 4

HEADER

INSULATION

2 x 4 BLOCKING

2 x 4

Add 2 x 4's at Both Top and Bottom of Stud Wall.

G.   Two-by-four blocking in the center helps prevent twisting and also gives added wall support (Figure 3).

H.   Building paper or #15 roofing felt is sometimes added as a protection against moisture. If added, use a staple gun and apply it horizontally, overlapping each course about 6″.

I.   On new walls, staple 1½″ or 2″ thick rock wool insulating bats between studs before applying paneling (Figure 3).

5. ARRANGING PANELS FOR INSTALLATION

A.   Stand panels up against the wall loosely, then back off for a study of the wood grain, color match, groove blend, etc.

B.   Rearrange to achieve the most attractive sequence of grain and tone.

C.   Use care in rearranging panels. Don't rub them against each other. This mars the finish.

D.   If panels have been stored in another area, leave in the room at least 24 hours before installing.

FIGURE 9-2

Instructions with Graphics and Checklist (Continued)

6. MEASURING AND CUTTING PANELS

A. Measure the floor-to-ceiling height at several points to detect any variations.

B. If height varies no more than ¼″, take the shortest height, subtract ¼″, and cut all panels to this measurement.

C. If height varies more than ¼″, cut each panel separately.

D. In either case, cut each panel ¼″ shorter than the actual ceiling height. Ceiling and floor moldings will cover height variations of up to 2″.

E. Start the panel installation in a corner (Figure 4). Be sure the panel is plumb. Use a level of sufficient length to get an accurate reading.

Figure 5

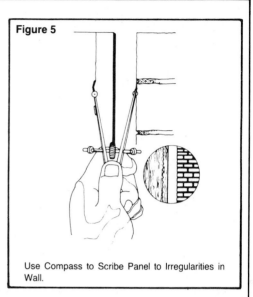

Use Compass to Scribe Panel to Irregularities in Wall.

Figure 4

Start Panels in a Corner. Use level to Plumb.

F. The slightest irregularity in the wall can cause trouble. To allow for such irregularities, place the first panel in a plumb position in such a way that the panel and the wall can be spanned by your scribing compass (Figure 5). Draw a scribed line from top to bottom. Use a china marking pencil for scribing.

G. Cut to scribed line with coping saw where irregular cuts are necessary.

H. Double check all measurements before sawing.

I. Use a fine tooth, finish saw. Never cut paneling with a coarse saw.

J. If cutting paneling with a hand saw, always cut from finished side (Figure 6). If using a power saw, always cut from back of panel (Figure 7).

Figure 6

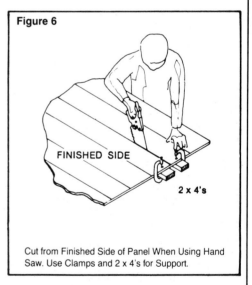

FINISHED SIDE

2 x 4's

Cut from Finished Side of Panel When Using Hand Saw. Use Clamps and 2 x 4's for Support.

K. After first corner panel is straight, proceed with installation by putting each panel firmly against the one already installed.

FIGURE 9-2

Instructions with Graphics and Checklist (Continued)

Figure 7

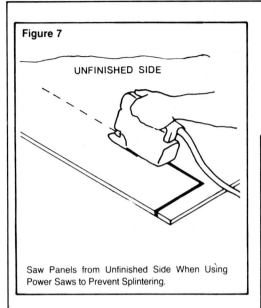

UNFINISHED SIDE

Saw Panels from Unfinished Side When Using Power Saws to Prevent Splintering.

7. APPLYING PANELS WITH NAILS

A.   When nailing panels to furring strips, use 1" nails.

B.   If nailing directly to old wall, use 1½" nails.

C.   Start at one side and nail straight across to opposite wall.

D.   Nail every 6" along panel edges.

E.   Space the nails 16" apart across face of panels. This will place each nail on a stud.

F.   Use a nailset to countersink all nails approximately $\frac{1}{32}$".

G.   Fill countersink holes with matching-color putty stick.

H.   If you use matching-color head nails, countersinking is not necessary.

8. APPLYING PANELS WITH ADHESIVES

A.   If wall is papered, remove all paper before applying adhesive and paneling.

B.   Do not use contact cement unless skilled in its application. Use the recommended cement—either rubber-base or plastic cement—and follow instructions carefully.

C.   Apply a 3" ribbon of adhesive along the furring strip lines and across bottom and top. Use serrated spreader or glue gun.

D.   Nail loosely at top to hinge panel in proper location. Use scrap of lumber about 8" long (Figure 8) to prop bottom of panel out from wall for 2 or 3 minutes, or until the adhesive gets "tacky."

E.   Remove block after about 3 minutes, press the panel into position, and nail about every 6" along base.

Figure 8

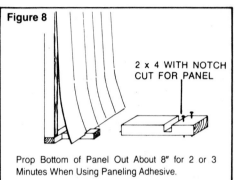

2 x 4 WITH NOTCH CUT FOR PANEL

Prop Bottom of Panel Out About 8" for 2 or 3 Minutes When Using Paneling Adhesive.

9. HANDLING SPECIAL PANELING PROBLEMS

A.   If a switch receptacle box requires a cutout in a panel, place the panel in its normal position over the box. Lay a soft wood block over the approximate location and tap the block soundly.

B.   The outlet box will make its imprint on back of the panel. Drill a small pilot hole and cut the box outlet with a keyhole saw (Figure 9).

C.   When a cut must be made for a door or window, measure the distance from the edge of the last panel installed to the door or window facing.

Figure 9

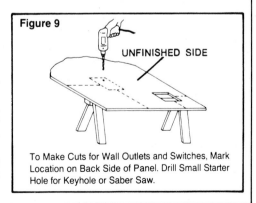

UNFINISHED SIDE

To Make Cuts for Wall Outlets and Switches, Mark Location on Back Side of Panel. Drill Small Starter Hole for Keyhole or Saber Saw.

FIGURE 9-2

Instructions with Graphics and Checklist (Continued)

D. Next, place the panel in the correct position, plumb it, and mark that same measurement on the face of the panel with a china marking pencil. For example (Figure 10), suppose the distance from the edge of the last panel to the door facing was 24″. When positioned and plumbed, that measurement should be marked on the panel (A). Section C can then be cut.

E. Be sure to measure from the floor to the top of the door (B), and transcribe this measurement to the panel to be cut. This should give you a perfect fit around the door.

F. To cut around a fireplace, scribe around the fireplace as in the method shown in Figure 5, marking the irregular cut on the paneling. You can use quarter round to conceal small variations.

10. APPLYING THAT FINISHING TOUCH

A. Apply proper moldings at top and bottom of panels.

B. Cut the wood moldings with a fine-tooth saw and a miter box. Apply with small nails.

C. Countersink and fill holes with matching-color putty stick. Clean with a dry rag if prefinished. Paint, stain, or varnish if unfinished.

D. Metal moldings are available if desired.

11. IF TILE OR CARPETING IS TO BE LAID

A. Install base molding first.

B. Lay tile or carpet up to the base molding.

C. Nail shoe molding over flooring tile.

Figure 10

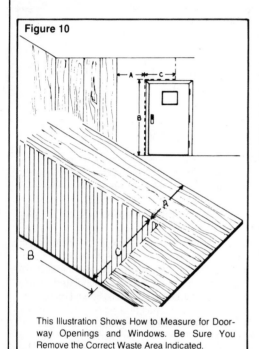

This Illustration Shows How to Measure for Doorway Openings and Windows. Be Sure You Remove the Correct Waste Area Indicated.

FIGURE 9-2

Instructions with Graphics and Checklist (Continued)

ning. A bigger investment is at stake for the organization, not only in the costs of writing and producing the manual but in the potential effects of the manual after it is published. A good manual will attract customers and reduce costs of doing business, for the organization will need fewer people to deal with questions in the field. A poor manual will be expensive to produce, will have to be revised more often, and will alienate customers.

THE PROCESS OF WRITING A MANUAL
The process of writing a manual, like the process of writing a book, is so complex that the following discussion can provide only an introduction. If you are involved in creating a manual, consult the books listed in the bibliography (Appendix E) for detailed advice.

In some respects, the process of writing a manual is the same as the general writing process: prewriting, drafting, and revising. However, the special requirements of manuals make the general stages somewhat different. The following discussion will explain some of the special requirements of manual writing.

The major difference between manual writing and other kinds of technical writing is that manuals are almost always collaborative projects. One reason is that a full-size manual is simply too big for one person to write quickly. A second reason is that no one person has all the skills necessary: the technical skills in the subject area, the writing skills, the artistic skills, and the knowledge of contracts and law that is required in an age of lawsuits. A third reason is that it is too risky to have one person performing such an ambitious task; any mistakes the person makes early on in the process will be compounded as the manual develops. Other perspectives are needed to prevent little problems from becoming big problems.

PREWRITING The most important stage of manual writing is prewriting. First, you must analyze your audience and purpose.

When you plan the documentation for a sophisticated procedure or system, you will almost always face a multiple audience. Often you will decide to produce a set of manuals—one for the user, one for the manager, one for the installer, one for the maintenance technicians, and so forth. Or you might decide on a main manual and some other, related documents, such as brochures, flyers, and workbooks.

Your primary purpose varies with the kind of manual. A user's manual helps a user carry out a process or operate a system. A reference manual enables a system designer to find detailed design information quickly. With most manuals, however, you also have a secondary purpose: to motivate your readers. The mere fact that the reader will be holding a 200-page manual means that you have to persuade that person to learn what has to be learned. In many cases the reader is uncomfortable with the new system. Your job is to make the task seem less overwhelming. A well-designed manual, with plenty of white space and graphic aids, will help a lot.

The next step is outlining the manual. As discussed in Chapter 3, outlining starts with brainstorming. For manuals, brainstorming almost always takes more than the fifteen or twenty minutes required for small reports. Often, several brainstorming sessions will be needed to plan the dozens of discussions and graphics. In addition, the brainstorming sessions usually involve all the people on the documentation team.

DRAFTING The drafting stage of manual writing is much the same as it is with any other kind of technical writing. Sometimes one person collects information from all the people on the documentation team and creates a draft from it. Sometimes different subject-area specialists write their own drafts, which are then revised or rewritten by one person or, for very large manuals, the writing team.

During the drafting stage the graphic aids are created or assembled.

REVISING Because manuals are bigger and more complex than most other kinds of technical writing, revision is a more substantial process. In fact, revision is a series of different kinds of checks: from revisions of technical information down to revisions of structure, organization, emphasis, and style.

Revising manuals usually involves field testing. Because a manual is, above all, a practical teaching document, the documentation team wants to make sure that it works well. The general approach is to try out the manual on people with backgrounds similar to those of the eventual readers. A number of computer manufacturers, for example, hire people and have them sit before the computer and try to perform the tasks described in the draft of the manual. Everything is recorded during the test. Test subjects are videotaped so that the documentation team can see what they are doing. Every action the test subjects make—every keystroke or command—is electronically recorded for analysis.

This sort of field testing, if performed carefully with enough test subjects, can be an extremely valuable way to find out which parts of the manual are effective and which aren't. The ineffective portions are revised, and then the manual is retested. This process is continued until the manual is as effective as it can be or until time or money runs out.

THE
STRUCTURE
OF THE
MANUAL

There is no single way to structure a manual, of course. However, many organizations have found through experience that certain components make the manual easier to use and therefore more effective. In general, a manual consists of the following three components:

1. the front matter
2. the body
3. the back matter

THE FRONT MATTER   The front matter consists of everything before the body of the manual. All manuals have a cover or title page and a table of contents. In addition, most manuals contain a preface and a how-to-use-this-manual section. Sometimes, however, these last two items are simply combined in an introduction.

Whether to use a cover or just a title page is determined by the manual's intended use and size. Manuals that will receive some wear and tear—such as maintenance manuals for oil-rigging equipment—will be given a hard cover, usually made of a water-resistant material. Manuals used around an office usually don't need hard covers, unless they are so big that extra strength is needed.

The title page, which in most cases is designed by a graphic artist, contains at least the title of the manual. Some examples follow:

How to Fill Out Credit Application Forms

Filer: An Electronic Database Program

The Murphy 4330 Bottling Machine: Maintenance Manual

In addition, any other identifying information that will help the readers see if they have got the right manual is included. For example:

Filer: An Electronic Database Program

For the IBM PC and All Compatible Models

Or:

The Murphy 4330 Bottling Machine: Maintenance Manual

For 1981-1986 Models

Sometimes a title page includes the names of the authors of the manual. For more information on title pages, see Chapter 11.

The table of contents is especially important in a manual because readers need to find particular information quickly. Manuals are usually referred to for specific information more than they are read straight through; therefore, an extensive table of contents is crucial. The headings in an effective table of contents are easy to understand and focus on the task the reader wants to accomplish. For example, notice in the following excerpt how the writer has used -*ing* verb phrases to highlight the action:

```
DELETING, INSERTING, AND MOVING CELLS    31
    Deleting Cells                       31
    Inserting Cells                      32
    Inserting Blank Columns and Rows     32
    Sorting Rows                         33
```

Another effective way to phrase headings is to use infinitive phrases preceded by "how":

```
HOW TO DELETE, INSERT, AND MOVE CELLS    31
    How to Delete Cells                  31
    How to Insert Cells                  32
    How to Insert Blank Columns and Rows 32
    How to Sort Rows                     33
```

For more information on headings, see Chapter 5.

The other introductory information, which sometimes precedes and sometimes follows the table of contents, takes a number of forms. The first page of text in the manual, for instance, can be called "Introduction" or have no heading at all. Or the first page of text can be called "Preface," which is a brief statement describing the manual. A manual with a preface often has a section called "How to Use This Manual" or some similar phrase.

Regardless of what you call this front matter, you want to answer several basic questions for your reader:

1. Who should use this manual?
2. What product, procedure, or system does it describe?
3. What is its purpose?
4. What are its major components?
5. How should it be used?

These questions need not be answered in separate paragraphs. Sometimes several of the answers can be combined in a single sentence or paragraph.

Following is an example of an introduction that answers all these questions. The manual is entitled "Travel Procedures for De-John Nursing Service Employees":

As a DeJohn Nursing Service representative, you will spend approximately 40 percent of each year on business-related overnight travel. Like any other organization, DeJohn has a specific set of procedures and regulations that its representatives are expected to follow when preparing for, and after returning from, business trips.

This manual has been designed to familiarize you with DeJohn travel procedures. It is divided into two parts: (1) a detailed discussion of travel regulations and procedures that apply to all business trips and (2) a detailed discussion of additional procedures that apply to trips that include training seminars.

In addition, a glossary and samples of filled-out travel forms are included in the appendices. The manual concludes with an index.

Read the entire manual before your first trip for DeJohn. The manual contains valuable information that will help you make your trip pleasant and productive.

If you have any questions, please call Ms. Calkins in Personnel, x3229.

The following example, from the tutorial manual for a software program, divides the same kinds of information into a preface and a how-to-use-this-manual section.

PREFACE

Congratulations. You are now the owner of Data-Ease, a powerful data-base program for the Apple IIe® computer. With Data-Ease, you will soon be able to manipulate data in ways you never thought possible. You will be able to enter, list, categorize, summarize, calculate, and report data.

This is the Tutorial Manual. It will show you how to use those functions of Data-Ease that can be accessed by menu selections. Most of you will find that menu selections will fulfill almost all your data-base management needs.

The other manual, the Reference Manual, discusses some additional features of Data-Ease: the text editor and the E-ZEE programming language.

Before you turn to the next page, How to Use the Data-Ease Tutorial Manual, fill out the attached License Agreement. When we receive your License Agreement, you are a member of the Data-Ease Users Group, which will entitle you to receive information and updates on this software package and our other new releases.

Good luck!

HOW TO USE THE DATA-EASE TUTORIAL MANUAL

The Data-Ease Tutorial Manual is structured so that you can read it like a book or refer to it like a reference source.

Everyone should read "Data-Ease in a Snap," the introductory chapter that provides an overview of the program. Even if you have used other data-bases in the past, you'll find powerful new features described here— and you'll find out how to read more about them.

The Tutorial Manual is divided into three major sections:

Section 1. Learning the Elementary Commands

Section 2. Learning the More Advanced Commands

Section 3. Using More than One Relation

If you want to skip from section to section, here is a brief summary of each chapter:

Chapter 1. Data-Ease in a Snap. Tells you how to start the program, use the basic commands, add a record, and print a report.

Chapter 2. Creating a Report.

Chapter 3. Printing a Report

(NOTE: Chapters 9 and 10, on how to edit reports and on the advanced features of the screens, tell you more about the report generator.)

If at this point you want to create your own applications, read the following three chapters:

Chapter 8. Using the Structure of Your Relations

Chapter 9. Modifying the Shape of Your Relations

Chapter 10. Creating Your Own Relations

Chapter 4. Modifying the Content of Your Relation. Tells you how to manipulate records within a single relation.

For information on how to use the Query function, read the following two chapters:

Chapter 5. Using the Query

Chapter 7. Using Small Groups of Records

Chapter 6. Entering Data. Tells you how to enter data into more than one relation at the same time.

To learn how to create simple and complex screens, read the following three chapters:

Chapter 12. Formatting a Data Entry Screen

A FEW NOTES ABOUT PUNCTUATION

We use single quotation marks to set off commands. When you see double quotation marks in this manual, type them.

One other point: Standard punctuation puts commas and periods within quotation marks. For clarity, we put them outside. So if you see a comma or a period as part of a command, type it.

THE BODY The body of a manual might look like the body of a traditional report, or it might look radically different. Its structure, style, and use of graphics will depend on its purpose and audience. The body of a manual might have summaries and diagnostic tests to help the readers determine whether they have understood the discussion.

The body of a manual is structured according to the way the reader will use it. For example, if the manual describes a process that the reader is supposed to carry out, the manual will probably be structured chronologically. If the manual describes a system that the reader is supposed to understand, the manual might be structured from more important elements to less important elements. The various organizational patterns discussed in Chapter 3 are appropriate, but, as always, the writing situation might call for another pattern.

The writing style in a manual must be clear. Simple, short sentences and common vocabulary are necessary. Use the imperative when giving instructions. For some kinds of manuals, especially those used by readers untrained in the subject being discussed, a rather informal style is appropriate. Contractions, casual vocabulary, and a sense of humor are common. A software manual, for example, might have a section titled "The Lazy Person's Guide to Spreadsheets." However, the reference manual for a computer system probably would be more formal.

Graphic aids are as important in manuals as they are in instructions. They break up the text and, in many situations, are easier to understand than words. Whenever you want your readers to perform some action with their hands, include a diagram or photograph showing the action being performed. Whenever possible, use appropriate tables and figures.

Following are two samples from the bodies of manuals. The first sample is from the manual describing travel procedures used at DeJohn Nursing Services. The excerpt is from the section on how to make hotel arrangements.

2.1.1 How to Make Hotel Reservations

As with ground transportation, it is your responsibility to make hotel reservations for yourself for each stop on your trip. If you are not familiar with your destination, check with your co-workers or consult the Hotel Guide. If one of your co-workers has been to the area, he or she might be able to suggest an acceptable hotel close to the exhibit or seminar site. The Hotel Guide lists most of the hotels in the United States, Canada, Mexico, and the Caribbean. The Guide is divided alphabetically by country, then further divided by state or province, and then city or town.

DeJohn prefers that, whenever possible, you stay in a moderately priced hotel that charges between $60 and $100 per night for a single room. This is not, however, an unbreakable rule. In some areas, an average room will cost less; in other areas, it will cost more. DeJohn does not expect you to stay in an unacceptable room, or in one located far from the exhibit site just to save money.

This sample from a manual is similar to any other kind of technical writing. There are no graphic aids because the subject does not lend itself to them. The writer is trying to explain an idea that cannot be visualized.

Figure 9-3, an excerpt from the user's guide for Microsoft® Multiplan® for the Apple® Macintosh (Microsoft Corporation 1984: 49–50), shows a creative use of type, spacing, and graphics.

The excerpt begins with a chapter heading followed by the topic sentence. The excerpt covers the first third of the chapter: changing cell contents. Notice the use of the phrases in the left-hand margin: they serve as headings that clarify the text. The items preceded by the arrows are steps that the reader is to carry out in sequence. The items preceded by boxes represent alternative actions that the reader might want to carry out. The graphic shows what appears on the screen when the actions being described are actually carried out. This strategy, of course, is the best way to clarify the verbal description. The excerpt concludes with traditional paragraphs that furnish explanations and advice.

THE BACK MATTER   The audience and purpose of the manual will determine what makes up the back matter. However, two items are common: a glossary and an index.

A glossary, which is discussed in Chapter 11, is an alphabetized list of definitions of crucial terms used in the document. An index is common for most manuals of 20–30 pages or more. The index, along with the table of contents, is the principal accessing tool. If a person wants to read about a particular item—such as the query function in a software package—the index will be the easiest way to find the references to it.

Once you have entered information in the worksheet, you can change it, erase it, or copy it elsewhere in the worksheet.

Changing Cell Contents

You can change the information in a cell simply by selecting the cell and typing new information. Or, you can change part of a cell's contents without retyping.

First, make the formula bar active:

► Move the pointer anywhere in the formula bar.

► Click the mouse button. This makes the formula bar active.

Current cell address _ Cancel icon __ Cell contents

```
 é \File  Edit  Select  Format  Options  Calculate
```

| R2C1 | (X) | Advertising| |

Untitled

	1	2	3	4	5	6
1						
2	Advertising					

Character insertion point

To move the character insertion point, move the pointer and click the mouse button.

An Active Formula Bar

To edit the contents of the selected cell:

- Delete the character to the left of the character insertion point by pressing the Backspace key.
- Insert text to the left of the character insertion point by typing the text.
- Select characters by clicking and dragging the mouse across the characters. The selected characters will appear as white characters on a black background.
- Delete selected characters by pressing the Backspace key.

- Select an entire reference, name, number, or operator by double clicking the mouse button. "Double click" means to click the mouse button twice, very fast.
- Reselect the entire contents of the formula bar by pressing and dragging over the contents with the mouse.
- Cancel editing changes by clicking the Cancel icon.
- Put the edited contents back in the selected cell and recalculate the worksheet by pressing the Enter or Return key.
- If you clicked the Cancel icon to restore the cell's original contents, you don't have to press the Enter key.

You can also use the Cut, Copy, and Paste commands to edit the contents of a cell in the formula bar. These commands work only on the characters you have selected. For more information, see the descriptions of these commands in "Commands" in "Multiplan Reference."

When the formula bar is active, the Calculate Now command calculates the value of the formula in the formula bar, then replaces the formula with its result. Used this way, it is similar to Calculator from the Apple menu.

FIGURE 9-3

Excerpt from a Manual

221

When you make a lot of changes	Multiplan calculates the worksheet whenever you change any cell's contents and press the Enter key. If you have a large and complex worksheet, calculation could take several minutes. When calculation takes more than a few seconds, Multiplan displays the word *CALC* on the worksheet, and counts off the cells as they are recalculated. If you have a lot of changes to make, it is more convenient to suspend automatic calculation by choosing the Manual Calculation command from the Calculate menu. When you want Multiplan to recalculate the worksheet, you can choose the Calculate Now command. Or, you can turn calculation on again by choosing the Automatic Calculation command. See "Commands" in "Multiplan Reference" for more information.

FIGURE 9-3

Excerpt from a Manual (Continued)

Creating an index is a difficult task because it involves making decisions about content, such as whether an item is important enough to list or whether an item is a subset of another item. And making the entries consistent requires sustained concentration. Much of the labor involved in creating an effective index is being performed today by computers. Rather than having to read through the whole manual to find every reference to a term (such as "query function") the indexer can simply use the search function to find the references. When the writers have used vocabulary consistently, this technique is very effective.

Appendixes to manuals can contain many different kinds of information. Procedures manuals often have flow charts or other graphic aids that picture the processes described in the body of the manual. These graphics can be removed from the manual and taped to the office wall for frequent reference. Users' guides often have diagnostic tests. Reference materials—error messages and sample data lists for computer systems, troubleshooting guides, and the like—are common. For a further discussion of appendixes, see Chapter 11.

The following excerpt (Plum 1987) is from a procedures manual for student aides working at the information desk in the student center at a university. The manual was written by a student. Although the table of contents covers the entire manual, only the first section of the manual is included here. Also included here is the letter of transmittal that accompanies the manual. For a discussion of letters of transmittal, see Chapter 11.

Brief marginal comments have been added.

403 Franklin Street
Philadelphia, PA 19104

December 3, 19--

Mr. Thomas Williams
Director, Hamilton Student Center
Eastern University
Philadelphia, PA 19104

Dear Mr. Williams:

I am enclosing a copy of the first draft of the <u>Hamilton Student Center Student Aide Manual</u> that I have written.

The writer explains the purpose of the manual.

I wrote this manual because student aides who work at the Information Desk need a training and reference guide. At present, there is no such manual, and as a result staff members spend an inordinate amount of time training new student aides each quarter. This manual is designed to help shorten the training period.

This paragraph provides an overview of the manual.

The manual is divided into three sections: Procedures, Resources, and Policies. Section 1 deals with the daily routine of the Information Desk, explaining the procedures that aides will be asked to perform. Section 2 discusses the resources available to student aides to help them work more efficiently. Section 3 sets forth the policies of the Hamilton Student Center. I have also included appendices listing frequently called telephone numbers and answering frequently asked questions, along with floor plans of HSC, a map of campus, and a sample timesheet.

Here the writer explains why she has decided *not* to discuss a particular subject.

I have deliberately omitted a discussion of the duties specific to the Game Room, because Game Room aides do little besides make change and provide a physical presence to deter vandalism. However, because they are sometimes called upon to work at the Information Desk in the absence of the regular desk worker, I hope that this manual will be given to them as well as to the Information Desk aides.

The writer concludes politely and offers her assistance with future projects.

I have enjoyed writing this manual and welcome your suggestions for its revision. If you would like my help with other projects, such as a codification of the responsibilities of the Evening/Weekend Manager, please do not hesitate to contact me.

Sincerely,

Sandra D. Plum

Sandra D. Plum

Enclosure

HAMILTON STUDENT CENTER STUDENT AIDE MANUAL

Prepared for: Thomas Williams, Director

                Hamilton Student Center

     by: Sandra D. Plum

December 3, 1987

CONTENTS

Notice the writer's use of capitalization, underlining, and indention to clarify the hierarchical relationships among the various headings.

Notice too her use of leader dots to direct the reader's eye to the page number.

(The preface should start on a new page.)

Purpose of the manual

An overview of the organization of the manual

PREFACE

This manual is intended as a training and reference guide to help you in your work at the Hamilton Student Center Information Desk.

This manual is divided into three sections. The first section, "Procedures," discusses the daily routine of the Information Desk, those duties that you will be called upon to perform on a regular basis. The second section, "Resources," lists some of the publications, phone numbers, and people available to help you do your job. The list is not complete but may help you discover some different places to look for answers to your own and other people's questions. The final section, "Policies," explains some general guidelines established by the Hamilton Student Center.

The appendices should be quite useful to you and I hope that you will take time to look them over. Appendix A lists some frequently called phone numbers and Appendix B answers some frequently asked questions. Appendices C and D are plans of Hamilton Student Center and the Eastern campus, respectively. Finally, Appendix E shows a sample timesheet used for payroll.

I hope this manual proves helpful to you in your job here at the Hamilton Student Center.

(Each first-level heading should start on a new page.)

1.0 PROCEDURES

The writer introduces the topics she will be describing in detail.

Your job as a student aide at the Information Desk involves many different tasks, some routine and some not at all routine. The more predictable things you are asked to do daily include signing in each time you work, answering the telephone, operating the cash register, signing out equipment, and doing photocopying.

Notice how the writer uses a wider left margin to emphasize that this is a subsection.

1.1 SIGNING IN

You must sign in each time you work. The sign-in sheets are located on the center post inside the Information Desk and are used by the secretary for calculating the payroll. If you do not sign in, you will not be paid.

There are two sign-in sheets. If you usually work in the Game Room, sign in on the sheet marked "Game Room"; if you usually work at the Information Desk, sign in on the "Main Desk" sheet. Even if you are covering a different area than you normally do, sign in on the sheet for the location to which you are normally assigned. This system facilitates bookkeeping because Game Room and Information Desk aides are paid from different payroll accounts.

Be sure to sign in on the correct line and for the correct days and times. If you are late, sign in for the time you do come in, and if you work overtime, sign out for the time you actually leave. Times must be recorded by the quarter-hour, however, so you may need to estimate slightly. (See Appendix E for a sample timesheet.)

The writer clearly describes the reasons behind the procedures.

1.2 ANSWERING THE TELEPHONE

One of your primary duties is answering the telephone. Because many callers receive their first impression of Hamilton

Student Center, and often Eastern itself, from the manner in
which the Information Desk presents itself, it is very important
to be courteous and friendly when answering the telephone.

Greet the caller by saying either, "Good morning (afternoon, evening), Hamilton Student Center," or "Hamilton Student
Center. May I help you?" Do not merely state "Hamilton Student
Center." Some callers may be put off by the abruptness of this
approach.

Respond to the person warmly; he or she is calling an information number and may need encouragement to state question(s).
Always strive to leave the caller with the feeling that HSC
cares, and never hang up without either getting the information
that the caller needs or referring him or her to a source that can
help out (see Section 2.0, RESOURCES).

1.2.1   Taking Messages

When a caller wishes to speak with a staff member who
is unable to take the call, fill out a pink "WHILE YOU WERE
OUT" phone message form. Be certain to record a telephone
number where the caller can be reached, even if the person
has left previous messages. (This prevents the problem of
the recipient's needing to locate another message.) Put
the message on the staff member's desk. If the call is for
the Director or Assistant Director, and that person is
busy, give the message to the secretary, and he or she will
forward it when the recipient is available. Do not hold
messages until the end of your shift.

1.2.2   Transferring Calls

There are two ways to transfer calls--one method for
transferring calls within the office, and another for
transferring calls to other offices within the University.

1.2.2.1   Inside the Office

Transferring calls to the secretary is simple
because of the proximity of the telephones. Merely

inform the secretary that a call is waiting and identify the incoming line (3515, 3516, or 3517).

However, transferring calls to the Director or
Assistant Director is not so easy. First, try to as-
certain if he or she is already on another call. If
more than one button (line) on the telephone is lit,
and you are not sure who is on the other line, check
with the secretary before transferring the call. If
neither of the other lines is lit, and you are reasonably sure that the staff member is both in the office
and free to take a call, you may then transfer it by
"buzzing" the office.

There are two buttons that buzz the offices. The
button for the Assistant Director is on the telephone
at the extreme right and is marked with her name. The
button for the Director is mounted on the wall to the
right of the telephone and is also marked.

For step-by-step in-
structions, the
writer uses a num-
bered list in single-
spaced format.

To transfer a call, do the following:

1. Place the caller on hold by pressing in the
   red "Hold" button firmly. The light on the
   extension line will begin to blink.
2. Press the appropriate buzzer briefly, using
   the following code:
   (a) one buzz for 3515
   (b) two buzzes for 3516
   (c) three buzzes for 3517
3. Check that the recipient picks up the phone.
   (The extension light will stop blinking.) If
   he or she does not, press in the extension
   button again, express your regret to the
   caller that you are unable to put through the
   call, and take a message so that the call may
   be returned later.

1.2.2.2 Outside the Office

If the caller wants to be connected to another
office inside the University, first find out if he or
she is telephoning from inside Eastern. If so, tell
him or her the number of the extension desired--calls
cannot be transferred from within the university it-

self. If the caller is from outside the university and would like to be transferred, do the following:

1. Tell the caller the extension number (complete with the Eastern prefix 695) to which you are transferring the call. Doing this will enable the caller to dial directly in case you accidentally disconnect the call.
2. Press the receiver button down firmly but briefly, and listen for a sound similar to a dial tone. This step must be repeated if you do not hear the sound.
3. Dial the extension number. If it rings, you may hang up. If the extension is busy, however, press the receiver again (which will reconnect you to the caller), ask the person to try the call again later, and repeat the telephone number if necessary.

1.3 OPERATING THE CASH REGISTER

A major service of the Information Desk is handling money transactions: providing change for visitors, staff, and students, selling parking tokens, and collecting fees for parking tickets and bowling. The cash register is simple to operate. You need only remember a few guidelines.

The cash register can be used as a storage device that does not monitor the flow of money through it, or it can be used to record (ring up) those transactions in which money is actually received over the amount that the drawer normally contains.

To use the register as a simple holding device, open it by pushing the "total" key only. To ring up a sum, do the following:

1. Enter the amount of the transaction on the tens, ones, tens-of-cents, and cents keys.
2. Push the appropriate numerical key for the account to which the sum is being desposited.
3. Hit the "total" key.

For example, to ring up $56.82 on key #1, you press the keys for $50 and $6 and 80¢ and 2¢, then key #1, and finally "total." The register drawer will open and the transaction will be recorded on the cash register tape inside the machine. The receipt button is kept depressed unless a receipt is needed. If the customer needs a receipt of the transaction, flip the lever at the top right of the register before ringing up the transaction. Doing this will make the receipt button at the lower left of the register pop out, and the transaction will be recorded on a tape that will emerge from the receipt slot when you press the "total" key. If nothing has been rung up on a key, the receipt will list the amount as $0.00.

1.3.1 Making Change

When you make change, do not ring up the amount. Simply hit the "total" key, which will open the register, and substitute the denominations. If you need more change, ask the secretary or the desk clerk to get more from the safe. Always verify the amount you have received from the customer and count the change in front of the person.

1.3.2 Selling Tokens

The HSC Information Desk sells tokens for use in the Eastern Parking Garage. Each token costs $1.00, and two tokens are required for each stay at the facility. Do not ring up parking tokens; just open the register by hitting "total." The tokens are kept in a compartment at the far left side of the register drawer. Money for tokens is also put in that compartment--not with the other cash.

1.3.3 Validating Parking Tickets

Parking tickets issued by Eastern University may be paid at the Information Desk. Each ticket must be validated by placing it in a slot at the left side of the register while the transaction is being rung up. Parking tickets are rung up on key #10.

To accept payment for a ticket, do the following:

1. Check the ticket for the total amount. This amount is usually $3.00 but may be much higher, depending upon the number and type of the outstanding tickets.

2. Flip the lever at the top right of the cash register in order to obtain a receipt for the ticket amount.
3. Place the ticket in the slot under the cash register at the left side. Be sure to push the ticket firmly against the register so the ticket will be properly validated.
4. Enter the amount of the ticket. Next hit key #10, and finally hit the "total" key.

The register drawer will open and a receipt will pop up from its slot. Put the cash or check and the validated ticket in the left-hand compartment where the tokens are stored. Give the cash-register receipt to the person as proof of payment. If you forget to validate the ticket, do not ring it up a second time in order to validate it. Doing so will cause the amount recorded on the register tape to double. Instead, check the three-digit transaction number on the register receipt and give that number to a staff member for verification of the cash register tape.

Validated tickets are picked up weekly by a Parking Garage representative.

1.3.4   Accepting Bowling Fees

Fees for Physical Education bowling classes are taken at the HSC Information Desk. Students usually pay these fees during the first week of the term. Bowling classes currently cost $25.00 per term; ring up the fees on key #12. Students may pay in cash or by check made out to Eastern University. Do not accept credit cards or traveler's checks. Students paying bowling fees must receive a written receipt to give to their instructor as proof of payment.

The following information must be recorded on the receipt:

- student's name
- amount paid

- days and times of classes (e.g., M, W--1:30 p.m.)
- three-digit transaction number (from the register receipt)
- signature of person receiving the payment

1.4   SIGNING OUT EQUIPMENT

Because of the possibility of theft from the HSC Information Desk, it is necessary to check I.D.s and often request that I.D.s be left at the desk in exchange for certain borrowed items. I.D.s are placed inside the key box located on the center post inside the desk.

1.4.1   Magazines

HSC has a wide variety of popular magazines available for browsing. These magazines, stored in a vertical file next to the cash register, may be borrowed by any member of the Eastern community. You must obtain a valid I.D. from the person and place it in the key box until the magazine is returned.

1.4.2   Copier Key

The Information Desk holds the only key to the copier machine located next to the HSC main entrance. Members of organizations funded by the Student Activity Committee (SAC) are allowed to use the copier key to obtain free photocopies of organization-related documents. The names of the funded organizations are listed on a sheet of paper hanging on the Information Desk post. Only these organizations may use the copier key.

When a student identifies himself or herself as an organization member and requests the key, do the following:

1. Check the list to confirm that the organization is funded. Some student organizations are not SAC funded and therefore cannot use the copier key.

2. Ask the student to sign in on the log (kept next to the telephone) with his or her name, the organization's name, and the number of copies being made. Note that the total number of copies per day for any organization is limited to 25.

3. Take the student's I.D. and put it in the key box. If the student does not have an I.D., he or she must obtain a temporary one from the Off-Campus Housing office (room 222) before you give out the key.

4. Remove the copier key (#61) from the key box and give it to the student.

When the student returns the key, return the I.D.

1.4.3   Piano Room Keys

Four piano-practice rooms are located in the Performing Arts Department on the second floor of Phillips Hall. Both the Information Desk and the Performing Arts Department have keys to these rooms; however, to avoid confusion and possible double-booking of rooms, only one department issues keys at any one time. During Eastern working hours, keys to the practice rooms must be obtained from the Performing Arts Department. Beginning at 4:45 p.m., though, only the HSC Information Desk gives out keys. You must obtain an Eastern I.D. from the person and place it in the key box in exchange for the key. The piano rooms are designated A, B, C, and D, and the corresponding keys are numbered 63, 65, 67, and 69.

Do not give out a piano room key before 4:45 p.m. on a weekday without first checking with a staff member.

1.4.4   Typewriters

HSC has the only typewriters available for student use on the campus. These typewriters are located in the room at the rear of the secretary's office. Currently, HSC has three IBM electric and four Royal manual typewriters available for use. These machines may be used from 8:45 a.m. until one-half hour before the building closes on weekdays and from the time the building opens until one-half hour before it closes on weekends.

To use a typewriter, the person must show a valid Eastern I.D. and must sign in on the log kept at the secretary's desk. Do not ask the person to surrender the I.D.--merely check it for current validation.

Typewriters may only be used by Eastern-affiliated persons.

1.5   COPYING MATERIALS

You may be requested by staff members to photocopy official HSC materials. Two copiers are used; one is located on the second floor of Hamilton and the other on the fifth floor of Phillips. The copier you use will depend on the number of copies you are making.

1.5.1   Up to 20 Copies

If you are asked to make 20 copies or fewer, use the Hamilton copier located in room 245 across the hall from the office of the Vice-President for Student Affairs. The key to the room is kept in room 215, at the desk of the Dean of Student's secretary. Be sure to return the key immediately to the secretary's desk.

1.5.2   More Than 20 Copies

When you are making more than 20 copies, use the copier in room 5147 on the fifth floor of Phillips. This copier is operated by an auditron, which you must obtain from the secretary in the Dean of Humanities/Social Sciences' office. She will show you where the copier room is located if you are unfamiliar with the area and can instruct you in the use of the auditron.

The following questions cover instructions:

1. Does the introduction to the set of instructions
   a. State the purpose of carrying out the task?
   b. Describe safety measures or other concerns that the readers should understand?
   c. List necessary tools and materials?

2. Are the step-by-step instructions
   a. Numbered?
   b. Expressed in the imperative mood?
   c. Simple and direct?

3. Are appropriate graphic aids included?

4. Does the conclusion
   a. Include any necessary follow-up advice?
   b. Include, if appropriate, a troubleshooter's guide?

The following questions cover manuals:

1. Does the manual include, if appropriate, a cover?

2. Does the title page provide all the necessary information to help readers understand whether they are reading the appropriate manual?

3. Is the table of contents clear and explicit? Are the items phrased to indicate clearly the task the reader is to carry out?

4. Does the other front matter clearly indicate
   a. The product, procedure, or system the manual describes?
   b. The purpose of the manual?
   c. The major components of the manual?
   d. The best way to use the manual?

5. Is the body of the manual organized clearly?

6. Are appropriate graphic aids included?

7. Is a glossary included, if appropriate?

8. Is an index included, if appropriate?

9. Are all other appropriate appendix items included?

10. Is the writing style clear and simple throughout the manual?

EXERCISES

1. Write a set of instructions for one of the following activities or for a process used in your field. Include appropriate graphic aids. In a brief note preceding the instructions, indicate your audience and purpose.

   a. how to load film into a 35-mm camera
   b. how to change a bicycle tire
   c. how to parallel-park a car
   d. how to study a chapter in a text
   e. how to light a fire in a fireplace
   f. how to make a cassette-tape copy of a record
   g. how to tune up a car
   h. how to read a corporate annual report
   i. how to tune a guitar
   j. how to take notes in a lecture class

2. In an essay, evaluate the effectiveness of the following set of installation instructions for a set of outdoor shutters.

   OUTDOOR SHUTTER INSTALLATIONS

   1. Draw a 4″ to 6″ light vertical line from the top and bottom, 1″ from the sides of the shutter. Locate screw locations by positioning shutter next to window and marking screw holes. If six or eight screws are to be used, make sure they are equidistant.
   2. Drill ³⁄₁₆″ holes at marked locations. DO NOT DRILL INTO HOUSE. Drilling should be done on the ground or a workbench.
   3. Mount the shutter screws.
      Notes: 1. For drilling into cement, use a cement bit.
      2. Always wear eye protection when drilling.
   4. Make sure shutters are mounted with the top side up.
   5. If you are going to paint shutters, do so before mounting them.

3. In an essay, evaluate the effectiveness of the set of installation instructions for a sliding door for a shower found on p. 238.

4. Write a brief manual for some process or system you are familiar with. In looking for topics, consider any activities you carry out at school. If, for instance, you are the business manager of the newspaper staff, you could write a manual describing how to perform the duties of that position. Off-campus activities, such as participation in civic groups, are another source of ideas. Part-time jobs are a source of ideas for procedures manuals.

5. In an essay, evaluate the effectiveness of the following excerpt from a manual. The manual is titled "A Reference Manual for the Swimming Instructor." Presented here are the preface, table of contents, and one section from the body: the discussion on teaching the backstroke.

INSTALLATION INSTRUCTIONS

CAUTION: SEE BOX NO. 1 BEFORE CUTTING ALUMINUM HEADER OR SILL

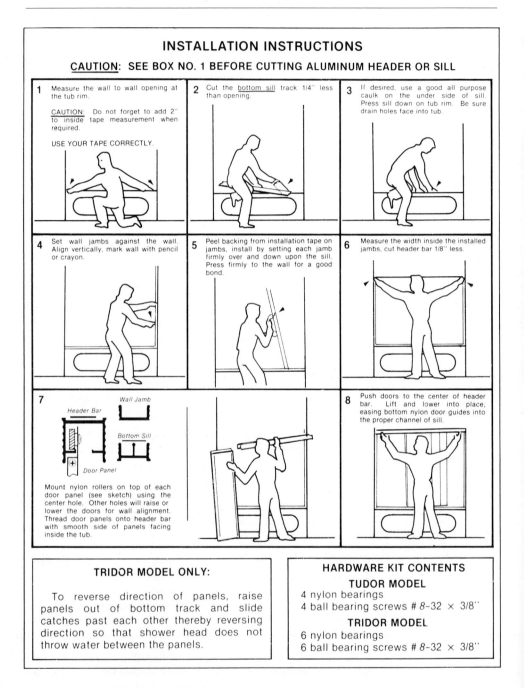

1 Measure the wall to wall opening at the tub rim.

CAUTION: Do not forget to add 2" to inside tape measurement when required.

USE YOUR TAPE CORRECTLY.

2 Cut the bottom sill track 1/4" less than opening.

3 If desired, use a good all purpose caulk on the under side of sill. Press sill down on tub rim. Be sure drain holes face into tub.

4 Set wall jambs against the wall. Align vertically, mark wall with pencil or crayon.

5 Peel backing from installation tape on jambs, install by setting each jamb firmly over and down upon the sill. Press firmly to the wall for a good bond.

6 Measure the width inside the installed jambs, cut header bar 1/8" less.

7

Wall Jamb

Header Bar

Bottom Sill

Door Panel

Mount nylon rollers on top of each door panel (see sketch) using the center hole. Other holes will raise or lower the doors for wall alignment. Thread door panels onto header bar with smooth side of panels facing inside the tub.

8 Push doors to the center of header bar. Lift and lower into place, easing bottom nylon door guides into the proper channel of sill.

TRIDOR MODEL ONLY:

To reverse direction of panels, raise panels out of bottom track and slide catches past each other thereby reversing direction so that shower head does not throw water between the panels.

HARDWARE KIT CONTENTS

TUDOR MODEL
4 nylon bearings
4 ball bearing screws # 8-32 × 3/8"

TRIDOR MODEL
6 nylon bearings
6 ball bearing screws # 8-32 × 3/8"

Preface

A Reference Manual for the Swimming Instructor is designed to help the swimming instructor teach the different strokes required for Red Cross certification. Because the manual is written in plain language, it is also a useful tool for beginning swimmers.

The manual covers the different strokes, teaching methods, games, workouts, and other subjects.

Over the course of the years, I have gained much valuable information from different instructors, to all of whom I am very grateful.

CONTENTS

Elementary Backstroke

Beginning swimmers are often afraid to put their heads in the water. Because the elementary backstroke involves keeping the swimmer's head out of the water at all times, it is a good stroke to teach first.

Initial Body Position

The swimmer starts in a horizontal, supine position. The back is kept almost straight, with the face always clear of the water. The arms are placed alongside the body, with the palms touching the thighs. The legs are together with the toes pointed forward. Keep the hips near the surface at all times.

Arm Stroke

The arm stroke for this stroke is continuous. At the start of the recovery phase, the palms are drawn upward alongside the body until they reach the armpits. The wrists are then rotated outward so that the fingers point away from the shoulders and the palms face the feet. A breath is taken at this point. Then the arms are put out straight toward the sides. The swimmer's body will look like a T. The student should not extend the arms or hands above the shoulder level. Then the palms and arms press back and down toward the feet. The student then recovers to the glide position.

Kick

There are two kicks that can be used with the elementary back-stroke. They are the frog kick and the inverted wedge.

In the wedge kick, which is the favorite for all students except those with limited rotational ability in the hips or knees, the legs recover from the glide position by bending and separating at the knees. The heels move toward the bottom of the pool and toward the rear. The feet are relaxed with the heels almost touching. At the end of the recovery, the knees should be a little wider than the hips. The ankles should be dorsiflexed with the feet rotated outward.

The next step is propulsion. The feet slide sideways. When they are under the knees they rotate slightly inwards. Then with a rounded, outward, backward, and then inward movement the feet and legs are brought back together. This movement starts slowly and then accelerates. The glide is then resumed. It is imperative that the emphasis be placed on the backward movement and not on the sideways movement.

The frog kick must be performed in a plane horizontal to the surface. Recovery is begun by bending the legs and separating the knees laterally as the heels are drawn toward the rear. The heels are kept close to each other. Then the ankles are dorsally flexed and rotated outward so that the toes point away from and to the side of the body. The toes lead the legs to a fully extended position outside the hips. Without stopping, the legs press backward and inward with force. The glide position is then resumed.

REFERENCES   Microsoft Corporation. 1984. Microsoft® Multiplan® Electronic Worksheet Program for Apple® Macintosh: 49–50.

National Retail Hardware Association. 1976. How to apply panels to stud walls.

Plum, S. L. 1987. "Hamilton Student Center Student Aide Manual." Unpublished manual.

CHAPTER
TEN

GRAPHIC AIDS

Graphic aids are the "pictures" of technical writing: photographs, diagrams, charts, graphs, and tables. Few technical documents contain only text, because graphic aids offer several benefits that sentences and paragraphs alone cannot provide.

First, graphics are visually appealing. Watch people pick up and skim a report. They almost automatically stop at the graphic aids and begin to study them. Readers are intrigued by graphics; that in itself increases the effectiveness of your communication.

Second, effective graphic aids are easy to understand and remember. Try to convey in words what a simple hammer looks like. It's not easy to describe the head and get it to fit on the handle correctly. But in 10 seconds you could draw a simple diagram that would clearly show the hammer's design.

Third, graphic aids are almost indispensable in demonstrating relationships, which form the basis of most technical writing. For example, if you wanted to show the profits of the 10 largest corporations in a given year, a paragraph would be a complicated jumble of statistics. A simple graph, however, would provide a meaningful and memorable picture. Graphic aids are also effective in showing relationships over time: the number of nuclear power plants completed each year over the last decade, for example. And, of course, graphic aids can show the relationships among several variables over time, such as the numbers of four-cylinder, six-cylinder, and eight-cylinder cars manufactured in the United States during each of the last five years.

CHARACTERISTICS OF EFFECTIVE GRAPHIC AIDS

A graphic aid is like a paragraph: it has to be clear and understandable alone on the page, and it must be meaningfully related to the larger discussion. When you are writing a paragraph, it is easy to remember these two tasks: having to provide a transition to the paragraph that follows reminds you of the need for overall coherence. When you are creating a graphic aid, however, it is easy to put the rest of the document out of your mind temporarily as you concentrate on what you "draw."

The tendency, therefore, is to forget to relate the graphic aid to the text. Although it is true that some graphics merely illustrate points made in the text, others must be explained to the readers. Graphic aids illustrate facts, but they can't explain causes, results, and implications. The text must do that.

Effective graphic aids are

1. appropriate to the writing situation
2. labeled completely
3. placed in an appropriate location
4. integrated with the text

1. A graphic aid should be appropriate to the writing situation. The first question you have to answer once you have decided that some body of information would be conveyed more effectively as a graphic aid than as text is "What type of graphic aid would be most appropriate?" Some kinds of graphic aids are effective in showing deviations from a numerical norm, some are effective in showing statistical trends, and so forth. Some types of graphic aids are very easy to understand. Other, more sophisticated types contain considerable information, and the reader will need help to understand them. Think about your writing situation: the audience and purpose of the document. Can your readers handle a sophisticated graphic aid? Or will your purpose in writing be better served by a simpler graphic aid? A pie chart might be perfectly appropriate in a general discussion of how the federal government spends its money. However, so simple a graphic would probably be inappropriate in a technical article addressed to economists.

2. *A graphic aid should be labeled completely.* Every graphic aid (except a brief, informal one) should have a clear and informative title. The columns of a table, and the axes of a graph, should be labeled fully, complete with the units of measurement. The lines on a line graph should also be labeled. Your readers should not have to guess whether you are using meters or yards as your unit of measure, or whether you are including in your table statistics from last year or just this year. If the information in the graphic aid was not discovered or generated by you, the source of the information should be cited.

3. *A graphic aid should be placed in an appropriate location.* Graphic aids can be placed in any number of locations. If your readers need the information to understand the discussion, put the graphic aid directly after the pertinent point in the text—or as soon after that point as possible. If the information functions merely as support for a point that is already clear, or as elaboration of it, the appendix is probably the best location for the graphic aid.

Understanding your audience and purpose is the key to deciding where to put a graphic aid. For example, if only a few of your readers will be interested in the graphic, it should be placed in an appendix.

Sometimes the way your readers will react to the graphic will help you determine the best place to put it. For instance, you might be trying to convince your supervisors that they should *not* put any more money into a project. You know the project is going to fail. However, you also know that your supervisors like the project: they think it will succeed, and they have supported it openly. In this case, the graphic aid conveying the crucial data should probably be placed in the body of the document, for you want to make sure your supervisors see it. If the situation were less controversial—for instance, if you knew your readers were already aware that the project is in trouble—the same graphic aid could probably be placed in an appendix. Only the "bottom line" data would be necessary for the discussion in the body.

4. *A graphic aid should be integrated with the text.* Integrating a graphic aid with the text involves two steps. First, introduce the graphic aid. Second, make sure the readers understand what the graphic aid means.

Whenever possible, refer to a graphic aid before it appears. The ideal situation is to place the graphic aid on the same page as the reference. If the graphic is included as an appendix at the end of a document, tell your readers where to find it: "For the complete details of the operating characteristics, see Appendix B, page 24."

Writers often fail to explain clearly the meaning of the graphic aid. When introducing a graphic aid, ask yourself whether a mere paraphrase of its title will be clear: "Figure 2 is a comparison of the costs of the three major types of coal gasification plants." If you want the graphic to make a point, don't just hope your readers will see what that point is. State it explicitly: "Figure 2 shows that a high-sulfur bituminous coal gasification plant is presently more expensive than either a low-sulfur bituminous or anthracite plant, but more than half of its cost is cleanup equipment. If these expenses could be eliminated, high-sulfur bituminous would be the least expensive of the three types of plants."

GRAPHIC AIDS AND COMPUTERS

Recent advances in computerized graphics packages have made it simple for anyone to create a number of graphic aids easily. Sophisticated modern graphics packages can do much more than create traditional graphics such as tables, pie charts, bar graphs, and line graphs. They can create all sorts of line drawings, from flow charts to maps, blueprints, and diagrams. They can turn a photograph into a computerized image, which can then be manipulated. In addition, modern software and printers let you print the graphic aids in different colors for emphasis. Over the next few years the new technology will make it much easier and cheaper to create clear, effective graphics.

One additional benefit is that writers will have greater control over their documents. With word processing and graphics software, writers will be able to control the entire writing process more completely. This means that they will be able to create whatever kind of graphic they want—and put it exactly where they want. Instead of having to give the text to the typist and then fit the graphic aids in at the end of the text, as often happens today, writers will be able to try out a number of organizational patterns to see which one works best.

Consult the bibliography (Appendix E) for a list of books about word processing in technical writing.

TYPES OF GRAPHIC AIDS

There are dozens of different types of graphic aids. Many organizations employ graphic artists to devise informative and attractive visuals. This discussion, however, will concentrate on the basic kinds of graphic aids that can be constructed by a writer who lacks special training or equipment.

The graphic aids used in technical documents can be classified into two basic categories: tables and figures. Tables are lists of data—usually numbers—arranged in columns. Figures are everything else: graphs, charts, diagrams, photographs, and the like. Generally, tables and figures have their own sets of numbers: the first table in a document is Table 1; the first figure is Figure 1. In documents of more than one chapter (as in this book), the graphic aids are usually numbered chapter by chapter. Figure 3–2, for example, would be the second figure in Chapter 3.

TABLES   Tables easily convey large amounts of information, especially quantitative data, and often provide the only means of showing several variables for a number of items. For example, if you want to show the numbers of people employed in 6 industries in 10 states, a table would probably be the best graphic aid to use. Tables lack the visual appeal of figures, but they can handle much more information with complete accuracy.

In general, tables should be structured so that they are read vertically, not horizontally. That is, the reader's eye should travel down the column, not along a row, to see the same parameter change. For example, if you are creating a table to show that the populations of three counties have changed at different rates over the last two years, the table should be structured like Table 1.

Table 1.

Population Changes (1985–1987) of A, B, and C Counties

	Population		
Date	County A	County B	County C
1985	35,912	46,983	53,572
1986	34,124	47,912	53,954
1987	29,876	47,321	62,876

This structure would help your readers see that the population of County A has fallen over the three-year period, whereas that of County C increased dramatically in 1987. Notice that the title of this table reflects a vertical reading of the data.

If, however, you wanted to compare the populations of the three counties year by year, the following structure would be more appropriate:

Table 1.

Population (1985, 1986, 1987) of A, B, and C Counties

	Year		
County	1985	1986	1987
A	35,912	34,124	29,876
B	46,983	47,912	47,321
C	53,572	53,954	62,876

This structure enables your readers to glance down the date columns to compare the populations of the three counties. Notice how the title of the table has changed to reflect the arrangement of the data.

Figure 10-1 shows the standard parts of a table. Notice that it identifies the table with both a number ("Table 1") and a substantive title. The title should encompass the items being compared, as well as the criteria of comparison.

Mallard Population in Rangeley, 1982–1986

Grain Sales by the United States to the Soviet Union in 1985

The Growth of the Robotics Industry in Japan, 1980–1985

Multiple Births in the Industrialized Nations in 1986

Note that most tables are enumerated and titled above the data. The number and title are centered horizontally.

If all the data in the table are expressed in the same unit, indicate that unit under the title:

Farm Size in the Midwestern States

(in Hectares)

If the data in the different columns are expressed in different units, indicate the unit in the column heading.

Population	Per Capita Income
(in millions)	(in thousands of U.S. dollars)

FIGURE 10-1

Parts of a Table

Table 1

Title
(subtitle)

Stub Heading	Column Heading 1	Column Heading 2	Column Heading 3
Stub Category 1			
Item A . . .	data[a]	data	data
Item B . . .	data	data	data
Item C . . .	data	data[b]	data
Item D . . .	data	data	data
Stub Category 2			
Item E . . .	data	data	data[c]
Item F . . .	data	data	data
Item G . . .	data	data	data

Notes: [a]Footnote
       [b]Footnote
       [c]Footnote
Source:

Provide footnotes for any information that needs to be clarified. Also at the bottom of the table, below any footnotes, add the source of your information (if you did not generate it yourself).

The stub is the left-hand column, in which you list the items being compared in the table. Arrange the items in the stub in some logical order: big to small, important to unimportant, alphabetical, chronological, and so forth. If the items fall into several categories, you can include the names of the categories in the stub:

```
Sunbelt States ....................................................
   Arizona .......................................................
   California ....................................................
   New Mexico ....................................................

Snowbelt States ...................................................
   Connecticut ...................................................
   New York ......................................................
   Vermont .......................................................
```

If the items in the stub are not grouped in logical categories, skip a line every four or five items to help the reader follow the rows across the table. Leader dots, a row of dots that links the stub and the next column, are also useful.

The columns are the heart of the table. Within the columns, arrange the data as clearly and logically as you can. Line up the numbers consistently.

```
3,147
  365
46,803
```

In general, don't change units. If you use meters for one of your quantities, don't use feet for another. If, however, the quantities are so dissimilar that your readers would have a difficult time understanding them as expressed in the same units, inconsistent units might be more effective.

```
 3 hr
12 min
 4 sec
```

This listing would probably be easier for most readers to understand than a listing in which all quantities were expressed in hours, minutes, or seconds.

Use leader dots if a column contains a "blank" spot: a place where there are no appropriate data:

```
3,147
 . . .
46,803
```

Make sure, however, that you don't substitute leader dots for a quantity of zero.

```
3,147
   0
46,803
```

Figure 10-2 is an example of an effective table.

FIGURES   Every graphic aid that is not a table is a figure. The discussion that follows covers the principal types of figures: bar graphs, line graphs, charts, diagrams, and photographs.

FIGURE 10-2

Table

Table 6

Test Results for Valves #1 and #2

Valve Readings	Maximum Bypass Cv	Minimum Recirc. Flow (GPM)	Pilot Threads Exposed	Main ΔP at Rated Flow (psid)
Valve #1				
Initial	43.1	955	+3	4.5
Final	43.1	955	+3	. . .
Valve #2				
Initial	48.1	930	+3	4.5
Final	48.1	950	+2	. . .

BAR GRAPHS Bar graphs provide a simple, effective way of representing different quantities so that they can be compared at a glance. The principle behind bar graphs is that the length of the bar represents the magnitude of the quantity. The bars can be drawn horizontally or vertically. Horizontal bars are generally preferred for showing different items at any given moment (such as quantities of different products sold during a single year), whereas vertical bars show how the same item varies over time (such as month-by-month sales of a single product). These distinctions are not ironclad, however; as long as the axes are labeled carefully, your readers should have no trouble understanding you.

Figure 10-3 shows the structure of basic horizontal and vertical bar graphs. When you construct bar graphs, follow four basic guidelines.

1. *Make the proportions fair.* For all bar graphs, number the axes at regular intervals. Using a ruler or graph paper makes your job easier.

For vertical bar graphs, choose intervals that will make your vertical axis about three-quarters the length of the horizontal axis. If your vertical axis is much longer than that, the differences between the height of the bars will be exaggerated. If the horizontal axis is too long, the differences will be unfairly flattened. Figure 10-4 shows two poorly proportioned graphs.

Make all bars equally wide, and use the same amount of space between them. The space between the bars should be about half the width of the bars themselves.

2. *If at all possible, begin the quantity scale at zero.* This will

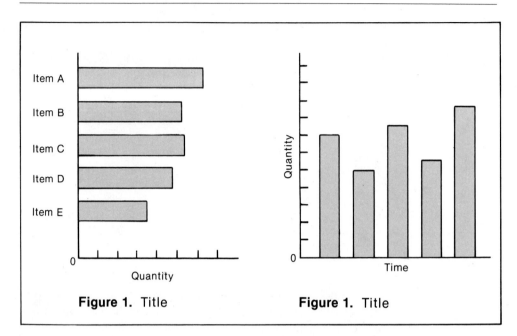

Figure 1. Title   **Figure 1.** Title

FIGURE 10-3

Structure of a Horizontal and a Vertical Bar Graph

ensure that the bars accurately represent the quantities. Notice in Figure 10-5 how misleading a graph is if the scale doesn't begin at zero. Version *a* is certainly more dramatic than version *b*. In version *a*, the difference in the lengths of the bars suggests that Item A is much greater than Item B, and that Item B is much greater than Item C.

If it is not practical to start the quantity scale at zero, break the quantity axis clearly, as in Figure 10-6.

3. *Use tick marks or grid lines to signal the amounts.* Ticks are the little marks drawn to the axis:

Grid lines are ticks extended through the bars:

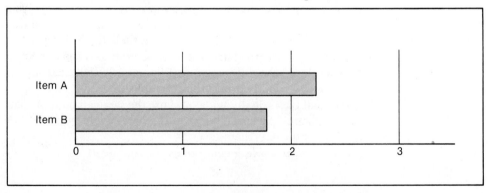

FIGURE 10-4

*Bar Graphs with
Excessively Long
Axes*

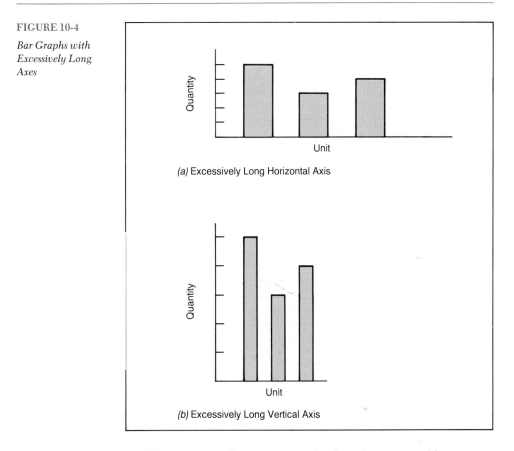

(a) Excessively Long Horizontal Axis

(b) Excessively Long Vertical Axis

Grid lines are usually necessary only if you have several bars, some of which would be too far away from tick marks for readers to gauge the quantity easily.

4. *Arrange the bars in a logical sequence.* In a vertical bar graph, chronology usually dictates the sequence. For a horizontal bar graph, arrange the bars in descending-size order beginning at the top of the graph, unless some other logical sequence seems more appropriate.

Figure 10-7 shows an effective bar graph.

Notice that most figures are titled underneath. Unlike tables, which are generally read from top to bottom, figures are usually read from the bottom up. If the graph displays information that you have gathered from an outside source, cite that source in a brief note near the bottom of the graph.

The basic bar graph can be varied easily to accommodate many different communication needs. Here are a few common variations.

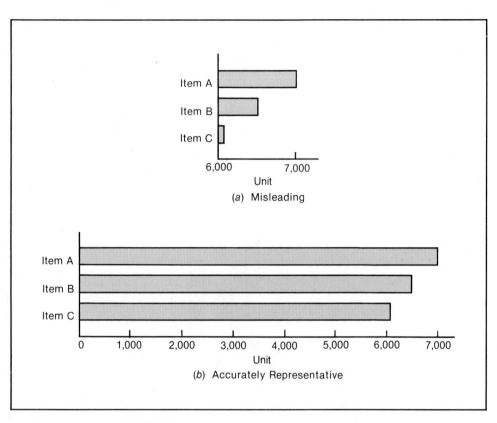

FIGURE 10-5

Misleading and Accurately Representative Bar Graphs

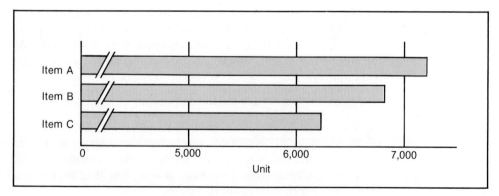

FIGURE 10-6

A Bar Graph with the Quantity Axis Clearly Broken

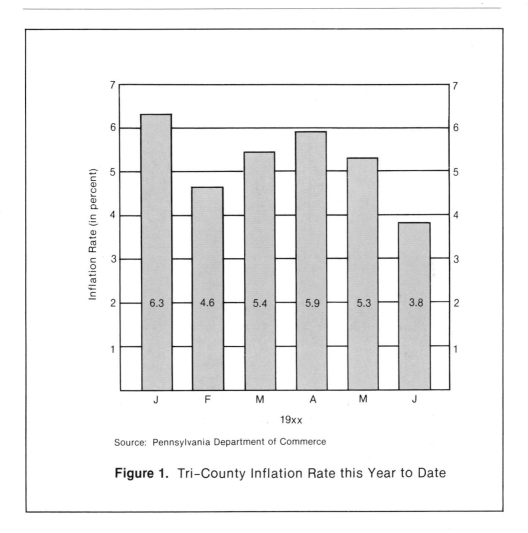

Source: Pennsylvania Department of Commerce

Figure 1. Tri–County Inflation Rate this Year to Date

FIGURE 10-7

Bar Graph

The *grouped bar graph*, such as that in Figure 10-8, lets you show two or three quantities for each item you are representing. Grouped bar graphs are useful for showing information such as the numbers of full-time and part-time students at several universities. One kind of bar represents the full-time students; the other, the part-time. To distinguish the bars from each other, use hatching (striping) or shading, and label one set of bars or provide a key. Leave at least one bar's width between sets of bars.

Another way to show this kind of information is through the *subdivided bar graph*, shown in Figure 10-9. A subdivided bar graph adds Aspect I to Aspect II, just as wooden blocks are placed on one another. Although the totals are easy to compare in a sub-

FIGURE 10-8

Grouped Bar Graph

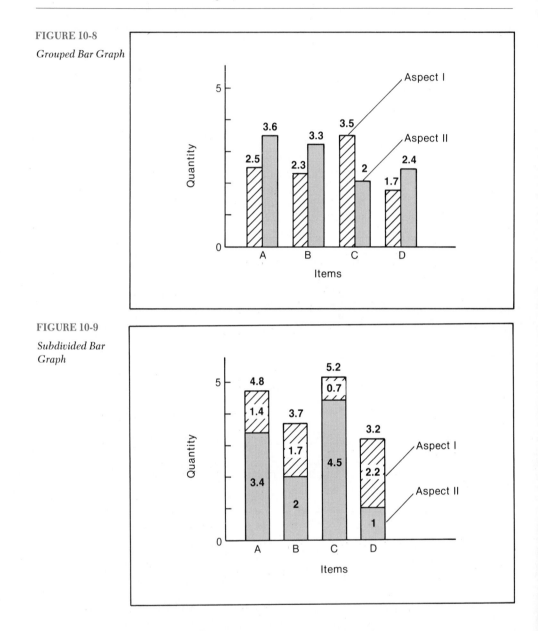

FIGURE 10-9

Subdivided Bar Graph

divided bar graph, the individual quantities (except those that begin on the horizontal axis) are not.

Related to the subdivided bar graph is the *100-percent bar
graph*, which enables you to show the relative proportions of the
elements that make up several items. Figure 10-10 shows a 100-
percent bar graph. This kind of graph is useful in portraying, for

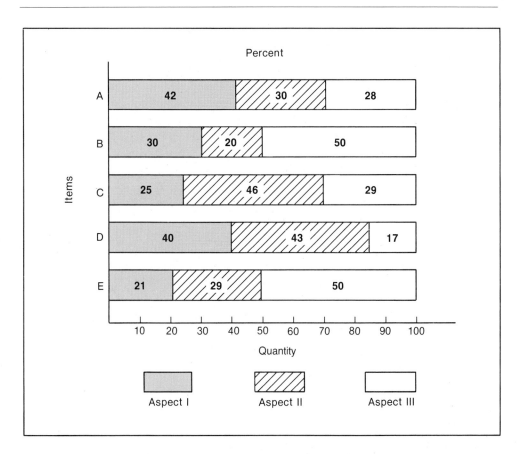

Percent

FIGURE 10-10

100-Percent Bar Graph

example, the proportion of full-scholarship, partial-scholarship, and no-scholarship students at a number of colleges.

The *deviation bar graph*, shown in Figure 10-11, lets you show how various quantities deviate from a norm. Deviation bar graphs are often used when the information contains both positive and negative values, as with profits and losses. Bars on the positive side of the norm line represent profits; on the negative side, losses.

Pictographs are simple graphs in which the bars are replaced by series of symbols that represent the items (see Figure 10-12). Pictographs are generally used only to enliven statistical information for the general reader. The quantity scale is usually replaced by a statement that indicates the numerical value of each symbol.

Pictographs are arranged horizontally rather than vertically: symbols sitting on top of each other would look foolish.

A related kind of error is to use a vertical format and make each symbol tall enough to represent the quantity desired. As Figure 10-13 shows, the larger symbol, if it is drawn proportionally,

FIGURE 10-11

Deviation Bar Graph

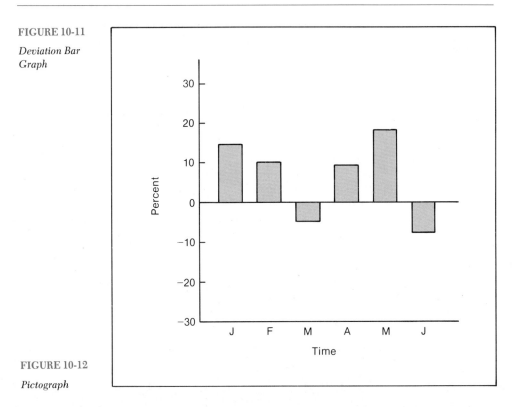

FIGURE 10-12

Pictograph

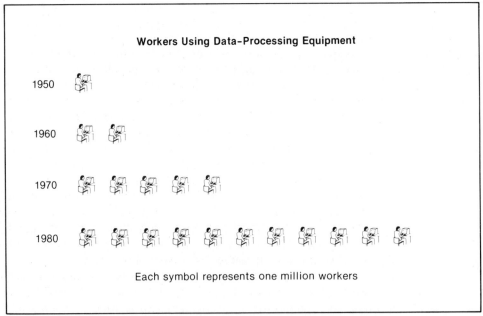

Workers Using Data–Processing Equipment

Each symbol represents one million workers

FIGURE 10-13

Misleading Picto-graph

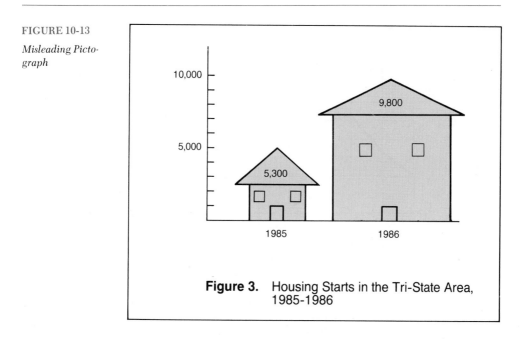

10,000

5,000

9,800

5,300

1985   1986

Figure 3. Housing Starts in the Tri-State Area,
1985-1986

looks many times larger than it should: the reader sees the total area of the symbol rather than its height.

LINE GRAPHS Line graphs are like vertical bar graphs, except that in line graphs the quantities are represented not by bars but by points linked by a line. This line traces a pattern that in a bar graph would be formed by the highest point of each bar. Line graphs are used almost exclusively to show how the quantity of an item changes over time. Some typical applications of a line graph would be to portray the month-by-month sales figures or production figures for a product or the annual rainfall for a region over a given number of years. A line graph focuses the reader's attention on the change in quantity, whereas a bar graph emphasizes the actual quantities themselves. Figure 10-14 shows a typical line graph.

An additional advantage of the line graph for demonstrating change is that it can accommodate much more data. Because three or four lines can be plotted on the same graph, you can compare trends conveniently. Figure 10-15 shows a multiple-line graph. However, if the lines intersect each other often, the graph will be unclear. If this is the case, draw separate graphs.

The principles of constructing a line graph are similar to those used for a vertical bar graph. The vertical axis, which charts the quantity, should begin at zero; if it is impractical to begin at zero

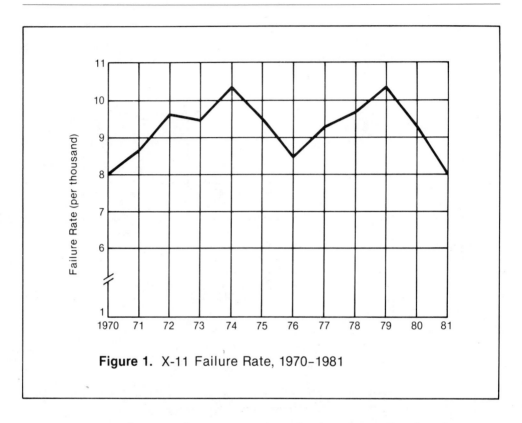

Figure 1. X-11 Failure Rate, 1970–1981

FIGURE 10-14

Line Graph with a
Truncated Axis

because of space restrictions, clearly indicate a break in the axis, as Figure 10-14 does. Where precision is required, use grid lines—horizontal, vertical, or both—rather than tick marks.

Two common variations on the line graph—the *stratum graph* and the *ratio graph*—deserve mention. A stratum graph shows an overall change, and then breaks down the total change into its constituent parts. Figure 10-16 shows a stratum graph. It's easy to read—with a little practice. For example, in Figure 10-16, in 1973 the total world military expenditures were approximately $570 billion. NATO countries accounted for $240 billion; Warsaw pact countries, $230 billion; other countries, $100 billion.

A *ratio graph* is a line graph used to emphasize percentages of change rather than the change in real numbers. A ratio graph makes it possible to chart data that could not be represented fairly on a standard line graph. For example, you might wish to compare the month-by-month sales of a large corporation with those of a small one. You would have great trouble making a vertical axis that would accommodate a company with sales of $20,000 a month and one with sales of $20,000,000. Even if you had a giant piece of paper, the graph could not reflect a true relation between

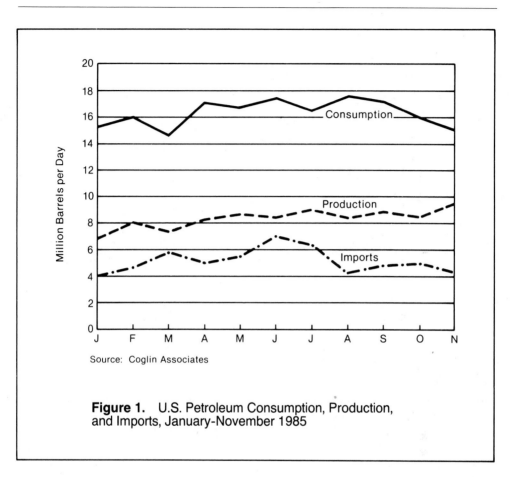

Figure 1. U.S. Petroleum Consumption, Production, and Imports, January–November 1985

FIGURE 10-15

Multiple-Line Graph

the companies. If both companies increased their sales at the same rate (such as 2 percent per month), the small company's line would appear relatively flat, whereas the big company's line would shoot upward, just because of the large quantities involved.

To solve this problem, the ratio graph compresses the vertical axis more and more as the quantities increase. Figure 10-17 shows how ratio graphs work. On the bottom graph, for example, the distance on the vertical axis from 50 to 100 is the same as the distance between 100 and 200. On an ordinary graph, the distance between 100 and 200 would of course be twice as great. Therefore, on a ratio graph a quantity that rose from 50 to 100—that is, rose 100 percent—would have the same slope as one that rose from 100 to 200—also 100 percent.

CHARTS Whereas tables and graphs present statistical information, most charts convey relationships that are more abstract,

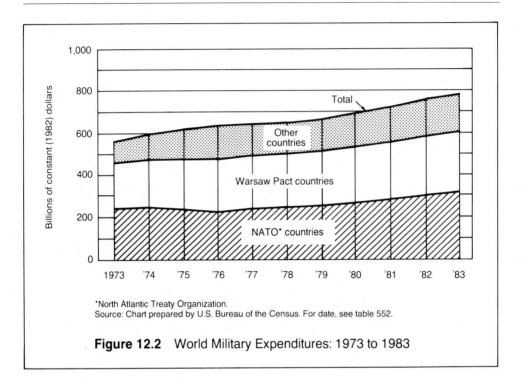

Figure 12.2   World Military Expenditures: 1973 to 1983

FIGURE 10-16

Stratum Graph
(from Statistical
Abstract of the
United States 1986,
p. 330)

such as causality or hierarchy. (The pie chart, which is really just a circular rendition of the 100-percent bar graph, is the major exception.) Many forms of tables and graphs are well known and fairly standard. By contrast, only a few kinds of charts—such as the organization chart and flow chart—follow established patterns. Most charts reflect original concepts and are created to meet specific communication needs.

The *pie chart* is a simple but limited design used for showing the relative size of the parts of a whole. Pie charts can be instantly recognized and understood by the untrained reader: everyone remembers the perennial "where-your-tax-dollar-goes" pie chart. The circular design is effective in showing the relative size of as many as five or six parts of the whole, but it cannot easily handle more parts because, as the slices get smaller, judging their sizes becomes more difficult. (Very small quantities that would make a pie chart unclear can be grouped under the heading "Miscellaneous" and explained in a footnote. This "miscellaneous" section, sometimes called "other," appears after the other sections as you work in a clockwise direction.)

To create a pie chart, begin with the largest slice at the top of the pie and work clockwise in decreasing-size order, unless you have a good reason for arranging the slices in a different order.

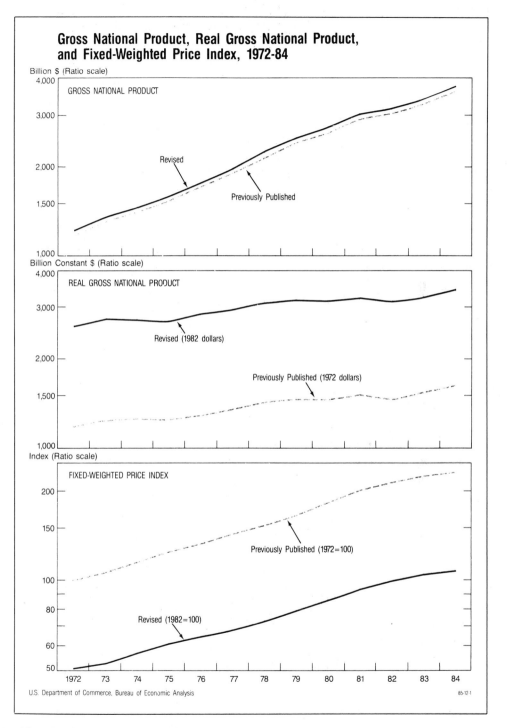

Gross National Product, Real Gross National Product, and Fixed-Weighted Price Index, 1972-84

Billion $ (Ratio scale)

GROSS NATIONAL PRODUCT

Revised

Previously Published

Billion Constant $ (Ratio scale)

REAL GROSS NATIONAL PRODUCT

Revised (1982 dollars)

Previously Published (1972 dollars)

Index (Ratio scale)

FIXED-WEIGHTED PRICE INDEX

Previously Published (1972=100)

Revised (1982=100)

U.S. Department of Commerce, Bureau of Economic Analysis

85-12-1

FIGURE 10-17

Ratio Graphs (from Survey of Current Business, *December 1985, p. 1)*

Label the slices (horizontally, not radially) inside the slice, if space permits. It is customary to include the percentage that each slice represents. Sometimes, the absolute quantity is added. To emphasize one of the slices—for example, to introduce a discussion of the item represented by that slice—separate it from the pie. Make sure your math is accurate as you convert percentages into degrees in dividing the circle. A percentage circle guide—a template with the circle already converted into percentages—is a useful tool.

Figure 10-18 shows two styles of pie charts. The chart on the left is the standard version with shading added to simulate a third dimension. The version on the right shows a slice removed from the pie for emphasis.

A *flow chart*, as its name suggests, traces the stages of a procedure or a process. A flow chart might be used, for example, to show the steps involved in transforming lumber into paper or in synthesizing an antibody. Flow charts are useful, too, for summarizing instructions that a reader is to carry out. The basic flow chart portrays stages with labeled rectangles or circles. To make it

FIGURE 10-18

Pie Charts (from Statistical Abstract of the United States 1986, *p. 222)*

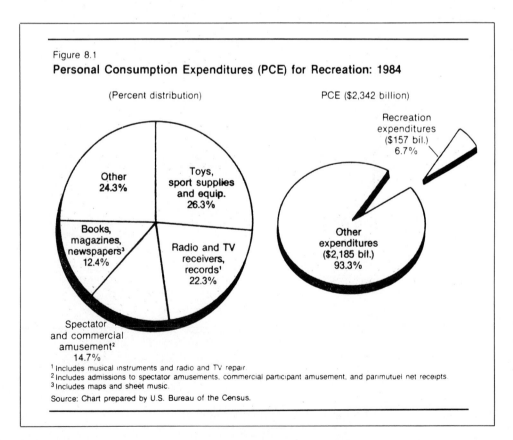

Figure 8.1
Personal Consumption Expenditures (PCE) for Recreation: 1984

(Percent distribution)

PCE ($2,342 billion)

Other
24.3%

Toys,
sport supplies
and equip.
26.3%

Books,
magazines,
newspapers[3]
12.4%

Radio and TV
receivers,
records[1]
22.3%

Spectator
and commercial
amusement[2]
14.7%

Recreation
expenditures
($157 bil.)
6.7%

Other
expenditures
($2,185 bil.)
93.3%

[1] Includes musical instruments and radio and TV repair
[2] Includes admissions to spectator amusements. commercial participant amusement. and parimutuel net receipts
[3] Includes maps and sheet music.
Source: Chart prepared by U.S. Bureau of the Census.

visually more interesting, use pictorial symbols instead of geometric shapes. If the process involves quantities (for example, the process of paper manufacturing might "waste" 30 percent of the lumber), they can be listed or merely suggested by the size of the line used to connect the stages. Flow charts can portray open systems (those that have a "start" and a "finish") or closed systems (those that end where they began). A special kind of flow chart, called a decision chart (in which the flow follows different routes depending on yes/no answers to questions), is used frequently in computer science.

Figure 10-19 shows an open-system flow chart whose bars correspond in width to the magnitude of the quantity they represent. The subject of the flow chart is the percentage of solar energy that reaches the Earth's surface. Figure 10-20 shows a typical flow chart from computer science. This chart incorporates the concept of the decision tree: the diamond-shaped stages of the process are questions that can be answered either yes or no. The process branches off at each diamond.

An *organization chart* is a type of flow chart that portrays the flow of authority and responsibility in a structured organization. In most cases, the positions are represented by rectangles. The

FIGURE 10-19

Flow Chart

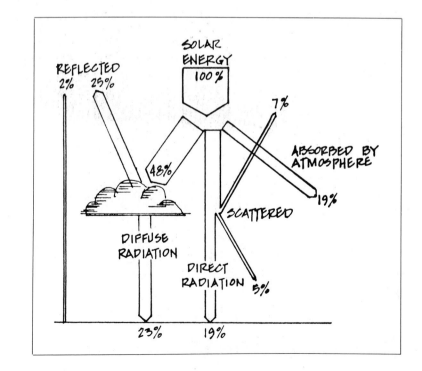

FIGURE 10-20

Flow Chart with Branches

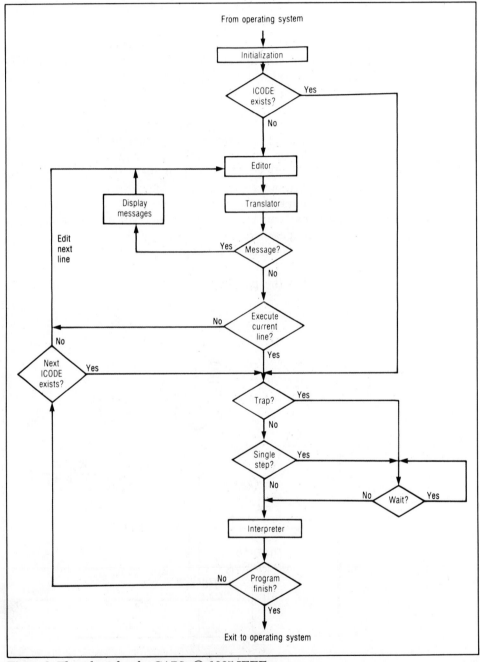

Figure 2. Flowchart for the CAPS. © 1985 IEEE.

more important positions can be emphasized through the size of the boxes, the width of the lines that form the boxes, the typeface, or the use of color. If space permits, the boxes themselves can include brief descriptions of the positions, duties, or responsibilities. Figure 10-21 is a typical organization chart. Unlike most other figures, organization charts are generally titled *above* the chart.

DIAGRAMS AND PHOTOGRAPHS To portray physical relationships, such as those in pieces of equipment or machinery, diagrams and photographs are often the most effective graphic aids. Photographs are unmatched, of course, for reproducing realistic images. Recent advances in specialized kinds of photography—especially in internal medicine and biology—are expanding the possibilities of the art. Diagrams drawn by hand are used to portray perspectives that cannot be photographed. Cutaways, for example, let you "remove" a part of the surface to expose what is underneath. "Exploded" diagrams separate components from each other while maintaining their physical relationship. Figure 10-22 shows a cutaway diagram; Figure 10-23, an exploded diagram. Notice how effectively the diagrams have been labeled.

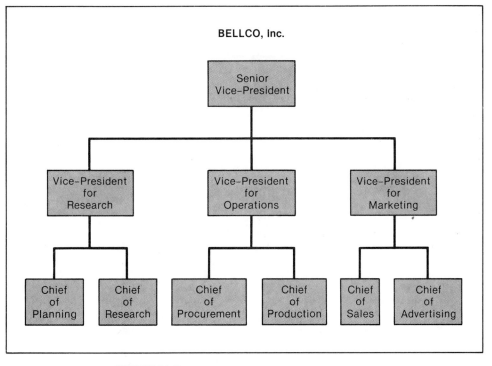

FIGURE 10-21

Organization Chart

FIGURE 10-22

Cutaway Diagram

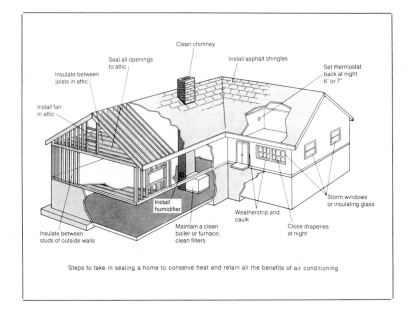

Clean chimney

Seal all openings
to attic

Install asphalt shingles

Set thermostat
back at night
6° or 7°

Insulate between
joists in attic

Install fan
in attic

Install
humidifier

Storm windows
or insulating glass

Weatherstrip and
caulk

Close draperies
at night

Maintain a clean
boiler or furnace;
clean filters

Insulate between
studs of outside walls

Steps to take in sealing a home to conserve heat and retain all the benefits of air conditioning

FIGURE 10-23

Exploded Diagram

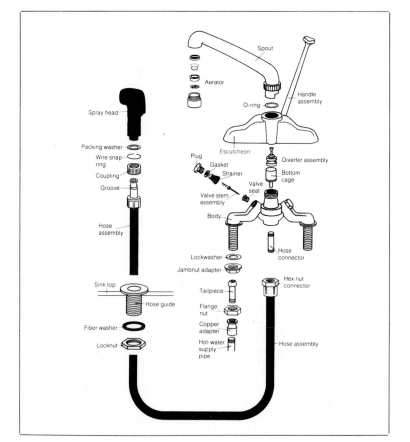

Spout

Aerator

O-ring

Handle
assembly

Spray head

Packing washer

Wire snap
ring

Coupling

Groove

Escutcheon

Plug

Gasket

Strainer

Diverter assembly

Bottom
cage

Valve
seat

Valve stem
assembly

Hose
assembly

Body

Lockwasher

Jambnut adapter

Hose
connector

Sink top

Hose guide

Fiber washer

Locknut

Tailpiece

Flange
nut

Copper
adapter

Hot-water
supply
pipe

Hex nut
connector

Hose assembly

EXERCISES

1. For each of the following graphic aids, write a paragraph evaluating its effectiveness and describing how you would revise it.

   a.

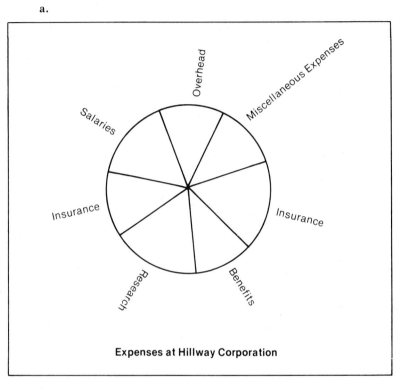

Expenses at Hillway Corporation

   b.

Engineering and Liberal Arts Graduate Enrollment

	1985	1986	1987
Civil Engineering	236	231	253
Chemical Engineering	126	134	142
Comparative Literature	97	86	74
Electrical Engineering	317	326	401
English	714	623	592
Fine Arts	112	96	72
Foreign Languages	608	584	566
Materials Engineering	213	227	241
Mechanical Engineering	196	203	201
Other	46	42	51
Philosophy	211	142	151
Religion	86	91	72

c.

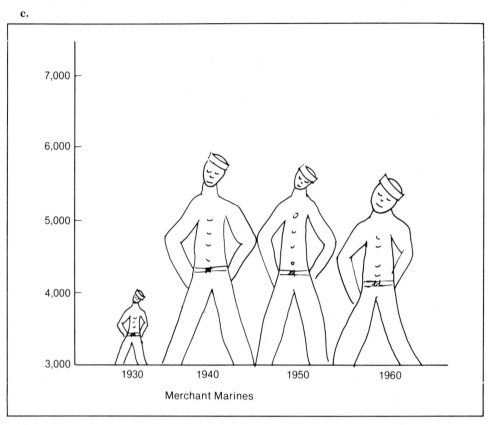

Merchant Marines

d.

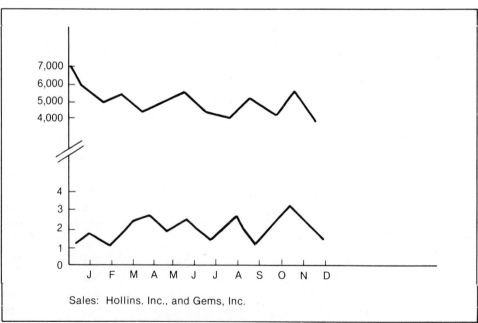

Sales: Hollins, Inc., and Gems, Inc.

e.

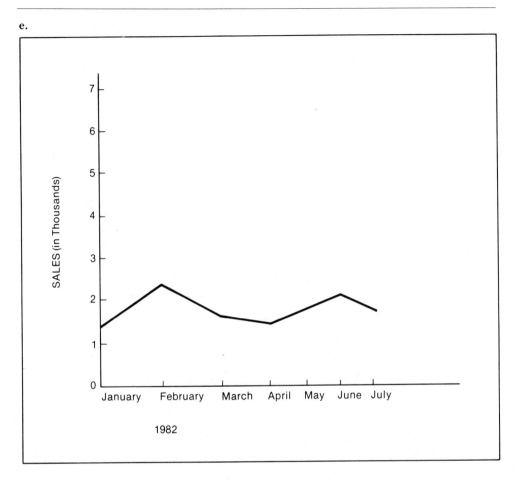

2. In each of the following exercises, translate the written information to at least two different kinds of graphic aids. For each exercise, which kinds work best? If one kind works well for one audience but not so well for another audience, be prepared to explain.

a. Following are the profit and loss figures for Pauley, Inc., in early 1982: January, a profit of 6.3 percent; February, a profit of 4.7 percent; March, a loss of 0.3 percent; April, a loss of 2.3 percent; May, a profit of 0.6 percent.

b. The prime interest rate had a major effect on our sales. In January, the rate was 11.5 percent. It went up a full point in February, and another half point in March. In April, it leveled off, and it dropped two full points each in May and June. Our sales figures were as follows for the Crusader 1: January, 5,700; February, 4,900; March, 4,650; April, 4,720; May, 6,200; June, 8,425.

c. Following is a list of our new products, showing for each the profit on the suggested retail price, the factory where produced, the date of introduction, and the suggested retail price.

THE TIMBERLINE

Profit 28%

Milwaukee

March 1984

$235.00

THE FOUR SEASON

Profit 32%

Milwaukee

October 1983

$185.00

THE FAMILY EXCURSION

Profit 19%

Brooklyn

October 1983

$165.00

THE DAY TRIPPER

Profit 17%

Brooklyn

May 1983

$135.00

d. This year, our student body can be broken down as follows: 45 percent from the tristate area; 15 percent from foreign countries; 30 percent from the other Middle Atlantic states; and 10 percent from the other states.

e. In January of this year we sold 50,000 units of the BG-1, of which 20,000 were purchased by the army. In February, the army purchased 15,000 of our 60,000 units sold. In March, it purchased 12,000 of the 65,000 we sold.

f. The normal rainfall figures for this region are as follows: January, 1.5 in.; February, 1.7 in.; March, 1.9 in.; April, 2.1 in.; May, 1.8 in.; June, 1.2 in.; July, 0.9 in.; August, 0.7 in.; September, 1.3 in.; October, 1.1 in.; November, 1.0 in.; December, 1.2 in. The following rainfall was recorded in this region: January, 2.3 in.; February, 2.6 in.; March, 2.9 in.; April, 2.0 in.; May, 1.6 in.; June, 0.7 in.; July, 0.1 in.; August, 0.4 in.; September, 1.3 in.; October, 1.2 in.; November, 1.4 in.; December, 1.8 in.

PART THREE

TECHNICAL REPORTS

CHAPTER ELEVEN

FORMAL ELEMENTS OF A REPORT

Because most reports today are read by many people with different backgrounds and needs, the writer's job is to make it as easy as possible for the various readers to find the information they seek. The structure and organization of a modern report are designed to make that information accessible. For example, a comprehensive table of contents makes it convenient to find a particular discussion. A well-written executive summary is a convenient way for a manager to understand the essentials of the project without having to read the whole report.

This chapter discusses the formal elements of a report—those components that are usually included in a formal report in business and industry. Few reports will have all the elements discussed here, in the order in which they are covered; most organizations have their own format preferences. You should therefore study the style guide used in your organization. If there is no style guide, study a few of the reports in your organization's files. Successful reports are the best teaching guides; ask a colleague to suggest some samples.

The following elements will be discussed here:

1. letter of transmittal
2. title page
3. abstract

4. table of contents
5. list of illustrations
6. executive summary
7. glossary and list of symbols
8. appendix

Two crucial elements of a report are not discussed in this chapter: the body, which varies according to the type of report, and the documentation, which can also vary. For discussions of the body, refer to the individual chapters on the various types of reports: proposals (Chapter 13), progress reports (Chapter 14), and completion reports (Chapter 15). Documentation—the system of citing sources of information used in a report—is discussed in Appendix B.

FORMAL ELEMENTS AND THE WRITING PROCESS

Chapter 3 discusses the writing process: the sequence of activities that begins with narrowing a topic and then proceeds to brainstorming, outlining, drafting, and finally revising. An important point made in that chapter is that most technical writers do not write a document straight through from the first sentence to the last. Instead, they work through the whole document in stages. For example, they brainstorm to try to determine what information will eventually be included. Then they outline to organize that information.

The same concept applies to the formal elements. The components of a report are not written in the same order in which they appear. The letter of transmittal, for instance, is the first thing the principal reader sees, but it was probably the last to be created. The reason is simple: the transmittal letter cannot be written until the document to which it is attached has been written. Many writers like to include in the first paragraph of the transmittal letter the title of the document; consequently, even the title has to be decided upon before the letter can be written. In the same way, the title cannot be chosen until the body of the report is complete, because the title must reflect accurately the contents of the report.

The body of the report is written before any of the other formal elements. After that, the sequence makes little difference. Many writers create the two summaries—the executive summary and the abstract—then the appendixes, the glossary and list of symbols, and finally the table of contents, title page, and letter of transmittal.

FORMAL ELEMENTS AND THE WORD PROCESSOR

Assembling the formal elements of a report is much simpler with a word processor than without one.

For one thing, with a word processor you have a much better sense of the length of the various elements because your draft is typed, not handwritten. A number of the formal elements have length restrictions. A word processor lets you see exactly how long each element is so that you can easily expand or contract it to meet your requirements.

In addition, the copy function on a word processor helps you create the different summaries and transmittal letter. You can make a copy of the body of the report and then eliminate the details that do not belong in the particular formal element you are creating. For instance, if the problem statement in the body is two paragraphs long, you can reduce it to a few sentences for the exclusive summary by eliminating some of the technical details. If the methods section will not be treated in the transmittal letter, you can simply erase it.

Creating the formal elements by cutting material from the body is not only faster than writing them from scratch; it is also more accurate because you don't introduce any technical errors. If in the body of the report you say that the Library of Congress catalogs 180,000 books each year, it will remain 180,000—not 18,000 or 1,800,000—in the executive summary.

THE LETTER OF TRANSMITTAL

The letter of transmittal introduces the purpose and content of the report to the principal reader of the report. The letter is attached to the report or simply placed on top of it. Even though it might contain no information that is not included elsewhere in the report, the letter is important, because it is the first thing the reader sees. It establishes a courteous and graceful tone for the report. Letters of transmittal are customary even when the writer and the reader both work for the same organization and ordinarily communicate by memo.

The letter of transmittal gives you an opportunity to emphasize whatever you think your reader will find particularly important or interesting in the attached materials. It also enables you to point out any errors or omissions in the materials. For example, you might want to include some information that was gathered after the report was typed or printed.

Transmittal letters generally contain most of the following elements:

1. a statement of the title, and, if necessary, the purpose of the report
2. a statement of who authorized or commissioned the project and when
3. a statement of the methods used in the project (if they are noteworthy) or of the principal results, conclusions, and recommendations
4. an acknowledgment of any assistance you received in preparing the materials
5. a gracious offer to assist in interpreting the materials or in carrying out further projects.

Figure 11-1 provides an example of a transmittal letter. (For a discussion of letter format, see Chapter 18.)

THE TITLE PAGE

The only difficult task in creating the title page is to think of a good title. The other usual elements—the date of submission and the names and positions of the writer and the principal reader— are simply identifying information.

A good title is informative without being unwieldy. It answers two basic questions: What is the subject of the report? and What type of report is it? Several examples of effective titles follow:

Choosing a Microcomputer: A Recommendation

An Analysis of the Kelly 1013 Packager

Open Sea Pollution-Control Devices: A Summary

Note that a convenient way to define the type of report is to use a generic term—such as *analysis, recommendation, summary, review, guide,* or *instructions*—in a phrase following a colon.

If you are creating a simple title page, center the title (typed in full capital letters) about a third of the way down the page. Then add the reader's and the writer's positions, the organization's name, and the date. Figure 11-2 provides a sample of a simple title page.

FIGURE 11-1

Letter of Transmit-tal

ALTERNATIVE ENERGY, INC.
Bar Harbor, ME 00314

April 3, 19 – –

Rivers Power Company
15740 Green Tree Road
Gaithersburg, MD 20760

Attention: Mr. J. R. Hanson
            Project Engineering Manager

Subject: Project #619-103-823

Gentlemen:

We are pleased to submit "A Proposal for the
Riverfront Energy Project" in response to your
request of February 6, 19 – –.

The windmill design described in the attached
proposal uses the most advanced design and materials.
Of particular note is the state-of-the-art storage
facility described on pp. 14 – 17. As you know,
storage limitations are a crucial factor in the
performance of a generator such as this.

In preparing this proposal, we inadvertently omitted
one paragraph on p. 26 of the bound proposal. That
paragraph is now on the page labeled 26A. We regret
this inconvenience.

If you have any questions, please do not hesitate to
call us.

Yours very truly,

Ruth Jeffries
Project Manager

RJ/fj
Enclosures 2

Most organizations have their own formats for title pages. Often, an organization will have different formats for different kinds of reports. Figure 11-3 shows the complex title page used for research reports at one company.

This company wants so much information that two pages are required. Notice that on the first page, the title of the report would be followed by a statement of whether the report is a progress report or final report. On the second page, the word *distribution* would be followed by the list of those who are to receive the report. Like many companies, this one in fact sends the complete

FIGURE 11-2

Simple Title Page

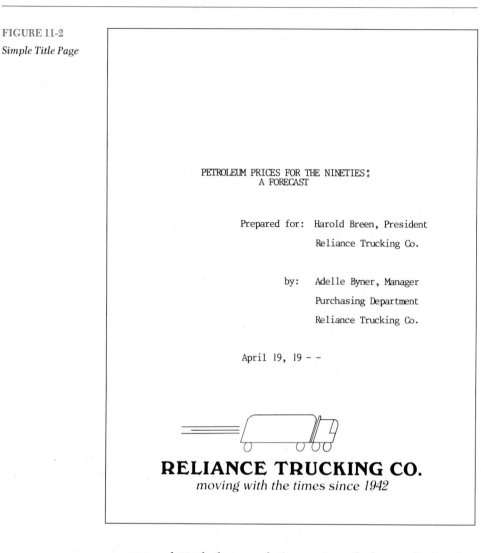

PETROLEUM PRICES FOR THE NINETIES:
A FORECAST

Prepared for: Harold Breen, President
                Reliance Trucking Co.

           by: Adelle Byner, Manager
                 Purchasing Department
                 Reliance Trucking Co.

April 19, 19 - -

RELIANCE TRUCKING CO.
moving with the times since 1942

report to relatively few people (approximately four or five) and to the files. The summary report (the executive summary) goes to many more readers (approximately twenty or thirty).

THE ABSTRACT

An abstract is a brief technical summary—usually no more than 200 words—of the report. Like the abstract that accompanies published articles (see Chapter 16), the abstract of a report is directed primarily to readers who are familiar with the technical

FIGURE 11-3

Complex Title Page

```
                    RESEARCH REPORT COVER PAGE

                      (Company Confidential)


                              Standard Technical Report No._____

                              Date Issued_____

                              Security (Check One)_____ RC_____ C

        Originating R & D Department_____

        Location (Facility, City, State)_____

        Group or Division_____



                       WILSON CHEMICALS, INC.



                 _____

                 _____

                              (Title)

                    _____
                    (Indicate Progress or Final Report)

        Work Done By:
        Report Written By:
        Supervisor:
        R & D Director:
        Previous Related Reports:
        Department Overhead Number:
        Project Number:
        Period Covered:
        Notebook Number(s)

           PROPRIETARY INFORMATION FOR AUTHORIZED COMPANY USE ONLY
```

subject and need to know whether to read the full report. Therefore, in writing an abstract you can use technical terminology freely and refer to advanced concepts in your field. (For managers who want a summary focusing on the managerial implications of a project, many reports that contain abstracts also contain executive summaries. See the discussion on pp. 292–296.)

Because abstracts can be useful before and after a report is read—and can even be read in place of the report—they are duplicated and kept on file in several locations within an organization: the division in which the report originated and many or all of the higher-level units of the organization. And, of course, a copy of

FIGURE 11-3

Complex Title Page
(Continued)

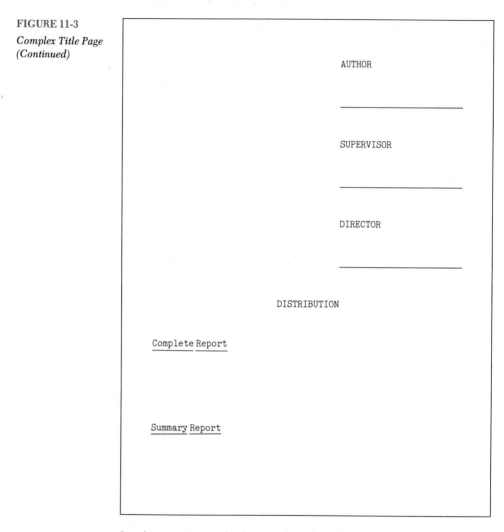

```
                                     AUTHOR


                             _____


                                     SUPERVISOR


                             _____


                                     DIRECTOR


                             _____


                        DISTRIBUTION

    Complete Report
    _____



    Summary Report
    _____
```

the abstract is attached to or placed within the report. It is not un-usual to find six or eight copies of an abstract somewhere in an or-ganization. To facilitate this wide distribution, some organiza-tions have special forms on which abstracts are typed.

The two basic types of abstracts are generally called descrip-tive and informative abstracts. The descriptive abstract is rapidly losing popularity, whereas the informative abstract is becoming the accepted standard.

THE
DESCRIPTIVE
ABSTRACT

The descriptive abstract, sometimes called the topical or table-of-contents abstract, does only what its name implies: it describes what the report is about. It does not provide the important results, conclusions, or recommendations. It simply lists the topics cov-

FIGURE 11-4
Descriptive Abstract

ABSTRACT

"Design of a Radio-based System for Distribution Automation"

by Brian D. Crowe

At this time, power utilities' major techniques of monitoring
their distribution systems are after-the-fact indicators such
as interruption reports, meter readings, and trouble alarms.
These techniques are inadequate in two ways. One, the information
fails to provide the utility with an accurate picture of the
dynamics of the distribution system. Two, after-the-fact indicators are expensive. Real-time load monitoring and load management would offer the utility both system reliability and long-
range cost savings. This report describes the design criteria we
used to design the radio-based system for a pilot program of
distribution automation. It then describes the hardware and
software of the system.

ered, giving equal coverage to each. Thus, the descriptive abstract simply duplicates the information included in the table of contents. Figure 11-4 provides an example of a descriptive abstract.

THE IN-
FORMATIVE
ABSTRACT
The informative abstract presents the major information that the report conveys. Rather than merely listing topics, it states the problem, the scope and methods (if appropriate), and the major results, conclusions, or recommendations.

The basic structure of the informative abstract includes three elements:

1. *The identifying information.* The name of the report, the writer, and perhaps the writer's department.

2. *The problem statement.* One or two sentences that define the problem or need that led to the project. Many writers mistakenly omit the problem statement, assuming that the reader knows what the problem is. The *writer* knows, being intimately involved with the project, but the readers are likely to be totally unfamiliar with it. Without an adequate problem statement to guide them, many readers will be unable to understand the abstract.

3. *The important findings.* The final three or four sentences—the biggest portion of the abstract—state the crucial information the report contains. Generally, this means some combination of results, conclusions, recommendations, and implications for further projects. Sometimes, however, the abstract presents other information. For instance, many technical projects focus on new or unusual methods for achieving results that have already been obtained through other means. In such a case, the abstract will focus on the methods, not the results.

Following are the introduction, conclusion, and recommendation from the report (Crowe 1985) whose descriptive abstract appears in Figure 11-4. The excerpts are annotated to show how the writer used these excerpts to create his descriptive abstract.

INTRODUCTION

The writer describes the problem, first in general terms and then in particular terms.

At this time, power utilities' major techniques of monitoring their distribution sytems are after-the-fact indicators such as interruption reports, meter readings, and trouble alarms. This system is inadequate in two ways.

One, the information fails to provide the utility with an accurate picture of the dynamics of the distribution system. To ensure enough energy for our customers, we have to overproduce. Last year we overproduced by 7 percent. This worked out to a loss of $273,000.

Two, after-the-fact indicators are expensive. Meter readings for our "easy-to-access" customers cost $20/year. Currently, 12,000 of our customers are classified as "difficult-to-access." Meter readings for each of these customers average $80/year, for an annual cost of $960,000. If we could reduce costs by installing radio-based monitoring systems on these 12,000 residences as a trial project, we could realize substantial savings while we perfect the system.

The writer describes the purpose of the report and of the project.

This report describes a project to design a radio-based system for a pilot project. If the radio-based units are technically and economically feasible, they will be installed on the 12,000 residences for a one-year study.

SYSTEM DESCRIPTION

[Here the writer describes the hardware and software in detail. The basic system, which uses packet-switching technology, consists of a base unit (built around a personal computer), a radio link, and a remote unit.]

The writer presents the four major conclusions.

CONCLUSIONS

The radio-based distribution monitoring system described in this report meets our four criteria: realiability, size, cost, and ease of use. It is more accurate than the currently used, after-the-fact indicators, it is small enough to replace the existing meters, it would pay for itself in 3.9 years (see Appendix C, p. 14), and it is simple to use.

RECOMMENDATION

The writer presents the major recommendation and links it to the long-range corporate goals.

We recommend that the pilot program begin with the purchase and installation of the 12,000 meters. This pilot program is the first step toward our long-range goal: total automation of the distribution monitoring system.

Total automation would increase the reliability of the readings while drastically reducing labor costs. In addition, total automation would enable us to save money by not overgenerating energy; we could base our generation on accurate, timely data, rather than on rough estimates derived from weather forecasts and precedent.

The informative abstract based on this excerpt appears in Figure 11-5.

The informative abstract resembles the descriptive abstract in that both contain identifying information and a problem statement. But whereas the descriptive abstract gives equal emphasis to most of the topics listed in the table of contents, the informative abstract concentrates on the important findings.

The distinction between descriptive and informative abstracts is not absolute. Sometimes you will have to combine elements of both in a single abstract. For instance, you are writing an informative abstract but the report includes 15 recommendations, far too many to list. You might decide to identify the major results and conclusions, as you would in any informative abstract, but add that the report contains numerous recommendations, as you would in a descriptive abstract.

THE TABLE OF CONTENTS

Far too often, good reports are ruined because the writer fails to create a useful table of contents. This element is crucial to the report, because it enables different readers to turn to specific pages to find the information they want. No matter how well organized the report itself may be, a table of contents that does not make the structure clear will be ineffective.

Most people will have read the abstract when they turn to the table of contents to find one or two items in the body of the report. Because a report usually has no index, the table of contents will provide the only guide to the report's structure, coverage, and pagination. The headings listed in the table of contents are the headings that appear in the report itself. To create an effective ta-

FIGURE 11-5

Informative Abstract

ABSTRACT

"Design of a Radio-based System for Distribution Automation"

by Brian D. Crowe

At this time, power utilities' major techniques of monitoring
their distribution systems are after-the-fact indicators such
as interruption reports, meter readings, and trouble alarms.
This system is inadequate in that it fails to provide the utility
with an accurate picture of the dynamics of the distribution
system, and it is expensive. This report describes a project to
design a radio-based system for a pilot project. The basic system, which uses packet-switching technology, consists of a base
unit (built around a personal computer), a radio link, and a
remote unit. The radio-based distribution monitoring system
described in this report is more accurate than the currently used
after-the-fact indicators, it is small enough to replace the
existing meters, it would pay for itself in 3.9 years, and it is
simple to use. We recommend installing the basic system on a trial
basis.

ble of contents, therefore, you first must make sure the report has
effective headings—and that it has enough of them. If the table of
contents shows no entry for five or six pages, the report could
probably be divided into additional subunits. In fact, some tables
of contents have a listing—or several listings—for every page in
the report.

Insufficiently specific tables of contents generally result from
the exclusive use of generic headings (those that describe entire
classes of items) in the report. Figure 11-6 shows how inadequate a
table of contents can become if it simply lists generic headings.

FIGURE 11-6

Ineffective Table of Contents

Table of Contents

Introduction	1
Materials	3
Methods	4
Results	19
Recommendations	23
References	26
Appendixes	28

To make the headings more informative, combine generic and specific items, as in the following examples:

Recommendations: Five Ways to Improve Information Retrieval

Materials Used in the Calcification Study

Results of the Commuting-Time Analysis

Then build more subheadings into the report. For example, in the "Recommendations" example, make a separate subheading for each of the five recommendations.

Once you have created a clear system of headings within the report, transfer them to the contents page. Use the same format—capitalization, underlining, indention, and outline-style or decimal headings (see the discussion of headings in Chapter 5)—that you use in the text.

If you are using a word processor, you have more format choices: boldface, italics, different sizes of type, and so forth. Regardless of whether you use a typewriter or a word processor, however, keep in mind some basic aspects of formatting:

1. *Size.* Larger letters are more emphatic than smaller letters. Uppercase letters are more emphatic than lowercase. Boldface and italics are more emphatic than regular type. And underlining is more emphatic than not underlining. Using a standard typewriter, you can create up to six hierarchical levels using capitalization and underlining:

CATHODE RAY TUBES

CATHODE RAY TUBES

Cathode Ray Tubes

```
Cathode Ray Tubes
cathode ray tubes
cathode ray tubes
```

Keep in mind, however, that most readers will become confused if you use more than four or five different levels.

2. *Indention.* In general, more emphatic items begin closer to the left-hand margin. Subheadings are therefore indented four or five spaces.

```
First-level heading
        Second-level heading
                Third-level heading
```

Many writers today make sure that they use indention not only in the table of contents but in the text as well. In other words, the text that follows a second-level heading, for instance, is indented the same five spaces that the heading itself is.

```
First-level heading
XXXXXXXXXXXXXXXXXXXXXXXXXXXXXXXXXXXXXXXXXXXXXXXXXXXXXXXXXXXXXXX
XXXXXXXXXXXXXXXXXXXXXXXXXXXXXXXXXXXXXXXXXXXXXXX.

        Second-level heading
        XXXXXXXXXXXXXXXXXXXXXXXXXXXXXXXXXXXXXXXXXXXXXXXXXXXXXXXXXXX
        XXXXXXXXXXXXXXXXXXXXXXXXXXXXXXXXXXXXXXXXXXXXXXXXXXXXX.
```

This indention reminds the reader of the level of the text.

One exception to the principle of indention regards first-level headings. Some writers like to center them:

```
                    First-level heading
Second-level heading
        Third-level heading
```

3. *Outline-style and decimal-style headings.* The outline-style or decimal-style headings that you use in the text of your report should be transferred intact to the table of contents.

```
I.   FIRST-LEVEL HEADING

     A.   Second-Level Heading

          1.   Third-Level Heading

          2.   Third-Level Heading

     B.   Second-Level Heading


1.0  FIRST-LEVEL HEADING

     1.1  Second-Level Heading

          1.1.1  Third-Level Heading

          1.1.2  Third-Level Heading

     1.2  Second-Level Heading
```

A word processor makes the task of transferring headings from the text of the report to the table of contents effortless. Some software programs will actually do the job for you automatically. But even the simplest software helps you make the table of contents quickly and easily—without introducing errors. Simply make a copy of the report. Then scroll through the copy, erasing the text. What you have left are the headings. If they are inconsistent you will notice it immediately and be able to fix any problems both in the table of contents and in the text of the report itself.

Figure 11-7 shows how you can combine generic and specific headings and use the resources of the typewriter to structure a report effectively. The report is titled "Methods of Computing the Effects of Inflation in Corporate Financial Statements: A Recommendation." This table of contents combines all three aspects of formatting discussed.

The table of contents in Figure 11-7 works well for several reasons. First, managers can find the executive summary quickly and easily. Second, all the other readers can find the information they are looking for, because each substantive section is listed separately. An additional advantage of a specific table of contents is that it gives the readers a clear idea of the scope and structure of the report before they start to read it.

A note about pagination is necessary. The abstract page of a report is generally not numbered, but other preliminary elements (for example, a preface or acknowledgments page) are numbered with lowercase roman numerals (i, ii, and so forth) centered at the bottom of the page. The report itself is generally numbered with Arabic numerals (1, 2, and so on) in the upper right-hand corner of the page. Some organizations include on each page the total

FIGURE 11-7

Effective Table of Contents

CONTENTS

number of pages (1 of 17, 2 of 17, and so forth) so the readers will be sure they have the whole document.

THE LIST OF ILLUSTRATIONS

A list of illustrations is a table of contents for the figures and tables of a report. (See Chapter 10 for a discussion of figures and tables.) If the report contains figures but not tables, the list is called a *list of figures*. If the report contains tables but not figures, the list is called a *list of tables*. If the report contains both figures and tables, figures are listed separately, before the list of tables, and the two lists together are called a *list of illustrations*.

Some writers will begin the list of illustrations on the same page as the table of contents; others prefer a separate page for the list of illustrations. If the list of illustrations begins on a separate page, it is listed in the table of contents. Figure 11-8 provides an example of a list of illustrations.

FIGURE 11-8

List of Illustrations

LIST OF ILLUSTRATIONS

ii

THE EXECUTIVE SUMMARY

The executive summary (sometimes called the *epitome*, the *executive overview*, the *management summary*, or the *management overview*) is a one-page condensation of the report. Its audience is made up, of course, of managers, who rely on executive summaries to cope with the tremendous amount of paper crossing their desks every day. The reason for the executive summary is that managers do not need or want a detailed and deep understanding of the various projects undertaken in their organizations; this kind of understanding would in fact be impossible for them, because of limitations in time and specialization. What managers *do* need is a broad understanding of the projects and how they fit together into a coherent whole. Consequently, a one-page (double-spaced) maximum for the executive summary has become almost an unwritten standard.

The special needs of managers dictate a two-part structure for the executive summary:

1. *Background.* Because managers are not necessarily technically competent in the writer's field, the background of the project is discussed clearly. The specific problem or opportunity is stated explicitly—what was not working or not working effectively or efficiently, or what potential modification of a procedure or product had to be analyzed.

2. *Major findings and implications.* Managers are not interested in the details of the project, so the methods—often the largest portion of the report—rarely receive more than one or two sentences. The conclusions and recommendations, however, are discussed in a full paragraph.

For instance, if the research and development division at an automobile manufacturer has created a composite material that can replace steel in car axles, the technical details of the report might deal with the following kinds of questions:

How was the composite devised?

What are its chemical and mechanical structures?

What are its properties?

The managerial implications, on the other hand, involve other kinds of questions:

Why is this composite better than steel?

How much do the raw materials cost? Are they readily available?

How difficult is it to make the composite?

Are there physical limitations to the amount we can make?

Is the composite sufficiently different from similar materials to prevent any legal problems?

Does the composite have other possible uses in cars?

The executives don't care about chemistry; they want to know how this project can help them make a better automobile for less money.

For several reasons, the executive summary poses a great challenge to writers. First, the brevity of the executive summary requires an almost unnatural restraint. Having spent some weeks or even months collecting data on a complex subject, writers find it difficult to reduce all that information to one page.

Second, the executive summary usually is written specifically for a nontechnical audience. Most writers are trained to write to specialists in their field. Beginning with their training in high-school science courses, writers were asked to address an audience (their teachers) that knew at least as much about the subject as they did. Learning and using the technical vocabulary was an important part of this training. In communicating with nonspecialists, however, writers must avoid or downplay specialized vocabulary and explain the subject without referring to advanced concepts.

Third, and most important, writers are so used to thinking in terms of supporting their claims with hard evidence that the notion of simply making a claim—and then not substantiating it—seems almost sacrilegious to them. Although some managers would probably like to read the full report, they simply don't have the time. They must trust the writer's technical accuracy.

Unfortunately, it is easier to define why executive summaries are a challenge to write than it is to suggest ways to write them easily. Implicit in this discussion is an obvious point: you must try to ignore most of the technical details of the project and think, instead, of the manager's needs. The expression *the bottom line* is useful to keep in mind when you must focus on the managerial implications of the project. Keep in mind the following questions as you draft the executive summary:

1. *What was the background of the study: the problem or the opportunity?* In describing problems you studied, focus on specific evidence. For most managers, the best evidence includes costs and savings. Instead of writing that the equipment we are now using to cut metal foil is ineffective, write that the equipment jams on the average of once every 72 hours, and that every time it jams we lose $400 in materials and $2,000 in productivity as the workers have to stop the production line. Then add up these figures for a monthly or annual total.

In describing opportunities you researched, use the same strategy. Your company uses thermostats to control the heating and air conditioning. Research suggests that if you had a computerized energy-management system you could cut your energy costs by 20–25 percent. If your energy costs last year were $300,000, you could save $60,000–$75,000. With these figures in mind, the readers have a good understanding of what motivated the study.

2. *What methods did you use to carry out the research?* In most cases, your principal reader does not care how you did what you did. He or she assumes you did it competently and professionally.

However, if you think your reader is interested, include a brief description—no more than a sentence or two.

3. *What were the main findings?* The findings are the results, conclusions, and recommendations. Sometimes your readers understand your subject sufficiently and want to know your principal results—the data from your study. If so, provide them. Sometimes, however, your readers would not be able to understand the technical data or would not be interested. If that is the case, go directly to the conclusions—the inferences you draw from the data. For example, the results of the feasibility study on a computerized energy-management system might consist of a description of the different hardware and software that make up the system. If a brief description would be informative, include it. Otherwise, go directly to the conclusion. That is, answer the question, Would the system be cost effective? Finally, most managers want your recommendations—your suggestions for further action. Recommendations can range from the most conservative—do nothing—to the most ambitious—proceed with the project. Often, a recommendation will call for further study of an expensive alternative.

After you have drafted the executive summary, give it to someone who has had nothing to do with the project—preferably someone outside the field. That person should be able to read the page and understand the basics of what the project means to the future of the organization.

Placement of the executive summary can be important. The current practice in business and industry is to place it before the detailed discussion. To highlight the executive summary further, writers commonly make it equal in importance to the entire detailed discussion. This strategy is signaled in the table of contents, in which the report as a whole is divided into two units. The executive summary becomes unit one of the report, and the detailed discussion unit two. In the traditional organization of a report, the executive summary would be unit one and the detailed discussion units two through six or seven.

Figure 11-9 shows an effective executive summary. Notice the differences between this executive summary and the informative abstract (Figure 11-5). The abstract focuses on the technical subject: whether the new radio-based system can effectively monitor the energy usage. The executive summary concentrates on whether the system can improve operations *at this one company*. The executive summary describes the symptoms of the problem at the writer's organization in financial terms.

After a one-sentence paragraph describing the system design—

FIGURE 11-9

Executive Summary

EXECUTIVE SUMMARY

Presently, we monitor our distribution system using after-the-fact indicators such as interruption reports, meter readings, and trouble alarms. This system is inadequate in two respects. First, it fails to give us an accurate picture of the dynamics of the distribution system. To ensure enough energy for our customers, we must overproduce. Last year we overproduced by 7 percent, or a loss of $273,000. Second, it is expensive. Escalating labor costs for meter readers and the increased number of "difficult-to-access" residences have led to higher costs. Last year we spent $960,000 reading the meters of 12,000 such residences. This report describes a project to design a radio-based system for a pilot project on these 12,000 homes.

The basic system, which uses packet-switching technology, consists of a base unit (built around a personal computer), a radio link, and a remote unit.

The radio-based distribution monitoring system described in this report is feasible because it is small enough to replace the existing meters and because it is simple to use. It would provide a more accurate picture of our distribution system, and it would pay for itself in 3.9 years. We recommend installing the system on a trial basis. If the trial program proves successful, radio-based distribution-monitoring techniques will provide the best long-term solution to the current problems of inaccurate and expensive data collection.

the results of the study—the writer describes the findings in a final paragraph. Notice how this last paragraph clarifies how the pilot program relates to the overall problem described in the first paragraph.

THE GLOSSARY AND LIST OF SYMBOLS

A glossary is an alphabetical list of definitions. A glossary is particularly useful if you are addressing a multiple audience that includes readers who will not be familiar with the technical vocabulary used in your report.

Instead of slowing down the detailed discussion by defining technical terms as they appear, you can use an asterisk or some other form of notation to inform your readers that the term is defined in the glossary. A footnote at the bottom of the page on which the first asterisk appears serves to clarify this system for readers. For example, the first use of a term defined in the glossary might occur in the following sentence in the detailed discussion of the report: "Thus the positron* acts as the . . . ." At the bottom of the page, add

*This and all subsequent terms marked by an asterisk are defined in the Glossary, page 26.

Although the glossary is generally placed near the end of the report, right before the appendixes, it can also be placed right after the table of contents. This placement is appropriate if the glossary is brief (less than a page) and defines terms that are essential for managers likely to read the body of the report. Figure 11-10 provides an example of a glossary.

A list of symbols is structured like a glossary, but rather than defining words and phrases, it defines the symbols and abbreviations used in the report. Don't hesitate to include a list of symbols if you suspect that some of your readers will not know what your symbols and abbreviations mean or might misinterpret them. Like the glossary, the list of symbols may be placed before the appendixes or after the table of contents. Figure 11-11 provides an example of a list of symbols (in this case, abbreviations).

THE APPENDIX

An appendix is any section that follows the body of the report (and the list of references or bibliography, glossary, or list of symbols). Appendixes provide a convenient way to convey information that is too bulky to be presented in the body or that will interest only a small number of readers. For the sake of conciseness in the report proper, this information is separated from the body. Examples of the kinds of materials that are usually found in appendixes include maps, large technical diagrams or charts, computations, computer printouts, test data, and texts of supporting documents.

Appendixes, which are usually lettered rather than numbered (Appendix A, Appendix B, etc.), are listed in the table of contents

FIGURE 11-10

Glossary

GLOSSARY

byte: a binary character operated upon as a unit and,
                         generally, shorter than a computer word.

error message: an indication that an error has been detected by
                         the system.

hard copy: in computer graphics, a permanent copy of a dis-
                         play image that can be separated from a display
                         device. For example, a display image recorded on
                         paper.

parameter: a variable that is given a constant value for a
                         specified application and that may denote that
                         application.

record length: the number of words or characters forming a
                         record.

FIGURE 11-11

List of Symbols

LIST OF SYMBOLS

CRT   cathode-ray tube

H_z hertz

rcvr   receiver

SNR   signal-to-noise ratio

uhf   ultra high frequency

vhf   very high frequency

and are referred to at the appropriate points in the body of the report. Therefore, they are accessible to any reader who wants to
consult them.

   Remember that an item in an appendix is titled "Appendix."
The item is not called "Figure" or "Table," even if it would have
been so designated had it appeared in the body of the report.

WRITER'S
CHECKLIST

In the rush to compile a report, it is easy to forget important elements of
the format. The following checklist is intended to help you make sure you
have included the appropriate format elements and written them correctly.

1. Does the transmittal letter
   a. Clearly state the title and, if necessary, the subject and purpose of the report?
   b. State who authorized or commissioned the report?
   c. Briefly state the methods you used?
   d. Summarize your major results, conclusions, or recommendations?
   e. Acknowledge any assistance you received?
   f. Courteously offer further assistance?

2. Does the title page
   a. Include a title that both suggests the subject of the report and identifies the type of report it is?
   b. List the names and positions of both you and your principal reader?
   c. Include the date of submission of the report and any other identifying information?

3. Does the abstract
   a. List your name, the report title, and any other identifying information?
   b. Clearly define the problem or opportunity that led to the project?
   c. Briefly describe (when appropriate) the research methods?
   d. Summarize the major results, conclusions, or recommendations?

4. Does the table of contents
   a. Clearly identify the executive summary?
   b. Contain a sufficiently detailed breakdown of the major sections of the body of the report?
   c. Reproduce the headings as they appear in your report?
   d. Include page numbers?

5. Does the list of illustrations (or tables or figures) include all the graphic aids included in the body of the report?

6. Does the executive summary
   a. Clearly state the problem or opportunity that led to the project?
   b. Explain the major results, conclusions, recommendations, and/or managerial implications of your report?
   c. Avoid technical vocabulary and concepts that the managerial audience is not likely to know?

7. Does the glossary include definitions of all the technical terms your readers might not know?

8. Does your list of symbols include all the symbols and abbreviations your readers might not know?

9. Do your appendixes include the supporting materials that are too bulky to present in the report body or that will interest only a small number of your readers?

EXERCISES

1. For each of the following letters of transmittal, write a one-paragraph evaluation. Consider each letter in terms of clarity, comprehensiveness, and tone.

   a.

   Dear Mr. Smith:

   The enclosed report, "Robots and Machine Tools," discusses the relationship between robots and machine tools.

   Although loading and unloading machine tools was one of the first uses for industrial robots, this task has only recently become commonly feasible. Discussed in this report are concepts that are crucial to remember in using robots.

   If at any time you need help understanding this report, please let me know.

   Sincerely yours,

   b.

   Dear General Smith:

   Along with Milwaukee Diesel, we are pleased to submit our study on potential fuel savings and mission improvement capabilities that could result from retrofitting the Air Force's C-3 patrol aircraft with new, high-technology recuperative or intercooled-recuperative turboprop engines.

   Results show that significant benefits can be achieved, but because of weight and drag installation penalties, the new recuperative or intercooled-recuperative engines offer little additional savings relative to a new conventional engine.

   Sincerely,

   c.

   Dear Mr. Smith:

   Enclosed is the report you requested on the effects consumer preferences will have on the future use of natural flavors.

   The major object of this project was to assess the strength of the apparent consumer desire for naturally flavored food products stemming from the controversy surrounding artificial additives. The lesser objective was to examine the other factors affecting the development and production of natural flavors.

I have found that food purchasers seem to be more interested in the taste of their foods than they are in the source of their ingredients. Moreover, the use of natural flavors is complicated by problems such as high cost, lack of availability, and inconsistent quality. Therefore, I have recommended that Smith & Co. not enter the natural flavor market at this time.

If you require any additional information, please do not hesitate to contact me.

Sincerely yours,

2. For each of the following report titles, write a three- or four-sentence evaluation. How clear an idea does each title give you of the subject of the report? Of the type of report it is? On the basis of your analysis, rewrite each title.

   a. "Recommended Forecasting Techniques for Haldane Company"
   b. "Investigation of Computerized Tax Return Systems"
   c. "Analysis of Multigroup Processing Techniques for Pennco Billing System"
   d. "Site Study for Proposed Shopping Mall in Speonk, N.Y."
   e. "Robotics in Japan"
   f. "A Detailed Description of the Factors That Led to the Great Depression of 1929"
   g. "A Study of Disc Cameras"
   h. "Teleconferencing"
   i. "The Effect of Low-level Radiation on Farm Animals"
   j. "Synfuels—Fact or Hoax?"

3. For each of the following abstracts, write a one-paragraph evaluation. How well does each abstract define the problem, methods, and important results, conclusions, and recommendations involved?

   a.
   "Proposal to Implement an Amplifier Cooling System"

   In professional sound reinforcement applications, such as discotheques, the sustained use of high-powered amplifiers causes these units to operate at high temperatures. These high temperatures cause a loss in efficiency and ultimately audible distortion, which is undesirable at disco volume levels. A cooling system of forced air must be implemented to maintain stability. I propose the use of separate fans to cool the power amplifiers, and of the best components to eliminate any distorting signals.

   b.
   "Design of a New Computer Testing Device"

   The modular design of our new computer system warrants the development of a new type of testing device. The term modular design indicates that the overall computer system can be broken down into parts or modules, each of which performs a specific function. It would be both difficult and time consuming to test the complete system as a whole, for it consists of sixteen different modules. A more effective testing method would check out each module individually for design or construction errors prior to its installation into the system. This individual testing process can be accomplished by the use of our newly designed testing device.

The testing device can selectively call or "address" any of the logic modules. To test each module individually, the device can transmit data or command words to the module. Also, the device can display the status or condition of the module on a set of LED displays located on the front panel of the device. In addition, the device has been designed so that it can indicate when an error has been produced by the module being tested.

c.

"A Recommendation for a New System of Monitoring Patients with Implanted Cardiac Pacemakers"

The monitoring system used today to test pacemaker patients is inadequate at the data distribution and storage end. Often doctors receive 20 to 50 reports a week from individual tests on patients who, because of clerical and administrative difficulties, cannot be evaluated. This report recommends a digital time-sharing system for data storage and instant recall of test results at hospital locations throughout the country. With the current system and the recommendations in this report, pacemaker monitoring will reduce data storage procedures, extend the useful life of individual pacemakers, provide doctors with current and reliable information on individual heart conditions, and supply patients with personal support and professional care.

4. For each of the following tables of contents, write a one-paragraph evaluation. How effective is each table of contents in highlighting the executive summary, defining the overall structure of the report, and providing a detailed guide to the location of particular items?

a. from "An Analysis of Corporations v. Sole Proprietorships":

CONTENTS

b. from "Recommendation for a New Incentive Pay Plan: the Scanlon Plan":

CONTENTS

c. from "Initial Design of a Microprocessor-Controlled FM Generator":

CONTENTS

5. For each of the following executive summaries, write a one-paragraph evaluation. How well does each executive summary present concise and useful information to the managerial audience?

   a. from "Analysis of Large-Scale Municipal Sludge Composting as an Alternative to Ocean Sludge-Dumping":

   Coastal municipalities currently involved with ocean sludge-dumping face a complex and growing sludge management problem. Estimates suggest that treatment plants will have to handle 65 pecent more sludge in 1995 than in 1985, or approximately seven thousand additional tons of sludge per day. As the volume of sludge is increasing, traditional disposal methods are encountering severe economic and environmental restrictions. The EPA has banned all ocean sludge-dumping as of next January 1. For these reasons, we are considering sludge composting as a cost-effective sludge management alternative.

   Sludge composting is a 21-day biological process in which waste-water sludge is converted into organic fertilizer that is aesthetically acceptable, essentially pathogen-free, and easy to handle. Composted sludge can be used to improve soil structure, increase water retention, and provide nutrients for plant growth.

At $150 per dry ton, composting is currently almost three times as expensive as ocean dumping, but effective marketing of the resulting fertilizer could dramatically reduce the difference.

b. from "Recommendation for a New Incentive Pay Plan: the Scanlon Plan":

<u>Summary</u>

The incentive pay plan, Payment by Results, in operation at Cargo Corporation of America's Trenton plant, is not working effectively. Since the implementation of this plan, union and management have experienced increasing conflicts. The employee turnover ratio and absenteeism both have risen markedly, and productivity has increased only slightly.

An extensive research project was conducted with the objective of providing a solution to the plant's problem. Originally, four possible solutions were considered: (1) a revised Payment by Results, (2) Fixed Hourly Wage (no incentive plan at all), (3) Measured Daywork, (4) Scanlon Plan. After careful consideration and investigation, the best alternative was found to be the Scanlon Plan.

The Scanlon Plan is a companywide bonus system to reward increases in productivity and efficiency. The plan has two major parts: (1) a base productivity norm, and (2) a system of work committees. The base productivity norm is based on performance over a period compared to a base period. Any improvement over the base period is called a bonus pool to be shared by <u>all</u> employees. Many different formulas have been designed to calculate the norm, but the most common is total payroll divided by the sales value of production (sales-value of production method).

Most companies implement the Scanlon Plan with departmental production committees, all reporting to one central screening committee. The individual committees meet regularly and discuss suggestions for improvements. The production committee meets formally only once a month for an hour or so. At this meeting they review the suggestions that deal with operating improvements. Any issues that they do not have the authority to deal with--for example, union matters, wages, and bonuses--are forwarded to the central screening committee. This committee is composed of employees from a representative cross-section of the company, including union officials. The committee does not vote on suggestions but does thoroughly discuss all points of view when there is a disagreement. After careful consideration, management makes the final decision.

Managers who have implemented the plan have found that employees are much more compromising toward change, particularly technological change. Experience has shown that thorough discussion of planned changes enables employees to realize that the new technology is not going to bring with it all sorts of ills that cannot be solved satisfactorily.

Companies that have used the plan have not experienced any resentment from the presiding union, and actually relations have improved. In many union-management relationships, the most difficult problem for the union leader is to get top management to sit down and discuss their various problems. A basic prerequisite of the Scanlon approach is that management be "willing to listen."

The Scanlon Plan is not for companies that are seeking a gim-
mick that will solve their problems. It is not a substitute for good
management. The message of the plan is simple: operations improvement is an area in which management, the union, and employees can
work together without strife.

c. from "Applying Multigroup Processing to the Gangloff Billing Account":

Gangloff Accounting is divided into seven geographical areas. Previously, end-of-the-month billing was processed for each
group separately. Multigroup processing allows for processing all
the groups simultaneously.
Multigroup processing is beneficial to both the data processing department and the accounting department in many ways.
Running all the groups together will cut the number of job execu-
tions from 108 to 14, an 87 percent difference. This difference
accounts for a 60 percent saving in computer time and a 30 percent
saving in paper. With multigroup processing, all sense switches
would be eliminated. With the sense switches done away with and the
tremendous decrease in computer operator responsibilities, the
chance of human error in the execution will diminish significantly.

6. Following is the body of a report (Karody 1985) entitled "Determination of an Optimal Casein and Soy Protein Mixture for Wilson Labs
Rats." The writer was a student nutritionist employed by Wilson
Labs. The principal reader is Dr. George Breyer, Head of Nutritional
Sciences at Wilson labs.
On the basis of this material, write
a. a transmittal letter to Dr. Breyer
b. a title page
c. an informative abstract (for the body sections)
d. a table of contents
e. an executive summary

II. DETERMINATION OF AN OPTIMAL CASEIN AND SOY PROTEIN MIXTURE

A. INTRODUCTION
The purpose of this research was to determine an optimal ration of casein and soy protein for our rat feed. The
present formulation contains a 10 percent weight of a complete protein source, casein. This formula has been used
since the feed was designed in 1966. By December of 19--, a
25 percent price increase of casein is expected. This will
directly affect the production cost of the product. In order
to avoid a price increase an alternative formulation was
considered.
The nutritional quality of a protein depends on the
quantity, availability, and proportions of the essential
amino acids that make up the protein. A complete protein,
such as casein, can provide growth and a positive nitrogen
balance because it contains all the essential amino acids
for a growing animal. The Amino Acid Score or Amino Acid

Index method (Oser 1959) shows that casein provides a ratio
of essential amino acids similar to the requirement for a
growing rat. On the other hand, soy protein appears to limit
the sulfur-containing amino acids: L-cisteine and
methionine. Soy is an incomplete protein, unable to promote
growth by itself, and of inferior protein quality when compared to casein. The Amino Acid Index also shows that adding
casein to a soy-protein diet should complement the defi-
ciency of soy's limiting amino acids, increasing its protein quality. But the Amino Acid Index method makes an
unreliable prediction since it does not account for the
absorption, availability, and metabolic interactions of
the amino acids.

Bioassays, on the other hand, compensate for these
factors by measuring the efficiency of the biological use of
dietary proteins as sources of the essential amino acids
under standard conditions. Many of these biological methods are based on the effects of the quality and quantity of
protein on animal growth. Several biological assays have
been proposed to measure protein quality. The Protein Efficiency Ratio (PER) was chosen for this research because it
is widely used, is the official method in the United States
and in Canada, and is simple and inexpensive.

B.   METHODOLOGY

The PER method was first used in 1917 by Osborne and
Mendel in their studies to establish protein quality (Osborne et al. 1919). This method requires that energy intake
be adequate and the protein be fed at an adequate but not
excessive level in order to promote growth. This is compatible with both the 10 percent protein level and the caloric
content of our feed. PER is a measure of weight gain of a
growing animal divided by its protein consumption. The
official government procedure requires a number of factors
to be modified in order to standardize the reference casein
and test diets. These factors include the level of the protein intake; type and amount of dietary fat; fiber levels;
strain, age, and sex of rats; assay life-span; room conditions; etc. All the biological factors were followed in our
research, but instead of using the official casein diet we
used our current feed as our standard.

1.   Diet Preparation

Prior to the arrival of the rats, four kg of
each diet were prepared with the following protein
content:

Diet 1: 25% casein   75% soy protein
Diet 2: 50% casein   50% soy protein
Diet 3: 75% casein   25% soy protein
Diet 4: 100% casein   (our current feed)

With a fixed total-protein level of 10 percent, the
diets were isocaloric. All other factors and ingredients of the diets were fixed at the current
proportions of our current feed product. The fol-

lowing is the standard percentage weight formulation used for the reference product:

Ingredients	Percentage Weight
Protein (100% casein)	10%
Starch	72
Oil (corn-cottonseed)	9
Mineral Mixture*	5
Water	2
Cellulose	1
Vitamin Mixture*	1
	100%

*The composition of the salt and the vitamin
mixtures is given in Appendix A.

2.   Rat Bioassay
We used weanling male, Sprague-Dawly rats,
weighing 48 to 60 grams. The animals were housed in
the individual galvanized steel cages (7" width;
7" height; 15" length) in Room 505 of the Animal
Research Department. In Room 505 temperature (72-
75°F) and lighting (12 hrs light/12 hrs dark) were
controlled. Assay groups were assembled in lots of
5 rats per diet in a random manner so the weight
difference was minimized. Throughout a period of
28 days, feed and water were provided ad libitum,
after a fasting period of 36 hours. Food consumption and weight gains of individual rats were recorded weekly. Data were collected in an automatic
electronic balance. Food disappearance, considered as food intake, was determined as the difference between the weights of the food cup before and
after it was filled, with corrections made for food
spilled. Protein intake was calculated by multi-
plying the percentage of protein in the diet (10%)
times the total food intake, divided by 100. PER was
calculated as the weight gain divided by the amount
of protein consumed. The raw data for PER, weight
gain, food intake, and protein intake are given in
Appendix B.

C.   RESULTS
The PER, weight gain, food intake, and protein intake
values for the assay are shown in Table 1. Data in Table 1
were statistically manipulated using the MINITAB system. A
two-sample t-test was used to check for significant differences of the means ($\alpha = 0.05$). PER values range from 1.6 to
2.6. The PER values from the 50 percent casein, 50 percent
soy diet and our standard feed proved by this statistical
manipulation to have no significantly different means. The
other two diets have different PER values than our standard.
The food intake, and subsequently the protein intake, did
not vary in the standard or in the test diets.

Table 1

Weight Gain, Food Intake, Protein Intake,

and PER for Our Current Rat Feed and Test Diets

Diet	Weight Gain* (grs)	Food Intake (grs)	Protein Intake (grs)	PER
25% casein- 75% soy prot.	52.96 ± 7.62	331 ± 70.6	33.1 ± 7.06	1.6 ± 0.22
50% casein- 50% soy prot.	88.80 ± 5.56	370 ± 50.2	37.0 ± 5.25	2.4 ± 0.38
75% casein- 25% soy prot.	101.79 ± 9.84	352 ± 43.4	35.1 ± 4.34	2.9 ± 0.21
100% casein (our current feed)	93.86 ± 5.94	362 ± 32.3	36.1 ± 3.2	2.6 ± 0.45

*Mean ± Standard Deviation

D.   CONCLUSION

Our results show that a mixture of 50 percent casein and 50 percent soy protein has the same PER value (around 2.5) as our standard product. Therefore, it can be concluded that they are dietary protein sources with the same biological protein quality. Moreover, this optimal protein ratio could substitute for casein in the current feed formula without altering the expected nutritional performance of the product. At the same time--as described in the executive summary--modifying our current protein to 50 percent casein/50 percent soy will reduce by 27 percent, or $3,200 per year, our current rat feed costs.

E.   RECOMMENDATION

If the Production and Quality Control Managers have no objection, the change in formulation of our current feed might be a feasible solution to avoid the increase in its production cost.

REFERENCES   Crowe, B. 1985. Design of a Radio-based System for Distribution Automation. Unpublished manuscript.

Korody, E. 1985. Determination of an Optimal Casein and Soy Mixture for Wilson Labs Rats. Unpublished manuscript.

CHAPTER
TWELVE

MEMO REPORTS

The memorandum is the workhorse of technical writing. It is the basic means of written communication between two persons— or between individuals and groups—in an organization. Each day, the average employee is likely to receive as many as a half-dozen memos and send out another half-dozen. Whereas most memos convey routine news addressed to several readers, many organizations are turning to the memo as an efficient format for brief technical reports.

To facilitate the transfer of information, most organizations have distribution routes—lists of employees who automatically receive copies of memos that pertain to their area of expertise or responsibility. A technician, for example, might be on several different distribution routes: one that includes all employees, one that includes the technical staff, and one that includes all employees involved with a particular project.

This chapter concentrates not on the "FYI" memo—the simple communication addressed "For Your Information"—but on the more substantive kinds of brief technical reports that generally are written in memo form, such as directives, responses to inquiries, trip reports, and field reports.

Like all technical writing, the memo should be clear, accurate, comprehensive, concise, accessible, and correct. But unlike letters, memos need not be ceremonious. Memo writing is technical writing with its sleeves rolled up.

Clarity, conciseness, and accessibility are encouraged by a mechanical device: headings. The printed forms on which memos are written begin with a subject heading—a place for the writer to define the subject. If you use this space efficiently, you can define not only the subject but also the purpose of the memo and thus begin to communicate immediately.

In the body of the memo, headings help your readers understand the message you are conveying. For example, the simple word *Results* at the start of a paragraph tells your readers what you are describing and thus enables them to decide whether or not to read the discussion. (See Chapter 5 for a further discussion of headings.)

Conciseness is also encouraged by the size of the forms. Most organizations have at least two sizes of memo forms: a full page and a half page. Often, writers are reluctant to put down only two or three lines on a full page, so they start to pad the message. The smaller page reduces this temptation.

The following discussion describes in detail how to write effective memos.

THE STRUCTURE OF THE MEMO

The memo is made up of two components: the identifying information and the body.

THE
IDENTIFYING
INFORMATION
Basically, a memo looks like a streamlined letter. The salutation ("Dear Mr. Smith") and the complimentary close ("Sincerely yours") are eliminated. Most writers put their initials or signature next to their typed names or at the end of the memo. The inside address—the mailing address of the reader—is replaced by a department name or an office number, generally listed after the person's name. Sometimes no "address" at all is given.

Almost all memos have five elements at the top: the logo or a brief letterhead of the organization and the "To," "From," "Subject," and "Date" headings. Some organizations have a "copies" or "cc" (carbon copy) heading as well. "Memo," "Memorandum," or "Interoffice" might be printed on the forms.

Organizations sometimes have preferences about the ways in which the headings are filled out. Some organizations prefer that the full names of the writer and reader be listed. Others want only the first initials and the last names. Job titles are sometimes used.

It the organization for which you work does not object, include your own job title and that of your reader. In this way, the memo will be complete and informative for a reader who refers to it after you or your reader has moved on to a new position as well as for readers elsewhere in the organization who might not know you and your reader. The names of persons receiving copies of the memo (generally photocopies, not carbons) are listed either alphabetically or in descending order of organizational rank. In listing the date, write out the month (March 4, 19-- or 4 March 19--); do not use the all-numeral format (3/4/--). Foreign-born people could be confused by the numerals, because in many countries the first numeral identifies the day, not the month.

The subject heading deserves special mention. Don't be *too* concise. Avoid naming only the subject, such as "Tower Load Test"; rather, add a clue identifying what it is about the test you wish to address. For instance, "Tower Load Test Results" or "Results of Tower Load Test" would be much more informative than "Tower Load Test," which does not tell the reader whether the memo is about the date, the location, the methods, the results, or any number of other factors related to the test.

The top of a memo is designed to identify the writing situation as efficiently as possible. The writer names himself or herself, the audience, and the subject, ideally with some indication of purpose.

The following examples show several common styles of filling in the identifying information of a memo.

AMRO   MEMO


```
      To:    B. Pabst
    From:    J. Alonso
 Subject:    MIXER RECOMMENDATION FOR PHILLIPS
    Date:    11 June 19--
```

NORTHERN PETROLEUM COMPANY   INTERNAL CORRESPONDENCE


```
Date:      January 3, 19--
To:        William Weeks, Director of Operations
From:      Helen Cho, Chemical Engineering Dept.
Subject:   Trip Report--Conference on Improved Procedures
           for Chemical Analysis Laboratory
```

```
HARSON ELECTRONICS      MEMORANDUM

To:      John Rosser, Accounting  From:    Andrew Miller,
                                           Technical Services

                                  Subject:   Budget Revision
                                             for FY 19--

cc:      Dr. William De Leon,     Date:    March 11, 19--
         President
         John Grimes, Comptroller
```

```
                     INTEROFFICE

         To:    C. Cleveland         cc:    B. Aaron
       From:    H. Rainbow                  K. Lau
    Subject:    Shipment Date of Blueprints  J. Manuputra
                to Collier                  W. Williams
       Date:    2 October 19--
```

The second and all subsequent pages of memos are typed on plain paper. The following information is typed in the upper left-hand corner of each page:

1. the name of the recipient
2. the date of the memo
3. the page number

Often, writers will define the communication as a memo and repeat the primary names as well. A typical second page of such a memo begins like this:

```
    Memo to:    J. Alders          April 6, 19--
        from:   R. Rossini         Page 2
```

THE BODY OF THE MEMO   The average memo has a very brief "body." A message about the office closing early during heavy snow, for example, requires only one or two sentences. However, memos that convey complex technical information, and those that approach one page or longer (memos are always single-spaced), are most effective when they follow a basic structure. This structure gives you the same sense of direction that a full-scale outline does as you plan a formal report. A useful structure for most substantive memos includes four parts:

1. purpose statement
2. summary
3. discussion
4. action

As with any kind of document, the organization of the memo does not necessarily reflect the sequence in which the parts were written. After brainstorming and outlining the memo, start with the discussion—the most technical section. Until you have written the discussion, you cannot write the summary or the purpose statement, because both these parts depend on the discussion. For most writers, the most effective sequence of composition will be discussion, action, summary, and purpose.

Make sure to highlight the structure of the memo. Use headings and lists (see Chapter 5).

Headings make the memo easier to read, easier *not* to read, and easier to refer to later. Headings are not subtle; they define clearly what the discussion that follows is about and thus improve the reader's comprehension. In addition, headings represent a courtesy to executive readers who are not interested in the details of the memo. These readers can read the purpose and summary and stop if they need no further information. Finally, headings enable readers to isolate quickly the information they need after the memo has been filed away for a while. Rather than having to reread a three-page discussion, they can turn directly to the summary, for instance, or to a subsection of the discussion.

Lists help your reader understand the memo. If, for instance, you are making three points, you can list them consistently in the different sections of the memo. Point number one under the "Summary" heading will correspond to point number one under the "Recommendations" heading, and so forth.

PURPOSE STATEMENT Memos are reproduced very freely. The majority of those that you receive any given day might be only marginally relevant to you. Many readers, after starting to read their incoming memos, ask, "Why is he telling me this?" The first sentence of the body should answer that question. Following are a few examples of purpose statements.

I want to tell you about a problem we're having with the pressure on the main pump, because I think you might be able to help us.

The purpose of this memo is to request authorization to travel to the Brownsville plant Monday to meet with the other quality inspectors.

This memo presents the results of the internal audit of the Phoenix branch that you authorized March 13, 19--.

I want to congratulate you on the quarterly record of your division.

This memo confirms our phone call of Tuesday, June 10, 19--.

Notice that a purpose statement need not be long and complicated. In fact, the best purpose statements are concise and direct. Make sure your purpose statement has a verb that clearly communicates what you want the memo to accomplish, such as *to request*, *to explain*, or *to authorize*. (See Chapters 2 and 3 for a more detailed discussion of purpose.) Some students of the techniques of logical argument object to a direct statement of purpose—especially when the writer is asking for something, as in the example about requesting travel authorization. Rather than beginning with a direct statement of purpose, an argument for such a request would open with the reasons that the trip is necessary, the trip's potential benefits, and so forth. Then the writer would conclude the memo with the actual request: "For these reasons, I am requesting authorization to. . . ." Although some readers would rather have the reasons presented first, far more would prefer to know immediately why you have written. There are two basic problems with the standard argument structure: (1) some readers will toss the memo aside if you seem to be rambling on about the Brownsville plant without getting to the point; and (2) some readers will feel as if you are trying to manipulate them, to talk them into doing something they don't want to do. The purpose statement sacrifices subtlety for directness.

SUMMARY Along with the purpose statement, the summary forms the core of the memo. It helps all the readers follow the subsequent discussion, enables executive readers to skip the rest of the memo, and serves as a convenient reminder of the main points. Following are some examples of summaries:

The proposed revision of our bookkeeping system would reduce its errors by 80 percent and increase its speed by 20 percent. The revision would take two months and cost approximately $4,000. The payback period would be less than one year.

The conference was of great value. The lectures on new coolants suggested techniques that might be useful in our Omega line, and I met three potential customers who have since written inquiry letters.

The analysis shows that lateral stress was the cause of the failure.
We are now trying to determine why the beam could not sustain a lateral
stress weaker than that it was rated for.

In March, we completed Phase II (Design) on schedule. At this point,
we anticipate no delays that will jeopardize our projected completion
date.

The summary should reflect the length and complexity of the
memo. It might range in length from one simple sentence to a long
and technical paragraph. If possible, the summary should reflect
the structure of the memo. For example, the discussion following
the first sample summary should explain, first, the proposed revision of the bookkeeping system and, second, its two advantages:
fewer errors and increased speed. Next should come the discussion
of the costs and finally discussion of the payback period.

DISCUSSION   The discussion is the elaboration of the summary. It is
the most detailed, technical portion of the memo. Generally, the
discussion begins with a background paragraph. Even if you think
the reader will be familiar with the background, it is a good idea
to include a brief recap, just to be safe. Also, the background will
be valuable to a reader who refers to the memo later.

The background discussion is, of course, as individual as the
memo of which it is a part; however, some basic guidelines are
useful. If the memo defines a problem—for example, a flaw detected in a product line—the background might discuss how the
problem was discovered or present the basic facts about the product line: what the product is, how long it has been produced, and
in what quantities. If the memo reports the results of a field trip,
the background might discuss why the trip was undertaken, what
its objectives were, who participated, and so forth.

Following is a background paragraph from a memo requesting
authorization to have a piece of equipment retooled.

Background

The stamping machine, a Curtiss Model 6143, is used in the sheet-metal
shop. We bought it in 1976 and it is in excellent condition. However,
since we switched the size of the tin rolls last year, the stamping
machine no longer performs up to specifications.

After the background comes the detailed discussion of the message itself. Here you present whatever details are necessary to give
your readers a clear and complete idea of what you have to say.
The detailed discussion might be divided into the subsections of a
more formal report: materials, equipment, methods, results, con-

clusions, and recommendations. Or it might be made up of headings that pertain specifically to the subject you are discussing. Small tables or figures might also be included; larger ones should be attached as appendixes to the memo.

The discussion section of the memo can be developed according to any of the basic patterns for structuring technical documents:

1. chronological
2. spatial
3. general to specific
4. more important to less important
5. problem-methods-solutions

These patterns are discussed in detail in Chapter 3.

Following is the detailed discussion section from a memo written by a salesman working for the "XYZ Company," which makes electronic typewriters. The XYZ salesman met an IBM salesman by chance one day, and they talked about the XYZ typewriters. The XYZ salesman is writing to his supervisor, telling her what he learned from the IBM salesman and also what XYZ's research and development (R&D) department told him in response to the comments of the IBM salesman.

DISCUSSION

Salesman's Comments:

In our conversation, he talked about the strengths of our machines and then mentioned two problems: excessive ribbon consumption and excessive training time.

In general, he had high praise for the XYZ machines. In particular, he liked the idea of the rotary and linear stepping motor. Also, he liked having all the options within the confines of the typewriter. He said that although he knows we have some reliability problems, the machines worked well while he was training on them.

The major problem with the XYZ machines, he said, is excessive ribbon consumption. According to his customers who have XYZ machines, the $5 cartridge lasts only about two days. This adds up to about $650 a year, about a third the cost of our basic Model A machine.

The minor problem with the machines, he said, is that most customers are used to the IBM programming language. Since our language is very different, customers are spending more time learning our system than they had anticipated. He didn't offer any specifics on training-time differences.

R&D's Response:

I relayed these comments to Steve Brown in R&D. Here is what he told me.

>Ribbon Consumption: A recent 20 percent price reduction in the 4.1" cartridge should help. In addition, in a few days our 4.9" correctable cartridge--with a 40 percent increase in character output--should be ready for shipment. R&D is fully aware of the ribbon-consumption problem, and will work on further improvements, such as thinner ribbon, increased diameter, and new cartridges.

>Training Time: New software is being developed that should reduce the training time.

If I can answer any questions about the IBM salesman's comments, please call me at x1234.

The basic pattern of this discussion is chronological: the writer describes first his discussion with the IBM salesman and then the response from R&D. Within each of these two subsections the basic pattern is more-important-to-less-important: the ribbon-consumption problem is more serious than the training-time problem, so ribbon consumption is discussed first.

ACTION Most memo reports present information that will eventually be used in formulating or modifying major projects or policies. These memo reports will become parts of the files on these projects or policies. Some reports, however, require follow-up action more immediately, by either the writer or the readers. For example, a writer addressing a group of supervisors might define what he or she is going to do about a problem discussed in the memo. A supervisor might use the action component to delegate tasks for other employees to accomplish. In writing the action component of a memo, be sure to define clearly *who* is to do *what* and *when*. Following are two examples of action components.

Action:

I would appreciate it if you would work on the following tasks and have your results ready for the meeting on Monday, June 9.

1. Henderson to recalculate the flow rate.
2. Smith to set up meeting with the regional EPA representative for some time during the week of February 13.
3. Falvey to ask Armitra in Houston for his advice.

```
Action:

To follow up these leads, I will do the following this week:

1. Send the promotional package to the three companies.
2. Ask Customer Relations to work up a sample design to show the three
   companies.
3. Request interviews with the appropriate personnel at the three
   companies.
```

Notice in the first example that although the writer is the supervisor of his readers, he uses a polite tone in this introductory sentence.

TYPES OF MEMOS

Each memo is written by a specific writer to a specific audience for a specific purpose. Every memo is unique. Nevertheless, it is possible to define broad categories of memos according to the functions they fulfill. This section of the chapter discusses four basic types of memos: the directive, the response to an inquiry, the trip report, and the field/lab report.

Notice as you read about each type of memo how the purpose-summary-discussion-action strategy is tailored to the occasion. Pay particular attention to the headings, lists, and indention used to highlight the structure of the examples.

THE DIRECTIVE   In a directive memo, you define a policy or procedure you want your readers to follow. If possible, explain the reason for the directive; otherwise, it might seem like an irrational order rather than a thoughtful request. For short memos, to prevent the appearance of bluntness, the explanation should precede the directive. For longer memos, the actual directive might precede the detailed explanation. This strategy will ensure that the readers will not overlook the directive. Of course, the body of the memo should begin with a polite explanatory note, such as the following:

```
The purpose of this memo is to establish a uniform policy for dealing
with customers who fall more than sixty days behind in their accounts.
The policy is defined below under the heading "Policy Statement."
Following the statement is our rationale.
```

Figure 12-1 provides an example of a directive.

Notice in this example that the directive is stated as a request, not an order. Unless your readers have ignored previous requests, a polite tone is the most effective.

FIGURE 12-1

Directive

```
Quimby                                              Interoffice


        Date:    March 19, 19--
          To:    All supervisors and sales personnel
        From:    D. Bartown, Engineering
     Subject:    Avoiding customer exposure to sensitive
                 information outside Conference Room B.

     It has come to our attention that customers meeting in Conference
     Room B have been allowed to use the secretary's phone directly
     outside the room. This practice presents a problem: the proposals
     that the secretary is typing are in full view of the customers.
     Proprietary information such as pricing can be jeopardized unin-
     tentionally.

     In the future, would you please escort any customers or non-
     Quimby personnel needing to use a phone to the one outside the Es-
     timating Department? Your cooperation in this matter will be
     greatly appreciated.
```

Also note that in spite of its brevity and simplicity, this example follows, without headings, the purpose-summary-discussion-
action structure. Purpose is identified on the subject line, the first
paragraph is a combination of summary and discussion (extensive
discussion is hardly necessary in this situation), and the second
paragraph dictates the specific action to be taken.

THE
RESPONSE
TO AN
INQUIRY

Often you might be asked by a colleague to provide information
that cannot be communicated on the phone because of its complexity or importance. In responding to such an inquiry, the purpose-summary-discussion-action strategy is particularly useful.
The purpose of the memo is simple: to provide the reader with the
information he or she requested. The summary states the major
points of the subsequent discussion and calls the reader's attention
to any parts of it that might be of special importance. The action
section (if it is necessary) defines any relevant steps that you or
some other personnel are taking or will take. Figure 12-2 provides
an example of a response to an inquiry.

Notice in this sample memo how the numbered items in the
summary section correspond to the numbered items in the discussion section. This parallelism enables the reader to find quickly the
discussion he wants.

THE TRIP
REPORT

A trip report is a record of a business trip written after the employee returns to the office. Most often, a trip report takes the
form of a memo. The key to writing a good trip report is to re-

FIGURE 12-2

Response to an
Inquiry

NATIONAL INSURANCE COMPANY   **MEMO**

    To: J. M. Sosry, Vice President
    From: G. Lancasey, Accounting
    Subject: National's Compliance with the Federal Pay Stan-
            dards
    Date: February 2, 19--

    Purpose: This memo responds to your request for an assessment
              of our compliance with the Federal Non-Inflationary
              Pay and Price Behavior Standards.

    Summary: 1. We are in compliance except for a few minor vio-
                  lations.
              2. Legal Affairs feels we are exercising "good
                  faith," a measure of compliance with the Stan-
                  dards.
              3. Data Processing is currently studying the costs
                  and benefits of implementing data processing of
                  the computations.

              The following discussion elaborates on these three points.

    Discussion: 1. We are in compliance with the Standards except
                    for the following details related to fringe
                    benefits.

                a. The fringe benefits of terminated individ-
                    uals have not yet been eliminated from our
                    calculations. The salaries have been elim-
                    inated.

                b. The fringe benefits associated with promo-
                    tional increases have not yet been elimi-
                    nated from our calculations.

                c. The fringe benefits of employees paid
                    $9,800 or less have not yet been eliminated
                    from our calculations.

---

member that your reader is less interested in an hour-by-hour description of what happened than in a carefully structured discussion of what was important. If, for instance, you attended a
professional conference, don't list all the presentations—simply attach the agenda or program if you think your reader will be interested. Communicate the important information you learned—or
describe the important questions that didn't get answered. If you
traveled to meet with a client (or a potential client), don't describe
everything that happened. Focus on what your reader is interested
in: how to follow up on the trip and maintain a good business relationship with the client.

Memo to J. M. Sosry
February 2, 19xx
Page 2

2. I met with Joe Brighton of Legal Affairs last
Thursday to discuss the question of compliance.
Joe is aware of our minor violations. His research, including several calls to Washington,
suggests that the Standards define "good
faith" efforts to comply for various-size corporations, and that we are well within these
guidelines.

3. I talked with Ted Ashton of Data Processing last
Friday. They have been studying the costs/bene-
fits of implementing data processing for the
calculations. Their results won't be complete
until next week, but Ted predicts that it will
take up to two months to write the program in-
ternally. He is talking to representatives of
computer service companies this week, but he
doubts if they can provide an economical solution.

As things stand now--doing the calculations
manually--we will need three months to catch
up, and even then we will always be about two
weeks behind.

Action: I have asked Ted Ashton to send you the results
of the cost/benefits study when they are in. I
have also asked Joe Brighton to keep you informed of any new developments with the Standards.

If I can be of any further assistance, please let me know.

In most cases, the standard purpose-summary-discussion-
action structure is appropriate for this type of memo. It is a good
idea to mention briefly the purpose of the trip—even if your
reader might already know its purpose. By doing this, you will be
providing a complete record for future reference. In the action
section, list either the pertinent actions you have taken since the
trip or what you recommend that your reader do. Figure 12-3 provides an example of a typical trip report.

Notice in this example that the writer and reader appear to be
relatively equal in rank: the informal tone of the "Recommendation" section suggests that they have worked together before. De-

FIGURE 12-3

Trip Report

Dynacol Corporation

INTEROFFICE COMMUNICATION

To: G. Granby, R & D
From: P. Rabin, Technical Services
Subject: Trip Report--Computer Dynamics, Inc.
Date: September 20, 19--

Purpose:
This memo presents my impressions of the Computer
Dynamics technical seminar of September 18.
The purpose of the seminar was to introduce
their new PQ-500 line of computers.

Summary:
In general, I was not impressed with the new
line. The only hardware that might be of interest
to us is their graphics terminal, which I'd
like to talk to you about.

Discussion:
Computer Dynamics offers several models in
its 500 series, ranging in price from $11,000
to $45,000. The top model has a 1Mb memory.
Although it's very fast at matrix operations,
this feature would be of little value to us.
The other models offer nothing new.

I was disturbed by some of the answers offered
by the Computer Dynamics representatives, which
everyone agreed included misinformation.

---

spite this familiarity, however, the memo is clearly organized to make it easy for the reader to read and refer to later, or to pass on to another employee who might follow up on it.

FIELD AND LAB REPORTS

Many organizations use a memo format for reports on inspection and maintenance procedures carried out on systems and equipment. These memos, known as field or lab reports, include the same information that high-school lab reports do—the problem, methods, results, and conclusions—but they deemphasize the methods and add a recommendations section (if appropriate).

```
The most interesting item was the graphics
terminal.  It is user-friendly.  Integrating
their terminal with our system could cost $4,000
and some 4-5 person-months.  But I think that
we want to go in the direction of graphics
terminals, and this one looks very good.

Recommendation:
I'd like to talk to you, when you get a chance,
about our plans for the addition of graphics
terminals.  I think we should have McKinley
and Rossiter take a look at what's available.
Give me a call (x3442) and we'll talk.
```

A typical field or lab report, therefore, has the following structure:

1. purpose of the memo
2. problem leading to the decision to perform the procedure
3. summary
4. results
5. conclusions
6. recommendations
7. methods

FIGURE 12-4

Lab Report

Lobate Construction
MEMO

To:	C. Amalli
From:	W. Kabor
Subject:	Inspection of Chemopump after Run #9
Date:	6 January 19--

cc: A. Beren
      S. Dworkin
      N. Mancini

Purpose:
This memo presents the findings of my visual inspection of the
Chemopump after it was run for 30 days on Kentucky #10 coal and re-
quests authorization to carry out follow-up procedures.

Problem:
The inspection was designed to determine if the new Chemopump
were compatible with Kentucky #10, our lowest-grade coal. In
preparation for the 30-day test run, the following three modifications were made:

1. New front bearing housing buffer plates of tungsten car-
   bide were installed.
2. The pump casting volute liner was coated with tungsten car-
   bide.
3. New bearings were installed.

Summary:
A number of small problems with the pump were observed, but nothing serious and nothing surprising. Normal break-in accounts for
the wear. The pump accepted the Kentucky #10 well.

Findings:
The following problems were observed:

1. The outer lip of the front-end bell was chipped along two
   thirds of its circumference.

Sometimes several of these sections can be combined. Purpose and problem often are discussed together, as are results and conclusions.

The lab report shown in Figure 12-4 illustrates some of the possible variations on this standard report structure.

Notice the following points about this example:

1. A single word—"visual"—constitutes the discussion of the inspection procedure in the purpose section. Nothing else needs to be said.
2. In the "findings" section, the writer lists the "bad news"—the prob-

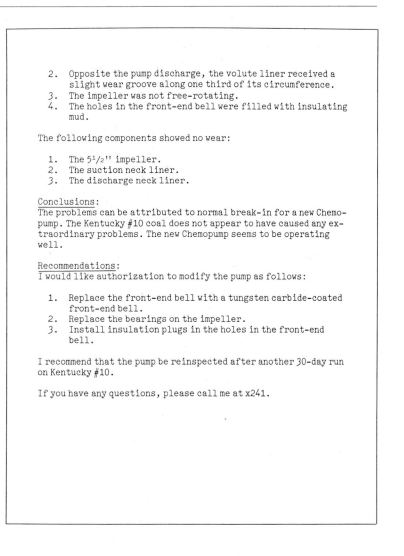

```
    2.  Opposite the pump discharge, the volute liner received a
        slight wear groove along one third of its circumference.
    3.  The impeller was not free-rotating.
    4.  The holes in the front-end bell were filled with insulating
        mud.

The following components showed no wear:

    1.  The 5¹/₂'' impeller.
    2.  The suction neck liner.
    3.  The discharge neck liner.

Conclusions:
The problems can be attributed to normal break-in for a new Chemo-
pump. The Kentucky #10 coal does not appear to have caused any ex-
traordinary problems. The new Chemopump seems to be operating
well.

Recommendations:
I would like authorization to modify the pump as follows:

    1.  Replace the front-end bell with a tungsten carbide-coated
        front-end bell.
    2.  Replace the bearings on the impeller.
    3.  Install insulation plugs in the holes in the front-end
        bell.

I recommend that the pump be reinspected after another 30-day run
on Kentucky #10.

If you have any questions, please call me at x241.
```

lems—before the "good news." This is a logical order, because the bad news is of more significance to the readers.

3. By including the last sentence, the writer makes it easy for the reader to get in touch with her to ask questions or authorize the recommended modifications.

WRITER'S CHECKLIST

The following checklist covers the basic formal elements included in most memo reports.

1. Does the identifying information
    a. Include the names and (if appropriate) the job positions of both you and your readers?

    b. Include a sufficiently informative subject heading?

    c. Include the date?

2. Does the purpose statement clearly tell the readers why you are asking them to read the memo?

3. Does the summary
   a. Briefly state the major points developed in the body of the memo?
   b. Reflect the structure of the memo?

4. Does the discussion section
   a. Include a background paragraph?
   b. Include headings to clarify the structure and content?

5. Does the action section clearly and politely identify tasks that you or your readers will carry out?

EXERCISES

1. As the manager of Lewis Auto Parts Store, you have noticed that some of your salespeople are smoking in the showroom. You have received several complaints from customers. Write a memo defining a new policy: salespeople may smoke in the employees' lounge but not in the showroom.

2. There are 20 secretaries in the six departments at your office. Although they are free to take their lunch hours whenever they wish, sometimes several departments have no secretarial coverage between 1:00 and 1:30 P.M. Write a memo to the secretaries, explaining why this lack of coverage is undesirable and asking for their cooperation in staggering their lunch hours.

3. You are a senior with an important position in a school organization, such as a technical society or the campus newspaper. The faculty adviser to the organization has asked you to explain, for your successors, how to carry out the responsibilities of the position. Write a memo in response to the request.

4. The boss at the company where you last worked has phoned you, asking for your opinions on how to improve the working conditions and productivity. Using your own experiences, write a memo responding to the boss's inquiry.

5. If you have attended a lecture or presentation in your area of academic concentration, write up a trip report memo to an appropriate instructor assessing its quality.

6. Write up a recent lab or field project in the form of a memo to the instructor of the course.

7. The following memos could be improved in tone, substance, and structure. Revise them to increase their effectiveness, adding any reasonable details.

a.

KLINE MEDICAL PRODUCTS

```
    Date:    1 September 19--
      To:    Mike Framson
    From:    Fran Sturdiven
 Subject:    Device Master Records
```

The safety and efficiency of a medical device depends on the adequacy of its design and the entire manufacturing process. To ensure that safety and effectiveness are manufactured into a device, all design and manufacturing requirements must be properly defined and documented. This documentation package is called by the FDA a "Device Master Record."

The FDA's specific definition of a "Device Master Record" has already been distributed.

Paragraph 3.2 of the definition requires that a company define the "compilation of records" that makes up a "Device Master Record." But we have no such index or reference for our records.

Paragraph 6.4 says that any changes in the DMR must be authorized in writing by the signature of a designated individual. We have no such procedure.

These problems are to be solved by 15 September 19--.

b.

```
                          Southwestern Gas          Interoffice

      To:    John Harlan
    From:    Robert Kalinowski
 Subject:    Official vehicle
    Date:    July 31, 19--
```

Someone has told us that our vehicles have been seen at recreational sites, such as beaches. I don't have to remind you that the vehicles are for official use only. This practice must stop.

c.

```
                          Viking National Bank        Memorandum

      To:    George Delmore, Expense Recording
    From:    David Derahl, Internal Audit
 Subject:    Escheat Procedures
    Date:    Dec. 6, 19--
```

Undeliverable checks received by the originating department should be voided immediately. The originating department should obtain a copy of the voided check and forward the original to Expense Recording. Efforts to determine a valid mailing address should be initiated by the originating department. If the check is less than seven months old and remains undeliverable, the liability should be recorded in the originating department's operating expense account.

This memo should define our recommended policy on escheat procedures. I hope it answers your question.

d.

CAPITAL ENERGY, INC.   MEMO

        To: L. Abrams, Technical Staff
        From: M. Cornish, R&D
    Subject: Separation by Gravity
        Date: May 3, 19--

Here is the information you requested last week.

At 120°F the difference in densities is only about 0.01 g/ml. At higher temperatures, up to 160°, the difference decreases. The best temperature was 80°, but still I got some oil mixed with the water. The separation at 120° was very slight.

From this limited data, I don't think gravity will work at Reynoldsville.

e.

Diversified Chemicals, Inc.
Memo

Date: August 27, 19--
To: R. Martins
From: J. Speletz
Subject: Charles Research Conference on Corrosion

The subject of the conference was high-temperature dry corrosion. Some of the topics discussed were

1 - thin film formation and growth on metal surfaces. The lectures focused on the study of oxidation and corrosion by spectroscopy.

2 - the use of microscopy to study the microstructure of thick film formation on metals and alloys. The speakers were from the University of Colorado and MIT.

3 - one of the most interesting topics was hot corrosion and erosion. The speakers were from Penn State and Westinghouse.

4 - future research directions for high-temperature dry corrosion were discussed from five viewpoints.

    1 - university research
    2 - government research
    3 - industry research
    4 - European industry research
    5 - European government research

5 - corrosion of ceramics, especially the oxidation of Si_3N_4. One paper dealt with the formation of Si ALON, which could be an inexpensive substitute for Si_3N_4. This topic should be pursued.

f.

Korvon Laboratories--Memo

To:	Ralph Eric
From:	Walt Kavon
Subject:	"Computers in the Laboratory"
Date:	May 1, 19--

The seminar on "Computers in the Laboratory" was held in New York on April 22, 19--. Approximately 30 managers of labs of various sizes attended.

The leader of the seminar, Mr. Daniel Moore, presented a program that included the following topics:

Modern analytical instrumentation
Maintaining quality in quality assurance
Harnessing the power of computers
Capital investments: justifying costs to management

The subjects of minicomputers versus terminals and how to increase reliability and reduce costs were discussed by several computer manufacturers' representatives.

My major criticism of the session was the ineffective leadership of Dr. Moore. Frequently, he read long passages from published articles. Often, he was very disorganized. I did enjoy, however, meeting the other lab managers.

g.

Technical Maintenance, Inc.   Memo

TO:	Rich Abelson
FROM:	Tom Donovan
SUBJECT:	Dialysis Equipment
DATE:	10/24/--

The Clinic that sent us the dialysis equipment (two MC-311's) reported that it could not regulate the temperature precisely enough.

I found that in both 311's, the heater element did not turn off. The temperature control circuit has an internal trim potentiometer that required adjustment. It is working correctly now.

I checked out the temperature control system's independent backup alarm system that will alarm and shut down the system if the temperature reaches 40°C. It is working properly.

The equipment has been returned to the client. After phoning them, I learned that they have had no more problems with it.

h.

FREEMAN, INC.   INTEROFFICE

```
     To:   C. F. Grant
   From:   R. C. Nedden
Subject:   Testing of Continuous Solder Strip Alternative for
           Large-Scale Integrated Terminals
     cc:   J. A. Jones
           M. H. Miller
```

We ordered samples of continuous solder strips in three thick-
nesses for our testing: 1.5 mils, 2.5 mils, and 4.0 mils. Then we
manufactured each thickness into terminals to test for pull
strength.

The 1.5 mil material had a test strength of 1.62 pounds, which is
above our goal of 1.5 pounds. But 30 percent of these terminals did
not meet the goal. The 2.5 mil material had an average pull strength
of 2.4 pounds, with a minimum force of 1.65 pounds. The 4.0 mil
material had an average pull strength of 2.6 pounds per terminal,
with a minimum of 1.9. Even though there was 60 percent more solder
available than with the 2.4 mils material, the average pull
strength increased by only 8 percent.

We concluded from this that the limit to the pull strength of the
terminal is dependent on the geometry of the terminal, not on the
amount of solder.

For this reason, we believe that the 2.5 mils material would be the
most cost-effective solution to the problem of inadequate pull
strength in our LSI terminals.

Please let me know if you have any questions.

CHAPTER
THIRTEEN

PROPOSALS

$\mathbf{M}$ost projects undertaken by organizations, as well as most changes made within organizations, begin with proposals. The document that persuades Mayer Contractors, for example, to have Technical Documentation, Inc., write a set of procedures manuals is an *external proposal* prepared by Technical Documentation. Similarly, when a local police department wishes to purchase a new fleet of police cars, automobile manufacturers and customizers interested in securing the contract submit external proposals that detail the cost, specifications, and delivery schedule of their products. In short, when one organization purchases goods or services from another, the decision to purchase is usually based on the strength of the supplier's external proposal.

When an employee suggests to his or her supervisor that the organization purchase a new word processor or restructure a department, a similar but generally less elaborate document—the *internal proposal*—is used. This chapter will discuss both types of proposals. Because the external proposal is the more formal and detailed of the two, it will be discussed at greater length and will serve as the model for the internal proposal, which borrows formal aspects of the external proposal according to the demands of the internal situation.

THE EXTERNAL PROPOSAL

No organization produces all the products or provides all the services it needs. Paper clips and company cars have to be purchased. Offices have to be cleaned and maintained. Sometimes, projects that require special expertise—such as sophisticated market analyses or feasibility studies—have to be carried out. With few exceptions, any number of manufacturers would love to provide the paper clips or the cars, and a few dozen consulting organizations would happily conduct the studies.

For those seeking the product or service, it is a buyer's market. In order to get the best deal, most organizations require that their potential suppliers compete for the business. By submitting a proposal, the supplier attempts to make the case that it deserves the contract.

A vast network of contracts spans the working world. The United States government, the world's biggest customer, spent about $250 billion in 1986 on work farmed out to organizations that submitted proposals. The defense and aerospace industries, for example, are almost totally dependent on government contracts. But proposal writing is by no means limited to government contractors. One auto manufacturer buys engines from another, and a company that makes spark plugs buys its steel from another company. In fact, most products and services are purchased on a contractual basis.

External proposals are generally classified as either solicited or unsolicited. A solicited proposal is written in response to a request from a potential customer. An unsolicited proposal has not been requested; rather, it originates with the potential supplier.

When an organization wants to purchase a product or service, it publishes one of two basic kinds of statements. An IFB—"information for bid"—is used for standard products. When the federal government needs pencils, for instance, it lets suppliers know that it wants to purchase, say, one million no. 2 pencils with attached erasers. The supplier that offers the lowest bid wins the contract. The other kind of published statement is an RFP—"request for proposal." An RFP is issued when the product or service is customized rather than standard. The automobiles that a police department buys are likely to differ from the standard consumer model: they might have different engines, cooling systems, suspensions, and upholstery. The RFP that the police department offers might be a long and detailed document, a set of technical specifications. The supplier that can provide the automobile most closely resembling the specifications—at a reasonable price—will probably win the contract. Sometimes, the RFP is a more general statement of goals. The customer is in effect asking the suppliers to create their own designs or describe how they will achieve the specified goals. The supplier that offers the most persuasive proposal will probably be successful.

Most suppliers respond to RFPs and IFBs published in newspapers or received in the mail. Government RFPs and IFBs are published in the journal *Commerce Business Daily* (see Figure 13-1). The vast majority of proposals are solicited through these channels.

An unsolicited proposal looks essentially like a solicited proposal except, of course, that it does not refer to an RFP. Even though the potential customer never formally requested the proposal, in almost all cases the supplier was invited to submit the proposal after the two organizations met and discussed the project informally. Because proposals are expensive to write, suppliers are very reluctant to submit them without any assurances that the potential customer will study them carefully. Thus, the term *unsolicited* is only partially accurate.

External proposals—both solicited and unsolicited—can culminate in contracts of several different types: a flat fee for a product or a one-time service; a leasing agreement; or a "cost-plus" contract, under which the supplier is reimbursed for the actual costs plus a profit set at a fixed percentage of the costs.

THE ELEMENT OF PERSUASION
An electronics company that wants a government contract to build a sophisticated radar device for a new jet aircraft might submit a three-volume, 2,000-page proposal. A stationery supplier offering an unusual bargain, on the other hand, might submit to a medium-sized company a simple statement indicating the price for which it would deliver a large quantity of 20-pound, 8½- by-11-inch, 25 percent cotton-fiber bond paper, along with an explanation of why the deal ought to be irresistible. One factor links these two very different kinds of proposals: both will be analyzed carefully and skeptically. The government officials will worry about whether the supplier will be able to live up to its promise: to build, on schedule, the best radar device at the best price. With perhaps a dozen suppliers competing for the contract, the officials will know only that a number of companies want the work; they can never be sure—not even after the contract has been awarded—that they have made the best choice. The office manager would be reluctant to spend more than has been budgeted for the short term in order to save money over the long term and would no doubt be suspicious about the paper's quality.

Issue No. PSA-9053; Tuesday, March 25, 1986   **COMMERCE BUSINESS DAILY**

Navy Aviation Supply Office, Code PGS1, 700 Robbins Avenue, Philadelphia, PA 19111; To Order 215-697-5777. For Questions 215-697-5743. YOUR FSCM IS MANDATORY.

● 59 -- DUMMY LOAD, DIM 2965" x 2.9" x 2936", Matl - aluminum alloy, Func acts as a liquid cooled dummy load for microwave applications. NSN 5985-01-118-3428CW, P/N 86481B & P/N WLA64B, 13 ea, appl to APG-63 sys on F-15 acft. Del to Robins AFB GA approx May 87. This notice is for RFP F09603-86-R-7693 which will be issued by Herley Industries Inc, Lancaster PA 17603; Microlab/Fxr, Livingston NJ 07039; Air Tron, Morris Plains NJ 07950; Micronetics, Norwood NJ 07648 with an approx closing date of 19 Jun 86. All responsible sources may submit a bid, proposal or quotation which shall be considered. No tel requests. Request bid sets from WR-ALC/PMXOA. Include mfg code. For info contact Barbara Grantham, PMXDZ, 912/926-3309. (080)

Commander, Naval Air Development Center (Code 84563) Warminster, PA 18974-5000.

59 -- TWO ELECTRONIC PRECISION AUTOCOLIMATING THEODOLITES with a variable focal length and magnification telescope, measurement accuracy of one half sec mean in both horizontal and vertical axis having an angular resolution of one tenth of an arc second with LCD digital displays. Theodolites shall have built in capability to interface with data processing recording and data transmission equipment not a part of this Sol. Also to be provided are four ea battery supply packs for the Theodolite, two ea battery chargers for battery packs, and one ea optical plumbing device with autocollimating eye piece. Del will be 60 days after award. RFP N62269-86-R0071. Closing date o/a 9 May 86. For info call 215/441-2683. (080)

Warner Robins ALC Directorate of Contracting and Manufacturing, Robins AFB, GA 31098

● 59 -- CORRECTION: COAX CABLE ASSY, part of target identification system. Electro-optical, consisting of ribbon type cable with two connectors on each end, NSN 5995-00-172-7799AY, PN 301202, 20 ea, appl to F4E acft. Del to Robins AFB GA approx Jun 86. This notice is for RFP F09603-86-R-3848 with an approx closing date of 12 May 86. All responsible sources may submit a bid, proposal or quotation which shall be considered. No tel requests. Request bid sets from WR-ALC, PMXOA. Include mfg code. For info contact Joann Hastings, PMZBN, 912/926-2533, x2300. See Notes 3, 24, 43 & 95. (079)

TVA Div of Purchasing 475I Commerce Union Bank Bldg Chattanooga TN 37401

59 -- CAPACITOR BANKS IFB/RFP RA-121248. IFB/RFP will issue fifteen FOB days after date of CBD publ. Purchasing agent RN Cutcher 615/751-7194. TVA intends to purchase the following goods or services by formal advertisement: 161-kV capacitor banks, outdoor, stack-type. 84,000 KVAR, 3-phase, 60 hz IAW TVA dwgs and specs. Shop inspection and dwg submittal required. Tech data will be furnished w/Sol. Del 1 Aug 87. FOB Tullahoma TN 37388. For copies of IBS or RFPs contact invitations clerk 615/751-5610. All responsible sources may submit a bid/proposal/quotation (as appropriate) which will be considered by TVA. For further info contact purchasing agent. (080)

U.S. Army Armament Munitions & Chemical Command, Attn: AMSMC-PCM-MS, Rock Island, IL 61299, Tel: 309/782-4664 or 4166

59 -- ELECTRICAL SOLENOID that complies w/spec 12524007 and dwg 12524181 for type II solenoids. MIL-I-45208 is required. This is an unfunded FY 87 requirement. NSN 5945-01-089-7648, P/N 12524181, end use M242 gun, qty 389 ea. FOB destn. Issue is o/a 2 May 86. Req for Sol should include contractor size and FSCM to facilitate handling. Sol DAAA09-86-R-0722. BOD is o/a 2 Jul 86. See Note 57, 66. (080)

Contracting Div, PO Drawer 5800, Shaw AFB SC 29152-5320. Attn: Sgt Hunt; Contracting Officer, Linda J Guerra 803/668-2406

59 -- SWITCH, AUTOMATIC TRANSFER & BYPASS Isolation, Automatic Switch Co. P/N 962326049CX or equal. 1 ea. Warranty required. RFQ F38601-86-T2378, required. Issue date o/a 11 Apr 86. Quote due at close of business on 12 May 86. Request for RFQ pkgs must be in writing. See Note 32. (079)

Dept of the Army, Tulsa Dist, Corps of Engineers, POB 61, Tulsa, OK 74121-0061, Attn: Chief, Procurement and Supply Div, 918/581-7318

● 59 -- INSULATED CONDUCTOR & GROUND CABLE for Hugo Project, OK. RFQ DACW56-86-Q-0073 will be available Apr 3 86. Quotations due May 2 86. Request for RFQ will be honored until 15 days after date of this publication or until supply is exhausted, whichever occurs first. (079)

UNICOR Fedl Prison Ind, Inc, POB 2000, Lexington KY 40512, Attn Hazel Helvey, Contr Officer, 606/255-6812 x360

59 -- CONNECTOR, Texas Instruments P/N 532775-3 per spec control dwg 532775 or Omni spectra P/N 2037-5009-00. and Delta Electronics Mfg Corp. P/N 1307-043-G001 w/no exceptions or deviations to referenced dwg or P/N. Selected item w/all appl amendments and changes thereto. 1,469 ea. See Note 25. RFQ 123PI-1723-6. Closing 30 Apr 86. (080)

U.S. Army Missile Command Directorate for Procurement & Production, Redstone Arsenal, AL 35898-5280

★ 59 -- FILTER, LOW PASS, NSN 5915 01 125 6142. APN 11500442-3. 66 ea plus 400% options. Destns to be furnished. RFQ DAAH01-86-R-A444 closing 16 May 86. RFQ will be issued to RFI Corp. 100 Pine Air Dr. Bayshore LI NY 110706; Captor Corp. E Main St, Tipp City OH 45371. POC E Beck address above, attn AMSMI-PC-DBD 205/876-4639. See Notes 26, 27, 40, 46. (080)

Navy Aviation Supply Office, Code PGS1, 700 Robbins Avenue, Philadelphia, PA 19111; To Order 215-697-5777. For Questions 215-697-5743. YOUR FSCM IS MANDATORY.

59 -- SEMICONDUCTOR DEVICE, NSN 1RM5961-00-004-7084EX, P/N 39555-9-4. 271 ea. Issuing BOA order to Eaton Corp Ail Div, Deer Pk, NY. PR N00383-86-XB-H099. Due 7 May 86. See Notes 22 & 40. (079)

FIGURE 13-1

An Extract from Commerce Business Daily

The key to proposal writing, then, is persuasion. The writers must convince the readers that the *future benefits* will outweigh the *immediate and projected costs.* Basically, external proposal writers must clearly demonstrate that they

1. understand the reader's needs
2. are able to fulfill their own promises
3. are willing to fulfill their own promises

UNDERSTANDING THE READER'S NEEDS   The most crucial element of the proposal is the definition of the problem or opportunity the project is intended to respond to. This would seem to be mere common sense: how can you expect to write a successful proposal if you do not demonstrate that you understand the reader's needs? Yet the people who evaluate proposals—whether they be government readers, private foundation officials, or managers in small corporations—agree that an inadequate or inaccurate understanding of the problem or opportunity is the largest single weakness of the proposals they see.

Sometimes, bad definitions of the situation originate with the client: the writer of the RFP fails to convey the problem or opportunity. More often, however, the writer of the proposal is at fault. The supplier might not read the RFP carefully and simply assume that he or she understands the client's needs. Or perhaps the supplier, in response to a request, knows that he cannot satisfy a client's needs but nonetheless prepares a proposal detailing a project that he can complete, hoping either that the reader won't notice or that no other supplier will come closer to responding to the real problem. It is easy for suppliers to concentrate on the details of what *they* want to do rather than on what is really required. But most readers will toss the proposal aside as soon as they realize that it does not respond to their situation. If you are responding to an RFP, study it thoroughly. If there is something in it you don't understand, get in touch with the organization that issued it; the organization will be happy to try to clarify it, for a bad proposal wastes everybody's time. Your first job as a proposal writer is to demonstrate your grasp of the problem.

If you are writing an unsolicited proposal, analyze your audience carefully. How can you define the problem or opportunity so that your reader will understand it? Keep the reader's needs in mind (even if the reader is oblivious to them) and, if possible, the reader's background. Concentrate on how the problem has decreased productivity or quality or on how your ideas would create new opportunities. When you submit an unsolicited proposal, your task in many cases is to convince your readers that a need exists. Even when you have reached an understanding with some of your customer's representatives, your proposal will still have to be approved by other officials.

DESCRIBING WHAT YOU PLAN TO DO Once you have shown that you understand why something needs to be done, describe what you plan to do. Convince your readers that you can respond to the situation you have just described. Discuss your approach to the subject: indicate the procedures and equipment you would use. Create a complete picture of how you would get from the first day of the project to the last. Many inexperienced writers of proposals underestimate the importance of this description. They believe that they need only convince the reader of their enthusiasm and good faith. Unfortunately, few readers are satisfied with assurances—no matter how well intentioned. Most look for a detailed, comprehensive plan that shows that the writer has actually started to do the work.

Writing a proposal is a gamble. You might spend days or months putting it together, only to have it rejected. What can you do with an unsuccessful proposal? If the rejection was accompanied by an explanation, you might be able to learn something from it. In most cases, however, all you can do is file it away and absorb the loss. In a sense, the writer takes the first risk in working on a proposal, which, statistically, is likely to be rejected. The reader who accepts a proposal also takes a risk in authorizing the work, for he or she does not know whether the work will be satisfactory.

No proposal can anticipate all of your readers' questions about what you plan to do, of course. This situation would require that the project already be completed. But the more work you have done in planning the project before you submit the proposal, the greater the chances are that you will be able to do the project successfully if you are given the go-ahead. A full discussion of your plan is effective, too, for reasons of psychology: it suggests to your readers that you are interested in the project itself, not just in winning the contract or in receiving authorization.

DEMONSTRATING YOUR PROFESSIONALISM After showing that you understand the reader's needs and have a well-conceived plan of attack, demonstrate that you are the kind of person—or yours is the kind of organization—that *will* deliver what is promised.

Many other persons or organizations could probably carry out the project. You want to convince your readers that you have the pride, ingenuity, and perseverance to solve the problems that inevitably occur in any big undertaking. In short, you want to show that you are a professional.

One major element of this professionalism is a work schedule, sometimes called a task schedule. This schedule—which usually takes the form of a graph or chart—shows when the various phases of the project will be carried out. In one sense, the work schedule is a straightforward piece of information that enables your readers to see how you would apportion your time. But in another sense, it reveals more about your attitudes toward your work than about what you will actually be doing on any given day. Anyone with even the slightest experience with projects knows that things rarely proceed according to plan: some tasks take more time than anticipated, some take less. A careful and detailed work schedule is therefore really another way of showing that you have done your homework, that you have attempted to foresee any problems that might jeopardize the success of the project.

Related to the task schedule is generally some system of quality control. Your readers will want to see that you have established procedures to evaluate the effectiveness and efficiency of your work on the project. Sometimes, quality-control procedures consist of technical evaluations carried out periodically by the project staff. Sometimes, the writer will build into the proposal provisions for on-site evaluation by recognized authorities in the field or by representatives of the potential client.

Most proposals conclude with a budget—a formal statement of how much the project will cost.

THE STRUCTURE OF THE PROPOSAL Most proposals follow a basic structural pattern. If the authorizing agency provides an IFB, an RFP, or a set of guidelines, follow it to the letter. If guidelines have not been supplied, or you are writing an unsolicited proposal, use the following structure to write a clear and persuasive proposal:

1. summary
2. introduction
3. proposed program
4. qualifications and experience
5. budget
6. appendixes

As is usually the case with technical documents, the sequence of composition is not the same as the sequence of presentation. The first section to write is the proposed program. Until you know what you propose doing, you cannot write the summary or the introduction. After you have drafted the proposed program, proceed to the end: the qualifications and experience, appendixes, and finally the budget. Once you have completed all these items, introduce the proposal and draft the summary.

SUMMARY   For any proposal of more than a few pages, provide a brief summary. Many organizations impose a length limit—for example, 250 words—and ask the writer to type the summary, single-spaced, on the title page. The summary is crucial, because in many cases it will be the only item the readers study in their initial review of the proposal.

The summary covers the major elements of the proposal but devotes only a few sentences to each. To write an effective summary, first define the problem in a sentence or two. Next, describe the proposed program. Then provide a brief statement of your qualifications and experience. Some organizations wish to see the completion date and the final budget figure in the summary; others prefer that these data be displayed in a separate location on the title page along with other identifying information about the supplier and the proposed project.

Figure 13-2 shows an effective summary taken from a proposal (Breen 1985) submitted by a metal foundry to a college to do some modifications to a piece of equipment owned by the college, an air gun. A sample internal proposal is included at the end of this chapter.

INTRODUCTION   The body of the proposal begins with an introduction. Its function is to define the background and the problem or the opportunity.

In describing the background, you probably will not be telling your readers anything they don't already know (except, perhaps, if your proposal is unsolicited). Your goal here is to show them that you understand the context of the problem or opportunity: the circumstances that led to the discovery, the relationships or events that will affect the problem and its solution, and so forth.

In discussing the problem, for example, be as specific as possible. Whenever you can, *quantify* the problem. Describe it in monetary terms, because the proposal itself will include a budget of some sort and you want to be able to convince your readers that spending money on what you propose represents a wise invest-

FIGURE 13-2

Summary of a Proposal

Summary

American Metal Foundry, Inc., proposes to modify St. Thomas College's pneumatic high-pressure air gun (PHPAG) to increase its usefulness in destructive testing research. The project entails redesigning the barrel of the gun and designing and manufacturing the receiver tank, which would hold the test specimens and catch the flying projectiles.

These modifications would enable St. Thomas College to carry out sophisticated destructive testing research in its materials laboratory.

This project, which would take seventeen days, would cost $8,066. The unit would be guaranteed to meet specifications after it is delivered and installed.

ment. Don't say that a design problem is slowing down production; say that it is costing $4,500 a day in lost productivity.

Figure 13-3 is the introduction to the air gun proposal.

PROPOSED PROGRAM Once you have defined the problem or opportunity, you have to say what you are going to do about it. As noted earlier, the proposed program demonstrates clearly how much work you have already done. The goal here is, as usual, to be specific. You won't be persuasive by saying that you plan to "gather the data and analyze it." How will you gather it? What techniques will you use to analyze it? Every word you say—or

Introduction

American Metal Foundry, Inc., proposes to modify St. Thomas
College's pneumatic high-pressure air gun (PHPAG) to increase
its usefulness in destructive testing research. The project
entails redesigning the barrel of the gun and designing and
manufacturing the receiver tank, which would hold the test specimens and catch the flying projectiles.

St. Thomas has a 15 ft PHPAG capable of firing projectiles at
speeds in excess of 1,200 ft/sec. A specimen of a desired material
is placed in front of the barrel. A projectile is then loaded and
fired at the specimen. The effect of the impact is then measured.
The results of this testing yield valuable information about the
material.

The problems that this proposal addresses are as follows:
1. Although the barrel of the PHPAG has an inside diameter large
enough to accommodate a 2.5 in. diameter projectile, the barrel
has an insert that is used to fire .60 caliber shells. This insert
prevents the use of larger shells.
2. The PHPAG does not have a receiving tank. Therefore, it cannot
accommodate the larger projectiles necessary for more advanced
destructive testing.

don't say—will give your readers evidence on which to base their
decision. If you know your business, the proposed program will
show it. If you don't, you'll inevitably slip into meaningless gener-
alities or include erroneous information that undermines the
whole proposal.

If your project concerns a subject that has been written about
in the professional literature, show your readers that you are familiar with the scholarship by referring to the pertinent studies.
Don't, however, just toss a bunch of references onto the page. For
example, it is ineffective to write, "Carruthers (1985), Harding
(1986), and Vega (1986) have all researched the relationship between acid-rain levels and groundwater contamination." Rather,

Modifying the barrel of the PHPAG and building a receiving
tank will allow St. Thomas to carry out high-level destructive
testing in its materials laboratory.

use the recent research to sketch in the necessary background and
provide the justification for your proposed program. For instance:

Carruthers (1985), Harding (1986), and Vega (1986) have demonstrated
the relationship between acid-rain levels and groundwater contamination. None of these studies, however, included an analysis of the
long-term contamination of the aquifer. The current study will
consist of . . .

A proposed program might include just one reference to recent research. If the proposal concerns a topic that has been researched

thoroughly, the proposed program might devote several paragraphs or even several pages to a discussion of recent scholarship. Figure 13-4 shows the proposed program for the air-gun project. In this example, the discussion of the proposed project is a general description of the process that the writer's organization will perform (see Chapter 8 for a discussion of process descriptions). Notice that this proposed procedure does not discuss any research, for the project does not involve any original research.

QUALIFICATIONS AND EXPERIENCE After you have described how you would carry out the project, turn to the question of your ability to undertake it. Unless you can convince your readers that you have the expertise to turn an idea into action, your proposal will be interesting—but not persuasive.

The more elaborate the proposal, the more substantial the discussion of qualifications and experience has to be. For a small project, a few paragraphs describing your technical credentials and those of your coworkers will usually suffice. For larger projects, the résumés of the project leader—often called the principal investigator—and the other important participants should be included.

External proposals should also include a discussion of the qualifications of the supplier's organization. Essentially similar to a discussion of personnel, this section outlines the pertinent projects the supplier has completed successfully. For example, a company bidding for a contract to build a large suspension bridge should describe other suspension bridges it has built. The discussion also focuses on the necessary equipment and facilities the company already possesses, as well as the management structure that will assure successful completion of the project. Everyone knows that young, inexperienced persons and new firms can do excellent work. But when it comes to proposals, experience wins out almost every time. Figure 13-5 shows the qualifications-and-experience section of the plastics proposal.

BUDGET Good ideas aren't good unless they're affordable. The budget section of a proposal specifies how much the proposed program will cost.

Budgets vary greatly in scope and format. For simple internal proposals, the writer adds the budget request to the statement of the proposed program: "This study will take me about two days, at a cost—including secretarial time—of about $400" or "The variable-speed recorder currently costs $225, with a 10-percent discount on orders of five or more." For more complicated internal

FIGURE 13-4

*Proposed Program
of a Proposal*

Proposed Procedure

The project entails two major stages:

1. redesigning the barrel of the gun
2. designing and manufacturing the receiver tank, which
   would hold the test specimens and catch the flying projectiles.

1. Redesigning the Gun

The barrel of the gun has an inside diameter large enough to
accommodate a 2.5 in. diameter projectile. However, the barrel
has been fitted with an insert used to fire 0.60 caliber shells.
This insert prevents the use of larger shells. We propose to
remove this insert and remachine the barrel so that it can accommodate the larger projectiles.

2. Designing and Manufacturing the Receiver Tank

Adding a receiver tank involves a number of steps. We propose to design and manufacture a tank large enough to tolerate the
enormous impact involved in destructive testing. We would install the gun in one side of the tank. At the insertion point we

proposals and for all external proposals, a more explicit and complete budget is usually required.

Most budgets are divided into two parts: direct costs and indirect costs. Direct costs include such expenses as salaries and fringe
benefits of program personnel, travel costs, and any necessary
equipment, materials, and supplies. Indirect costs cover the intan-
gible expenses that are sometimes called overhead. General secretarial and clerical expenses not devoted exclusively to the proposed
program are part of overhead, as are other operating expenses
such as utilities and maintenance costs. Indirect costs are usually
expressed as a percentage—ranging from less than 20 percent to

FIGURE 13-4

Proposed Program of a Proposal (Continued)

would install gaskets and flanges to keep the gun and tank air-

tight during firing. Finally, after testing the unit we would

mount the tank to the floor in your laboratory. We would then

repeat the testing to ensure that the unit performs according to

specifications.

more than 100 percent—of the direct expenses. In many external proposals, the client imposes a limit on the percentage of indirect costs.

Figure 13-6 shows an example of a budget statement.

APPENDIXES Many different types of appendixes might accompany a proposal. The air-gun proposal would include, among other items, some examples of the other projects the supplier has already performed. Another popular kind of appendix is the supporting letter—a testimonial to the supplier's skill and integrity, written by a reputable and well-known person in the field. Two

FIGURE 13-5

*Qualifications-and-
Experience Section
of a Proposal*

Qualifications

American Metal Foundry, Inc., has been performing high-
quality metal machining work since 1956. Our engineering staff
consists of experts in stress analysis, mechanics, and machine
design and fabrications. Our technicians combine state-of-the-
art technical expertise and old-fashioned pride in their workmanship. In fact, we have won seven Certificates of Merit from the
Association of Metal Foundries for our products.

Our team leader, Dr. Karen Mair, has over 14 years' experience in designing and fabricating metals and other materials. She
is supported by a fully professional support staff of engineers,
technicians, and clerical personnel. Dr. Mair has successfully
completed projects for dozens of organizations, including Val-
ley View Hospital, Parker State University, Hawkins Ford-Mer-
cury, and Demling Nursery. For a compete listing of her credentials, see her résumé in Appendix B. Please see Appendix D for a
full description of similar projects we have undertaken.

other kinds of appendixes deserve special mention: the task schedule and the evaluation description.

The task schedule is almost always drawn in one of two graphi-
cal formats. The Gantt chart is a horizontal bar graph, with time
displayed on the horizontal axis and tasks shown on the vertical
axis. (See Chapter 10 for a discussion of bar graphs.) A milestone
chart is a horizontal line that represents time; the tasks are written
in under the line. Both Gantt and milestone charts can include
prose explanations, if necessary. Figure 13-7 shows that the Gantt
chart is more informative than the milestone chart in that only the
Gantt chart can indicate several tasks being performed simulta-

FIGURE 13-6

Budget Section of a Proposal

Budget Itemization

   for period August 1, 19-- to August 17, 19--.

Direct Costs

   1. Salaries and Wages

Personnel	Title	Time	Amount
Karen Mair	Design Engineer	2 weeks	$1,900
Ed Smith	Chief Machinist	2 weeks	$1,200
	Typist	2 days	$ 175
Total Salaries and Wages			$3,275

   2. Supplies and Materials

	Amount
Tank	$2,300
Miscellaneous Materials	$ 400
Total Supplies and Materials	$2,700
Subtotal	$5,975

Indirect Costs

	Amount
35% of $5,975	$2,091
Total Cost	$8,066

neously; the milestone chart can indicate only the dates on which the various tasks are to be completed.

Much less clear-cut than the task schedule is the description of evaluation techniques. In fact, the term *evaluation* means different things to different people, but in general an evaluation technique can be defined as any procedure for determining whether the proposed program is both effective and efficient. Evaluation techniques can range from simple progress reports to sophisticated statistical analyses. Some proposals provide for evaluation by an outside agent—a consultant, a testing laboratory, or a university. Other proposals describe evaluation techniques that the supplier itself will perform, such as cost/benefit analyses.

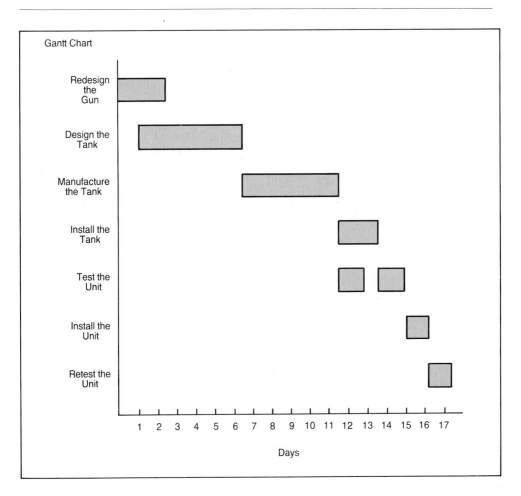

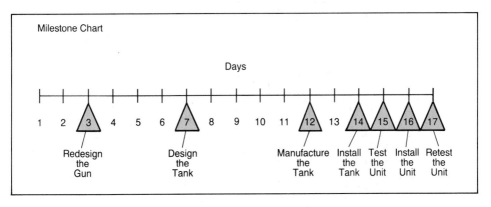

FIGURE 13-7

Gantt and Milestone Charts

The subject of evaluation techniques is complicated by the fact that some people think in terms of quantitative evaluations—tests to determine whether a proposed program is, for example, increasing production as much as had been hoped—whereas others think in terms of qualitative evaluations—tests of whether a proposed program is improving, say, the durability or workmanship of a product. And, of course, some people imply both qualitative and quantitative testing when they refer to evaluations. An additional complication is that projects can be tested both while they are being carried out (*formative evaluations*) and after they have been completed (*summative evaluations*).

When an RFP calls for "evaluation," experienced writers of proposals know it's a good idea to get in touch with the sponsoring agency's representatives to determine precisely what they mean. Figure 13-8 is a description of the evaluation techniques for the air-gun project.

THE STOP
TECHNIQUE

The STOP technique for writing long proposals and other kinds of technical documents is becoming very popular in business and industry. STOP (Strategic Thematic Organization of Proposals) was devised by Hughes Aircraft in 1962. The technique is based on brainstorming and a special organizational structure for the document itself.

Chapter 3 discusses brainstorming, the process of quickly generating a list of ideas and topics that might go into the document. With long proposals—those of 100 pages or more—brainstorming usually involves several people, including subject-area specialists, writers, editors, illustrators, and contract specialists, all working together in the same room. In the STOP technique, and in related techniques, brainstorming sessions are held frequently, not just once at the start of the project. Between sessions, the proposal-writing team creates storyboards. Often attributed to film director Alfred Hitchcock, a storyboard is a sketch with a brief prose statement or notes. Hitchcock used storyboards to coordinate his shots with his script before he actually began shooting the film. His goal was to avoid wasting time and film once the cast and crew were assembled on location.

The goal is the same in the STOP technique. The participants bring their storyboards to the brainstorming sessions, where they test them out on the other participants. The goal is to make sure the content and emphasis will be right—before the actual writing process begins. The storyboards, once they are approved, become the draft of the proposal.

The structure of a STOP proposal calls for two-page units of text and visuals, called topics. As Figure 13-9 shows, the left-hand

FIGURE 13-8

Evaluation-Techniques Section of a Proposal

Appendix A. Evaluation

We will submit to St. Thomas College two progress reports: one at the end of week one, and one at the end of week two. In addition, we would welcome your inspection of the progress on the project at any time. We will guarantee that the unit will work according to the agreed specifications once it is installed at your facility.

---

page of each topic consists of a heading, a thesis statement to unify the discussion, and the discussion itself, of 200–700 words. The right-hand page consists of any remaining text from the left-hand page and a graphic aid: a sketch, photograph, diagram, flow chart, etc. Longer topics that could not be reduced to a two-page topic are partitioned until they are concise enough. Topics are then gathered into sections: units of from one to seven topics.

The STOP technique forces the proposal team to use a systematic process in creating documents. The emphasis is on prewriting and on devising small, easy-to-read units with conveniently located graphic aids.

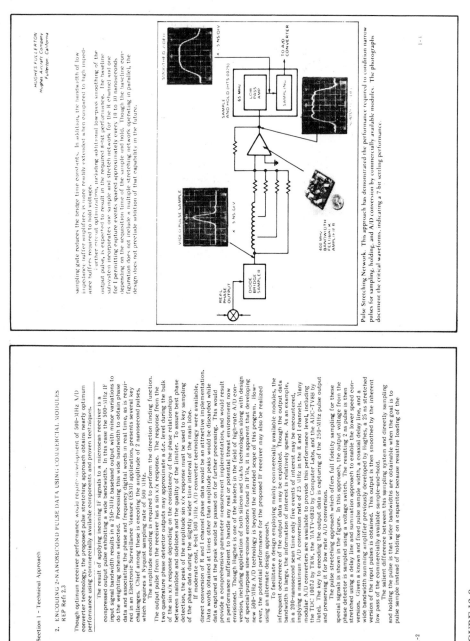

Pulse Stretching Network. This approach has demonstrated the performance required to condition narrow pulses for sampling, holding, and A/D conversion by commercially available modules. The photographs document the critical waveforms, indicating a 7-bit settling performance.

1-1

Section 1 - Technical Approach

2. ENCODING 2-NANOSECOND PULSE DATA USING COMMERCIAL MODULES
REP Ref: 2.3

Though optimum receiver performance would require development of 500-MHz A/D converter technology, the proposed pulse stretching approach obtains nearly optimum performance using commercially available components and proven techniques.

The result of applying incoming IF signals to a microscan receiver is a compressed output pulse exhibiting a wide bandwidth. In this case the 500-MHz IF input signal bandwidth results in a 250-MHz baseband output with minor variations to do the weighting scheme selected. Processing this wide bandwidth to obtain phase data and associating the phase and frequency digital words in real time, as is required in an Electronic Surveillance Measure (ESM) application, presents several key challenges. One of these is encoding the amplitude of 2 nanosecond pulses, which requires a Nyquist sampling rate of 500 MHz.

The amplitude encoding is required to perform the direction finding function. The output pulse from the receiver has a sin x/x response. The response from the two quadrature phase detector outputs may approximate a d.c. level during the bulk of the sin x/x response depending on the consistency of the phase relationships between mainlobe and sidelobes and the quality of the limiter. To assure best phase detection, the peak of the amplitude sin x/x response must be used to key sampling of the phase data during the slightly wider time interval of the main lobe.

If moderate cost, 8-bit 500 MHz A/D converter technology were available, direct conversion of R and I channel phase data would be an attractive implementation. Data words obtained at times other than amplitude peaks would be discarded while data captured at the peak would be passed on for further processing. This would provide a comprehensive parameter measurement implementation, and would result in performance sufficient to handle any real or potential threat environment now envisioned. Though Hughes is one of the leaders in the field of high-rate A/D conversion, including application of both silicon and GaAs technologies along with design of special-purpose sine-cosine encoders found in IFMs, it is apparent that developing new 500-MHz A/D technology is beyond the intended scope of this program. However, the potential performance for the proposed IF receiver may also be realized using an alternate design approach.

To facilitate a design employing mainly commercially available modules, the infrequent occurrence of the output data may be exploited. Though the output data bandwidth is large, the information of interest is relatively sparse. As an example, in a 200-nanosecond scan time only five events of interest may be encountered, implying a maximum A/D conversion rate of 25 MHz in the R and I channels. Many modular A/D converters are available to provide this performance level, including the TDC 1007J by TRW, the MATV-0820 by Computer Labs, and the ADC-TV8B by Datel. The key to encoding the output data is capturing of the 250-MHz pulse output and preserving it for low rate A/D conversion.

The pulse stretching approach, which offers full fidelity sampling for these special signals is shown in the figure. In this approach, the output voltage from the phase detector is sampled using a voltage switch. The resulting 2 ns pulse is then stretched using a delay line and summation approach to enable the lower speed conversion. Given a known and fixed pulse sample width, a coaxial delay line, and a wide bandwidth summing amplifier previously provided by Hughes, a 25 ns stretched version of the input pulses is obtained. This output is then smoothed by the inherent low pass of the commercially available sample-and-hold.

The salient difference between this sampling system and directly sampling and holding the pulse is that wider bandwidths are obtainable since the goal is to pulse sample instead of holding on a capacitor because resistive loading of the

1-2

FIGURE 13-9

STOP Topic © 1983 IEEE.

354

THE INTERNAL PROPOSAL

One day, while you're working on a project in the laboratory, you realize that if you had a new centrifuge you could do your job better and more quickly. The increased productivity would save your company the cost of the equipment in a few months. You call your supervisor and tell him about your idea. He tells you to send him a memo describing what you want, why you want it, what you're going to do with it, and what it costs; if your request seems reasonable, he'll try to get you the money.

The memo you write is an internal proposal—a persuasive argument, submitted within an organization, for carrying out an activity that will benefit the organization, generally by saving it money. An internal proposal is simply a suggestion, made by someone within an organization, about how to improve some aspect of that organization's operations. The suggestion can be simple—to purchase an inexpensive piece of office equipment—or complicated—to hire an additional employee or even add an additional department to the organization. The nature of the suggestion determines the format. A simple request might be conveyed orally, either in person or on the phone. A more ambitious request might require a brief memo. The most ambitious requests are generally conveyed in formal proposals. Often, organizations use dollar figures to determine the format of the proposal. A proposal that would cost less than $1,000 to implement, for instance, is communicated in a brief form, whereas a proposal of more than $1,000 requires a report similar to an external proposal.

The element of persuasion is just as important in the internal proposal as it is in the external proposal. The writer must show that he or she understands the organization's needs, has worked out a rational proposed program, and is a professional who would see that the job gets done.

A careful analysis of the writing situation is the best way to start, for as usual every aspect of the document is determined by your reader's needs and your purpose. Writing an internal proposal is both more simple and more complicated than writing an external proposal. It is simpler because you have more access to your readers than you would to external readers. And you can get more information, more easily. However, you might find it more difficult to get a true sense of the situation in your own organization. Some of your coworkers might not be willing to tell you directly if your proposal is a long shot. Another danger is that when you identify the problem you want to solve, you are often criticizing—directly or indirectly—someone at your organization who instituted the system that needs to be revised or who failed to take necessary actions earlier.

An unsuccessful external proposal does not linger long; it was a gamble and you lost. An unsuccessful internal proposal lingers longer because more of your colleagues know about it. Therefore, before you write an internal proposal, discuss your ideas thoroughly with as many potential readers as you can. In this way you will increase your chances of finding out what the organization really thinks of your idea before you commit it to paper.

SAMPLE INTERNAL PROPOSAL

Following is an internal proposal (Adams 1986). The author was a student working part-time in the sensory-evaluation department of a pharmaceutical corporation.

The progress report written after the project was under way is included at the end of Chapter 14. The completion report is in Chapter 15. Marginal notes have been added.

This proposal was written as a memo. If it had been longer, it probably would have taken the form of a report. For a discussion of memos, see Chapter 12.	DATE: February 10, 1986 TO: Laura Smith, Manager, Sensory Evaluation, Cavendish Laboratories, Inc. FROM: Cynthia Adams, Sensory Analyst, Sensory Evaluation, Cavendish Laboratories, Inc. SUBJECT: Proposal to Investigate Personal Computers for the Sensory Evaluation Department at Cavendish Laboratories, Inc.
This sentence describes the purpose of the memo and the purpose of the proposed investigation.	Purpose This memo describes a proposal to investigate the currently available personal computers, to determine whether any of them might be suitable for solving the problem with the keywords-filing system in Sensory Evaluation at Cavendish.
The problem. The relevant scholarship. The proposal.	Summary The keywords-filing system, which records all tests conducted in Sensory Evaluation, has not been kept up to date. No one in the department has enough time to manually enter into the filing system the tests that are conducted. Last year this resulted in missing information that cost the department $1,200 in lost time and products. If the system were kept current, this loss would not have occurred. Research in computer journals indicates that a personal computer can cut in half the time spent manually entering tests in the keywords-filing system. This memo proposes that major personal computers be investigated according to technical, usage/maintenance, and financial criteria, and that the results, conclusions, and recommendations of the investigation be reported. The investigation would cost $310.50.

Problem Definition

The background of
the problem.

Each time a sensory evaluation test is conducted, it is assigned a unique code of seven characters called the keyword code (such as LC05172). On the final test report, several keywords are listed. A keyword is the product tested (for example, lo-cal salad dressing), the variables in the product (for example, different amounts of vinegar in the salad dressing), the type of evaluation conducted (for example, triangle test, duo trio test, and qualitative evaluation), and the type of statistical analysis performed (for example, chi-square analysis or multidimensional scaling).

The keywords-filing system is a small box containing about 300 index cards. Each index card is labeled with a keyword, and the cards are stored in alphabetical order. When the system was established one year ago, the department planned that every time a report on a test was issued, someone would locate the cards for each of the keywords listed on the report. Then that person would write the keyword code of the test on each card.

The purposes of this keywords-filing system were listed as:

1. to maintain a permanent record of all tests conducted during the year
2. to record exactly how many tests are conducted of a certain type or with a particular variable
3. to enable sensory analysts to quickly find previous tests that have been conducted with a particular product or variable. This ease of access would prevent duplication of previous tests and would lead to the planning of more informative tests.

The problem is defined in specific terms.

Unfortunately, these purposes are not being fulfilled because reports written from sensory-evaluation tests have not been entered in the keywords-filing system from July 1985 through January 1986. That is, after one year, only the first six months of tests have been entered. No one in Sensory has enough time to enter reports in the keywords-filing system by manually searching for the proper card, writing the keyword code on the card, and then replacing it in the box. This requires approximately six minutes for each test. With 25 tests conducted each week, this task amounts to 130 hours a year.

Notice that the
writer defines the
problem in monetary terms.

The keywords-filing system therefore is ineffective. Within the last three months, five tests were conducted that had been conducted six months ago. Because Sensory did not have this information, this error cost the department $1,200 in lost time and products. If the system were kept current, this loss would not have occurred.

The writer places
her proposed procedure in context. She
has already done
some research on the
topic and describes a
professional approach to using the
scholarship.

Proposed Procedure

Two possible solutions to the problem of the ineffective keywords filing system are:

1. to employ someone for 2½ hours per week to manually record the reports in the filing system
2. to purchase a personal computer in which to enter and store the keywords-filing system

I would like to study reviews of the major personal computers (see References) to determine if one or more of them would fit our needs for the filing system. I would then compare the costs and benefits of the computer to those of hiring someone to enter the data manually.

Research in PC Magazine (Owen 1985) indicates that a personal computer can cut in half the time spent manually entering tests in the keywords-filing system. The PC's I will research have been reviewed in computer journals that are not affiliated with any computer company. In addition, the PC's are reviewed by independent consultants. To evaluate the PC's, I will consider reviews of them from at least two different journals, to avoid any bias the reviewer might have.

To determine if a PC would be effective, I have devised a set of criteria by which it might be evaluated.

In this discussion of the three sets of criteria, the writer shows that she has analyzed her department's situation and will be able to evaluate the different systems (if she receives authorization to carry out the investigation).

Technical Criteria:

1. The PC must have files to store approximately 250 keywords, with each keyword having 1 to 150 tests listed under it.
2. The PC software must handle word files.
3. The PC should be IBM compatible because the mainframes in Sensory Evaluation are IBM computers.
4. The PC must have a printer for hard copies of the files.

Usage/Maintenance Criteria:

1. The PC must be "user friendly" so that the sensory analysts do not need to understand how it operates in order to enter or retrieve data.
2. The PC must be easy to program because the sensory analysts have other responsibilities.
3. Maintenance and service personnel for the PC must be easily accessible.
4. The PC must be reliable, without bugs in the software or frequent downtime.

Financial Criteria:

1. The initial cost of the PC, including a printer, must not exceed $6,000.
2. The annual maintenance costs of the PC must not exceed $1,000.

Here the writer specifies the "product" of the investigation.

After evaluating the available personal computers against these criteria, I will write a report that contains my results, conclusions, and recommendations. This report will provide the information necessary for you to reach a decision on how to deal with the problem of the ineffective keywords-filing system.

Costs

This informal budget is appropriate because the investigation would require no unusual expenses.

This research project would require about 30 hours of my time and 5 hours of typing:

Cynthia Adams	30 hours at $9.25/hour=	$277.50
Typist	5 hours at $6.60/hour=	33.00
	Total	$310.50

Credentials

The writer describes her credentials for this investigation.

I have worked as a sensory analyst at Cavendish for 16 months and am quite familiar with the ineffectiveness of the keywords-filing system. Previously, I was a sensory analyst at Filmore Company for three years, where the sensory department maintained a record on a personal computer of the tests it conducted.

Task Schedule

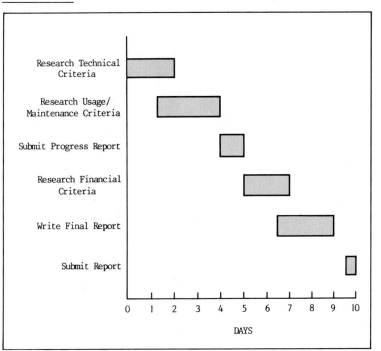

References

Archer, R. 1984. The IBM PCjr. BYTE 9, no. 8:254-68.

Bullard, B. 1983. Comparing the IBM PC and TI PC. BYTE 8, no. 11: 232-42.

Casella, P. 1985. NEC PC-8401A. InfoWorld 7, no. 11:42-43.

Casta, P., and J. Bernard 1984. Face off: The Apple IIc vs. IBM's new PCjr. Computers & Electronics 22, no. 11:50-53.

Finger, A. 1985. IBM PC AT. BYTE 10, no. 5:270-77.

Gilder, J. 1984. The Coleco Adam. BYTE 9, no. 4:206-20.

Hartmann, T. 1985. The Tandy 200: Tandy surpasses its Model 100 with an enhanced second-generation laptop computer. Popular Computing 4, no. 9:82-85.

Hass, M. 1984. The HP 150 Computer. BYTE 9, no. 12:262-75.

Krause, H. 1985. The Tandy 200 and the NEC PC-8401A. BYTE 10, no. 13:306-14.

Markoff, J. 1984. The Apple IIc Personal Computer. BYTE 9, no. 5:276-84.

The writer provides an unusually full bibliography because her evaluation of the PC's will depend on written evaluations.

Mazur, J. 1984. The Leading Edge Personal Computer. BYTE 9, no. 10:312-20.

Neudecker, T. 1984. Macintosh. InfoWorld 6, no. 13:82-90.

Owen, B. 1985. Fifteen real-life productivity solutions. PC Magazine 4, no. 24:111-35.

Satchell, S. 1985. AT&T 6300 Personal Computer. InfoWorld 7, nos. 1&2:49-54.

Schofield, J. 1984. IBM PC AT. Practical Computing 7, no. 12:72-74.

Talmy, S. 1984. The HP 150 Personal Computer. Creative Computing 10, no. 4:26-37.

Troiano, B. 1985. The AT&T PC 6300. BYTE 10, no. 13:294-302.

Vose, G.M., and R. Shuford 1984. A closer look at the IBM PCjr. BYTE 9, no. 3:320-32.

Webster, B. 1984. The Macintosh. BYTE 9, no. 8:238-47.

Williams, G. 1984. The Apple Macintosh Computer. BYTE 9, no. 2:30-54.

Wilson, D. 1984. Adam: Lowered price, some cured ills make a better machine. InfoWorld 6, no. 53:39-40.

Wood, L. 1985. The Leading Edge PC: Genuine enhancements make this computer stand out from the pack. Popular Computing 4, no. 3:104-7.

WRITER'S CHECKLIST

The following checklist covers the basic elements of a proposal. Any guidelines established by the recipient of the proposal should of course take precedence over these general suggestions.

1. Does the summary provide an overview of
   a. The problem or the opportunity?
   b. The proposed program?
   c. Your qualifications and experience?

2. Does the introduction define
   a. The background leading up to the problem or the opportunity?
   b. The problem or the opportunity itself?

3. Does the description of the proposed program
   a. Cite the relevant professional literature?
   b. Provide a clear and specific plan of action?

4. Does the description of qualifications and experience clearly outline
   a. Your relevant skills and past work?
   b. Those of the other participants?

c. Your department's (or organization's) relevant equipment, facilities, and experience?

5. Do the appendixes include the relevant supporting materials, such as a task schedule, a description of evaluation techniques, and evidence of other successful projects?

6. Is the budget
   a. Complete?
   b. Correct?

EXERCISES

1. Write a proposal for a research project that will constitute the major assignment in this course. Start by defining a technical subject that interests you. Using abstract journals and other bibliographic tools, create a bibliography of articles and books on the subject. (See Chapter 4 for a discussion of choosing a topic.) Then make up a reasonable real-world context: for example, you could pretend to be a young civil engineer whose company is considering the purchase of a new kind of earth-moving equipment. Address the proposal to your supervisor, requesting authorization to investigate the advantages and disadvantages of this new piece of equipment.

2. The following proposal was written by a student who had been a summer trainee with the Navy International Logistics Control Office (NAVILCO) (Fromnic 1981). In a short essay, evaluate the effectiveness of the proposal. Is it clear and persuasive?

TO: Commander William Haggerty, Commanding Officer, NAVILCO
FROM: Maureen Fromnic, Finance Department Trainee, NAVILCO
DATE: February 20, 19--

SUBJECT:
Proposal to Write a Terminology and Procedures Manual for All Incoming Summer Trainees

Context:
Each summer 150 trainees are hired to aid various departments in completing backlogged work as well as to temporarily replace vacationing workers. Of all incoming trainees, 98 percent have never had any previous government experience. The justification for hiring inexperienced clerks is that it enables the various departments to complete projects using inexpensive labor as well as to train prospective permanent employees.

Problem:
Since the trainees work only three months, there is very little time to train them and orient them to their respective departments. Three and a half weeks are required to familiarize the new employees completely with Navy terminology and work procedures. This nonproduc-

tive period represents approximately one-third of the trainee's term
and $75,150 ($501 per new employee) of the taxpayers' money. Further-
more, additional time and money ($25,100) is spent as permanent em-
ployees are "borrowed" from their regular jobs and assigned to train
the newcomers. By the end of the summer of 1980, $100,250 had been
spent and the amount of work completed was insufficient to justify the
cost of indoctrination.

Proposal:
I would like to assemble a brief but comprehensive manual defining
Navy terminology and describing NAVILCO work procedures. This manual
would serve as a directory for all incoming employees so that they
could become oriented to our routines more quickly and efficiently
than they now are.

Procedure:
The following is an outline to aid you in viewing the development of
the training manual.

Description and Purpose of the Summer Employment Program:
--The origin of the Summer Employment Program
--The function of the Summer Employment Program
--The advantages of the program for the Summer Employment Trainee
--The advantages of the program for NAVILCO

Trainee Orientation into the Program:
--Navy terminology
--Navy chain of command
--Explanation of leave time and pay periods
--Responsibilities of trainees
--Type of work assignments given during the employment period
--Guidelines for correspondence preparation

Extracurricular Activities Available During Employment:
--Equal Employment Opportunity Committee
--Special Events Committee
--Compound Sports Teams
--NAVILCO social events

Conclusion:
--Attitudes of supervisors and fellow workers toward trainees
--NAVILCO expectations for the summer employment trainees

In the first section of the manual, I shall explain how this particular program originated and why it was necessary to maintain this program. In addition, I shall explain the benefits of the program for
NAVILCO and for the trainee. Second, I shall define the various duties
and responsibilities the trainee must perform. In order to prevent a
language barrier, I shall discuss the various Navy terms that are
ordinarily familiar only to enlisted Navy personnel. Third, the manual will include the various organizations or groups available to
trainees within NAVILCO. I will point out the function of each group
and the role of its members. Lastly, I shall discuss how supervisors
and fellow workers feel about trainees. Also, I shall indicate to the
trainees the importance of fulfilling NAVILCO's expectations for
them, especially if they desire future employment with the Navy.

Budget:
Maureen Fromnic   45 hours @ $4.50/hr.: $202.50
Typist   7 hours @ $3.50/hr.:   24.50
                                              Total: $227.00

Credentials:
I was previously a summer employment trainee for two summers. I have
been a permanent employee of NAVILCO for two years. I have attended
several government seminars on how to train newly employeed trainees.
In addition, while attending Central University, I took courses that
covered government-sponsored programs and government spending for
special-education programs.

Task Schedule:
research the origin and function
of the program

interview supervisors, employees, and past trainees

compile information and write
the body of the manual

draw conclusions

proofread rough draft and make
necessary adjustments

submit manual

2/20   2/28   3/8   3/16   3/24   4/1
         Completion Dates

References:
1. Hills, C. A. 1980. Three ways to get a federal job. Working Woman
   (October): 66.
2. King, M. E. 1979. How government service can help your career. The
   Compound Chronicle (August): 22-24.
3. Miller, S. 1980. Career payoff in Washington, D.C. Working Woman
   (July): 53-56.
4. Rumaker, P. 1980. Profile of federal workers. U.S. News & World
   Report (16 August): 38-39.
5. U.S. Civil Service Commission. 1979. Summer Jobs. Announcement no.
   414. Washington, D.C.: Government Printing Office, 1-3.

3. The following proposal was written by a student who, for purposes of
   the assignment, assumed the role of a consultant (Tracy 1981). The
   proposal is addressed to a partner in an accounting firm. The student
   was requesting authorization to conduct a study to determine whether
   the firm could save money by converting from a manual system of preparing tax returns to a computerized system. In a brief essay, evaluate
   the effectiveness of the proposal. Is it clear and persuasive?

Proposal to Convert Manual Preparation
of Tax Returns to
Computerized Procedure

Submitted to: Mr. Saul O'Neill
Senior Partner
O'Neill and Bernstein

by: Terry Tracy, CPA

October 30, 19--

Background:

O'Neill & Bernstein is a medium-sized public accounting firm
consisting of a 40-person audit staff and an 18-person tax department
in the Philadelphia office. The volume of business has increased
substantially over the past two years, and therefore you have had to
hire 22 new audit personnel and 13 additional tax professionals.

Two years ago the tax department at O'Neill & Bernstein consisted of three staff members: one supervisor, one manager, and one
tax partner. Presently your tax department is made up of nine staff
members: three advanced staff, two supervisors, two managers, and two
partners. The reason for this drastic personnel increase has been the
tremendous gain in volume of tax clients. Two years ago you had 75 tax
clients; today you have 225 clients.

Your professionals work an average of 12 overtime hours a week
during the busy season (February 1-April 15, June 1-August 15) in
order to meet the deadline dates for client tax returns.

Problem:

Over the past year you have lost three $50,000-a-year corporate
tax customers and twenty individual and two partnership clients because of your failure to meet tax return deadlines. Because your work-
load diminishes substantially in the off season, it would not be economically feasible to hire additional tax people.

Proposal:

I believe your tax-return deadlines can be met without hiring
more personnel, by having all the returns processed through a computer. I have looked into a number of companies who have established
reputations in providing the computer personnel and facilities necessary to improve the efficiency, and reduce the cost of our tax department. As Avi Livenson states in his article on computer-processed
tax returns, "Accountants continue to make increasing use of computerized tax return preparation. Where large numbers of returns are
processed, computers provide a quick, efficient method for obtaining
accurate returns. Last year 6 million returns were prepared by computer."

Procedure:

My initial step in dealing with this proposal will be to research the implementation of computerized tax returns in other companies, and observe its effectiveness and efficiency.

The preliminary format I will use to determine whether this proposal will benefit your company is:
- --Prepare an introduction discussing our problem and recommended proposal.
- --Prepare a list of companies that have made the transition to computerized preparation of tax returns. This will list both the positive and negative reactions from these companies.
- --Calculate the amount of time and money saved through the method discussed in my proposal.
- --Prepare a list showing the alternatives of having your tax returns sent to the computer firm (through the mail) or having a computer terminal installed in your office and transmitting your information to the computer through the terminal.
- --Calculate the rate of return on your original investment and how long it will take to recoup your original investment.
- --Prepare a recommendation depicting the computer firm that I feel would best meet your needs, and also list the most favorable alternative companies.

Budget:

The charge for this analysis would be $3,000, payable in full within 30 days.

Credentials:

I am a graduate of Central University with a Bachelor of Science degree in accounting. I have a year-and-a-half's experience in the tax department at Carruthers and Higgins, a year's experience in the tax department with a "Big 8" CPA firm that prepared its returns through a computer, and I have worked for six months with computers that can process tax returns. I have been a freelance consultant for three years.

Task Schedule:

	11/02	11/10	11/13	11/16	11/20	11/22
Discuss Problem and Proposal	———					
Prepare List of Companies That Use Computerized Tax Preparation		———				
Prepare List Stating Our Alternative Methods			———			
Calculate Rate of Return on Investment				———		
Prepare Conclusion and Recommendation					———	
Submit Report						———

Bibliography:

1. Karter, N.A. 1976. Computer tax return preparation. Journal of Taxation (19 May): 2, 6, 8, 230.
2. Livenson, A. 1979. Computer processed tax returns. Taxation for Accountants (12 October): 144-74.
3. Meyers, B. A. 1977. How to computerize corporate tax operations. Tax Executive (6 June): 17-26.
4. Morson, D. M. 1977. Computer versus pencil for the tax professional. Tax Executive (October): 267-79.
5. Sonnichsen, G. W. 1980. Tax automation for administrative control. Tax Executive (January): 91-111.
6. Tanner, W. J. 1979. The computer: Its uses in the tax field. Journal of Taxation (19 October): 205-9.
7. Weiss, Richard. 1978. Computerized return preparation continues to be available in a wide range of services. Taxation for Accountants (September): 144-71.

REFERENCES

Adams, C. 1986. Proposal to investigate personal computers for the Sensory Evaluation Department at Cavendish Laboratories, Inc. Unpublished document.

Breen, D. 1985. Proposal to redesign and fabricate St. Thomas College's air gun. Unpublished document.

Fromnic, M. 1981. Proposal to write a terminology and procedures manual for all incoming summer trainees. Unpublished document.

Tracy, T. 1981. Proposal to convert manual preparation of tax returns to computerized procedure. Unpublished document.

CHAPTER FOURTEEN

PROGRESS REPORTS

A progress report communicates to a supervisor or sponsor the current status of a project that has been begun but is not yet completed. As its name suggests, a progress report is an intermediate communication—between the proposal (the argument that the project be undertaken) and the completion report (the comprehensive record of the completed project).

Although progress reports sometimes describe the entire range of operations of a department or division of an organization—such as the microcellular research unit of a pharmaceutical manufacturer—they usually describe a single, discrete project, such as the construction of a bridge or an investigation of excessive pollution levels in a factory's effluents.

Progress reports allow the persons working on a project to "check in" with their supervisors or sponsors. Supervisors are vitally interested in the progress of their projects, because they have to integrate them with other present and future commitments. Sponsors (or customers) have the same interest, plus an additional one: they want the projects to be done right—and on time—because they are paying for them.

The schedule for submitting progress reports is, of course, established by the supervisor or the sponsor. A relatively short-term project—one expected to take a month to complete, for example— might require a progress report after two weeks. A more extensive project usually will require a series of progress reports submitted on a fixed schedule, such as every month.

The format of progress reports varies widely. A small internal project might require only brief memos, or even phone calls. A small external project might be handled with letters. For a larger, more formal project—either internal or external—a formal report (see Chapter 11) generally is appropriate. Sometimes a combination of formats is used: for example, quarterly reports and memos for each of the other eight months. (See Chapter 12, "Memo Reports," and Chapter 18, "Correspondence," for discussions of these formats.) This chapter exemplifies both progress reports written as memos and those written as reports; the strategy discussed throughout applies to progress reports of every length and format.

The following example might help to clarify the role of progress reports. Suppose your car is old and not worth much. When you bought it for $600 three years ago, you knew you would be its third—and last—owner. Recently you've noticed some problems in shifting. You bring the car to your local mechanic, who calls within an hour to report that a small gasket has to be replaced, at a total cost of $25. Pleased with the mechanic's progress report, you tell him to do the work and that you'll pick up the car later in the afternoon.

If all projects in business and industry went this smoothly—with no major technical problems and no unanticipated costs—every progress report would be simple to write and a pleasure to read. The writer would merely photocopy the task schedule from the original proposal, check off the tasks completed, and send the page on to the supervisor or sponsor. The progress report would be a happy confirmation that the project personnel had indeed accomplished what they promised they would do by a certain date. The reader's task would be simple and pleasant: to congratulate the project personnel and tell them to continue.

Unfortunately, most projects in the real world don't go so smoothly. Suppose that you had received a different phone call from the mechanic. Your transmission is ruined, he tells you; a rebuilt transmission will cost about three hundred dollars; a new one, four hundred. "What do you want me to do?" he asks. You tell him you'll call back in a few minutes. You sort through your options:

1. bring your car to another mechanic for another estimate
2. have the rebuilt transmission installed
3. have the new transmission installed

4. take the bus to the garage, retrieve the license plates, and arrange to have the car sold for scrap

This version of the story illustrates why progress reports are crucial. Managers—you, in the case of the car—need to know the progress of their projects so that they can make informed decisions when things go wrong. The problem might not be what it was thought to be when the proposal was written, or unexpected difficulties might have hampered the working methods. Perhaps equipment failed; perhaps personnel changed; perhaps prices went up. An experienced manager could list a hundred typical difficulties.

The progress report is, to a large extent, the original proposal updated in the light of recent experience. If the project is proceeding smoothly, you simply report the team's accomplishments and future tasks. If the project has encountered difficulties—if the anticipated result, the cost, or the schedule has to be revised—you need to explain clearly and fully what happened and how it will affect the overall project. Your tone should be objective, neither defensive nor casual. Unless ineptitude or negligence caused the problem, you're not to blame. Regardless of what kind of news you are delivering—good, bad, or mixed—your job is the same: to provide a clear and complete account of your team's activities and to forecast the next stage of the project.

THE STRUCTURE OF THE PROGRESS REPORT

Progress reports vary considerably in structure, because of differences in format and length. Written as a one-page letter, a progress report is likely to be a series of traditional paragraphs. As a brief memo, it might also contain section headings. As a report of more than a few pages, it might contain the elements of a formal report.

Regardless of these differences, most progress reports share a basic structural progression. The writer

1. introduces the readers to the progress report by explaining the objectives of the project and providing an overview of the whole project
2. summarizes (if appropriate) the progress report
3. discusses the work already accomplished and the work that remains to be done, and speculates on the future promise and problems of the project
4. provides a conclusion that evaluates the progress of the project

If appropriate, appendixes are attached to the report.

In composing the progress report, leave the summary for last. As is the case with all technical documents, you cannot accurately summarize material that you have not yet written. However, the other elements of the standard progress report can be written in the sequence in which you will present them.

The best way to draft the progress report is to start with a copy of your proposal; many of the elements of the progress report are taken directly from it. If you are writing manually, photocopy the proposal and then cut and paste your changes. If you are working on a word processor, copy your proposal. The problem or opportunity that led to the project, for instance, can often be presented intact. The discussion of the past work is often a restatement of a portion of the proposed program, with the future tense changed to past. You might want to reproduce your task schedule—modified to show your current status—and your bibliography.

INTRODUCTION   The introduction provides background information that orients the reader to the report. First, of course, it identifies the document as a progress report and identifies the span of time the report covers. If more than one progress report has been (or will be) submitted, the introduction places the report in the proper sequence—for example, it might be the third quarterly progress report. Second, the introduction states the objectives of the project. These objectives have already been defined in the proposal, yet they are generally repeated in each progress report for the benefit of the readers, who are likely to be following the progress of a number of projects simultaneously. And third, the introduction provides a brief statement of the phases of the project. Figure 14-1 provides an example of an introduction to a progress report.

SUMMARY   Progress reports of more than a few pages are likely to contain a summary section, which, like all other summaries, provides a brief overview of the contents of the progress report for those readers who do not need to know all the technical details. Often, the summary enumerates the accomplishments achieved during the period covered in the report and then comments on the current work. Notice how the summary shown in Figure 14-2 calls the reader's attention to a problem described in greater detail later on in the progress report.

DISCUSSION   The points listed in the summary are elaborated in the discussion—the body of the progress report. The audience for the discussion includes those readers who want a complete picture of the team's activities during the period covered; many readers, how-

```
                          INTRODUCTION
      This is the first monthly progress report on the project to de-

   velop a new testing device for our computer system. The goal of

   the project is to create a device that can debug our current and

   future systems quickly and accurately.

      The project is divided into four phases:

          I.    define the desired capabilities of the device

         II.    design the device

        III.    manufacture the device

         IV.    test the device
```

ever, will not bother to read the discussion unless the summary highlights an unusual or unexpected development during the reporting period.

Of the several different methods of structuring the discussion section, perhaps the simplest is the past work/future work scheme. After describing the problem that motivated the project, the writer describes all the work that has been completed in the present reporting period and then sketches in the work that remains to be done. The advantage of this scheme is that it is easy to follow and easy to write.

Another common structure for the discussion is based on the tasks involved in the project. If the project requires that the researchers work on several of these tasks simultaneously, this structure is particularly effective, for it enables the writer to describe, in order, what has been accomplished on each task. Often, the task-oriented structure incorporates the past work/future work structure:

```
III. Discussion

     A. The Problem

     B. Task I

        1. past work

        2. future work

     C. Task II

        1. past work

        2. future work
```

FIGURE 14-2

*Summary of a
Progress Report*

```
                              SUMMARY
          In November, Phase I of the project--to define the desired
     capabilities of the testing device--was completed.
          The device should have two basic characteristics: versa-
     tility and simplicity of operation.
     1.   Versatility. The device should be able to test different
          kinds of logic modules. To do this, it must be able to "ad-
          dress" each module and ask it to perform the desired task.

     2.   Simplicity of Operation. The device should be able to display
          the response from the module being tested so that the opera-
          tor can tell easily whether the module is operating properly.
          Phase II of the project--to design the device--is now under-
          way.

     We would like to meet with Research and Development to discuss a
     problem we are having in designing the device so that it can "ad-
     dress" each module. Please see "Discussion" and "Conclusion,"
     below.
```

Figure 14-3 exemplifies the standard chronological progression—from the problem to past, present, and future work. Notice, however, that the writer uses a combination of generic and specific phrases in the headings.

CONCLUSION A progress report is, by definition, a description of the present status of a project. The reader will receive at least one additional communication—the completion report—on the same subject. The conclusion of a progress report, therefore, is more transitional than final.

```
                         DISCUSSION

The Problem
         Isolating and eliminating design and production errors
has always been a top priority of our company. Recently, the so-
phistication of new computer systems we are producing has out-
paced our testing capabilities. We have been asked to develop a
testing device that can quickly and accurately debug our current
and projected systems.
         Currently, we have no technique for testing the individual
logic modules of our new systems. Consequently, we cannot test
the system until all of the modules are in place. When we do dis-
cover a problem, such as a timing error in one of the signals, we
have to disassemble the system and analyze each of the modules
through which the signal flows. Even after we have discovered a
problem, we cannot know if that is the only problem until we reas-
semble the system and test it again. Although this testing method
is effective--we are well within acceptable quality standards--
it is very inefficient.
         The solution to this problem is to develop a device for
testing the logic modules individually before they are installed
in the system. That is the overall objective of the project.
```

In the conclusion of a progress report, your task is to convey to the reader your evaluation of how the project is proceeding. In the broadest sense, you have one of two messages:

1. Things are going well.
2. Things are not going so well as anticipated.

Through a careful use of language, try to communicate your evaluation accurately.

If the news is good, convey your optimism, but avoid overstatement.

FIGURE 14-3

*Discussion Section
of a Progress Report
(Continued)*

Phase I of the project involved our determining the desired
capabilities of this device.

Work Completed: Determining the Desired Characteristics of the
Testing Device

The first characteristic of the testing device must be
versatility. Over the next two years, we will be introducing
three new systems. This rate of introduction is expected to
continue at least through 1990. Although we cannot foresee the
specific components of these new systems, all are expected to
incorporate the latest large-scale integration techniques, in
conjunction with microprocessor control units. This basic
structure will enable us to employ separate logic modules, each
of which performs a specific function. The modules will be built
on separate logic cards that can be tested and replaced easily. To
achieve this versatility, the testing device should be able to
"address" each module automatically and ask it to perform the
desired task. Because the testing device will be asked to handle
many logic modules, it should be able to distinguish between the
different modules in order to ask them to perform their appropriate functions.

The second characteristic of the testing device should be

OVERSTATED

We are sure the device will do all that we ask of it, and more.

REALISTIC

We expect that the device will perform well and that, in addition, it
might offer some unanticipated advantages.

Beware, too, of promising that the project will be completed
early. Experienced writers know that such optimistic forecasts are
rarely accurate, and of course it is always embarrassing to have to
report a failure after you have promised success.

FIGURE 14-3

*Discussion Section
of a Progress Report
(Continued)*

---

simplicity of operation: the ability to display for the operator
the response from the module being tested. After the testing
device transmits different command and data lines to the module,
it should be able to receive a status or response word and communicate it to the operator.

Future Work

    We are now at work on Phase II--designing the device to
reflect these desired characteristics.

    We are analyzing ways to enable the device to "address"
automatically the various logic modules it will have to test. The
most promising approach appears to be to equip each logic module
with a uniform integrated circuit--such as the 45K58, a four-bit
magnitude comparitor--that can be wired to produce a unique word
that indicates the board address of that module.

    The display capability appears to be a simpler problem.
Once the module being tested has executed a command, it will
generate a status word. The testing device will receive this word
by sending an enable signal to the status enable pin on the unit
holding the module. Standard LCD indicators on the front panel of
the testing device will display the status word to the operator.

On the other hand, don't panic if the preliminary results are
not so promising as had been anticipated, or if the project is behind schedule. Readers are fully aware that the most sober and
conservative proposal writers cannot anticipate all the problems
that can—and generally do—arise. As long as the original proposal contained no wildly inaccurate computations or failed to
consider crucial factors, don't feel personally responsible. Just do
your best to explain what happened and the current status of the
work. If you suspect that the results will not match earlier predictions—or that the project will require more time, personnel, or
equipment—say so, clearly. Don't give your reader an unduly op-

timistic picture, in the hope that you eventually will be able to work out the problems on your own or make up the lost time. If your news is not good, at least give your reader as much time as possible to deal with it effectively.

Figure 14-4 shows the conclusion for the computer-testing-device progress report.

Because the design of the testing device will affect the future design of the new computer systems, the writer has wisely decided to ask for technical assistance—from the Research and Development Department.

APPENDIXES   In the appendixes to the report, include any supporting materials that you feel your reader might wish to consult: computations, printouts, schematics, diagrams, charts, tables, or a revised task schedule. Be sure to provide cross-references to these appendixes in the body of the report, so that the reader can consult them at the appropriate stage of the discussion.

Figure 14-5 shows an updated task schedule. The writer has taken the original task schedule from the proposal and added cross hatching to show the tasks that have already been completed.

SAMPLE PROGRESS REPORTS

Two sample progress reports are included here:

1. "Progress Report on the Investigation of Personal Computers for the Sensory Evaluation Department at Cavendish Laboratories, Inc." (Adams 1986).

FIGURE 14-4

Conclusion of a Progress Report

```
                        CONCLUSION
    Phase I of the project has been completed successfully and on
schedule.
    We hope to work out the basic schematic of the testing device
within two weeks. The one aspect of Phase II that is giving us
trouble is the question of versatility. Although we can equip our
future systems with a uniform integrated circuit that can be
wired to produce unique identifiers, this procedure will create
future headaches for R&D. We have arranged to meet with R&D next
week to discuss this problem.
```

FIGURE 14-5

Updated Task Schedule of a Progress Report

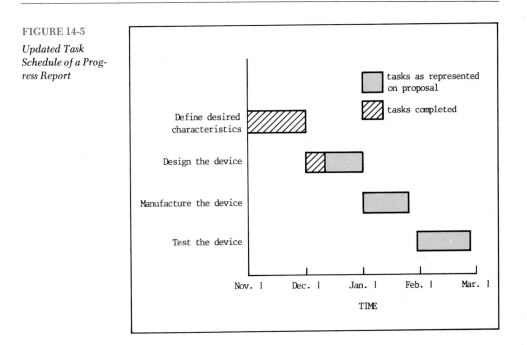

2. "Progress Report: Removing Miscible Organic Pollutants from Water Effluents" (Charles 1983).

The first progress report was written as a follow-up to the proposal included in Chapter 13. (The completion report on this study is included in Chapter 15.) The author was a student working part-time in the sensory-evaluation department of a pharmaceutical corporation. Her project was to determine whether any available personal computers would help her department better maintain its records.

The second progress report was written by a scientist employed by a chemical company. The subject of the report is a project to devise a method of removing miscible organic pollutants from water effluents.

Marginal notes have been added to both progress reports.

DATE: February 14, 1986

TO: Laura Smith, Manager, Sensory Evaluation, Cavendish
          Laboratories, Inc.

FROM: Cynthia Adams, Sensory Analyst, Sensory Evaluation,
          Cavendish Laboratories, Inc.

SUBJECT: Progress Report on the Investigation of Personal Com-
          puters for the Sensory Evaluation Department at Cavend-
          ish Laboratories, Inc.

This purpose state-
ment defines the
purpose of the
memo, identifies the
period the memo
covers, and states
the general subject
of the investigation.

The first sentence
suggests that the
memo will be struc-
tured according to
tasks. The second
sentence presents the
conclusions of the
first two parts of the
investigation and
sketches in how the
conclusion of the
third part will affect
the overall recom-
mendation. The fi-
nal sentence indi-
cates when the
completion report
will be submitted.

The problem is cop-
ied from the pro-
posal.

Purpose

This memo is the progress report covering two-thirds of my investigation of 11 personal computers for use by Sensory Evaluation.
The purpose of this investigation is to determine whether a PC would
solve the problem with the keywords-filing system. If one or more is
appropriate, it or they will be recommended.

Summary

I have completed my evaluations of 11 personal computers on the
basis of technical and usage/maintenance criteria. Five PC's--AT&T
PC 6300, HP 150, Leading Edge PC, IBM PC, and IMB PC AT--fully satisfy
both sets of criteria. Those PC's fulfilling the financial criteria
will be recommended; if no PC is feasible, then we will need to discuss
alternative solutions. You will receive the completion report on February 20.

The Problem

Each time a sensory evaluation test is conducted, it is assigned
a unique code of seven characters called the keyword code (for example, LC05172). On the final test report, several keywords are listed.
A keyword is the product tested (for example, lo-cal salad dressing),
the variables in the product (for example, different amounts of vinegar in the salad dressing), the type of evaluation conducted (for example, triangle test, duo trio test, and qualitative evaluation), and
the type of statistical analysis performed (for example, chi-square
analysis or multidimensional scaling). The keywords-filing system
is a small box containing about 300 index cards. Each index card is labeled with a keyword, and the cards are stored in alphabetical order.
When the system was established one year ago, the department planned
that every time a report on a test was issued, someone would locate the
cards for each of the keywords listed on the report. Then that person
would write the keyword code of the test on each card. The purposes of
this keywords-filing system were listed as:

1. to maintain a permanent record of all tests conducted during
   the year
2. to record exactly how many tests are conducted of a certain
   type or with a particular variable
3. to enable sensory analysts to quickly find previous tests
   that have been conducted with a particular product or variable. This would prevent duplication of previous tests and
   lead to the planning of more informative tests.

Unfortunately, these purposes are not being fulfilled because
reports written from sensory-evaluation tests have not been entered

in the keywords-filing system from July 1985 through January 1986. That is, after one year, only the first six months of tests have been entered. No one in Sensory has enough time to enter reports in the keywords-filing system by manually searching for the proper card, writing the keyword code on the card, and then replacing it in the box. This requires approximately six minutes for each test. With 25 tests conducted each week, this task amounts to 130 hours a year.

The keywords-filing system is therefore ineffective. Within the last three months, five tests were conducted that had been conducted six months ago. Because Sensory did not have this information, this error cost the department $1,200 in lost time and products. If the system were kept current, this would not have occurred.

The objective of this investigation is to determine whether a personal computer would be suitable for entering and storing the keywords-filing system.

Work Completed: Evaluating Personal Computers According to Technical and Usage/Maintenance Criteria

This section reports on the technical and usage/maintenance criteria applied to the 11 personal computers. The PC's chosen for the evaluation represent major domestic and foreign manufacturers. These 11 are the machines mentioned most often in the different articles I have researched.

1. Technical Criteria

In order to store approximately 300 keywords, the PC must have a memory storage of at least 256K bytes RAM. Of the 11 systems investigated, five have a memory capacity of at least 256K. All these PC's can use software that handles word files. In addition, these five PC's-- AT&T PC 6300, HP 150, Leading Edge PC, IBM PC, and IBM PC AT--are all IBM compatible and can accommodate a printer. These five personal computers fully satisfy the technical criteria.

2. Usage/Maintenance Criteria

After an evaluation of the 11 PC's based on usage/maintenance criteria, five PC's have been eliminated. These PC's either require a television set for the video display or have an LCD display that is difficult to read. Of the six remaining systems, all have documentation rated either "good" or "excellent." These PC's run most software, and in addition, two of them--AT&T PC 6300 and Leading Edge PC-- run two to three times faster than the IBM PC. The six remaining personal computers--AT&T PC 6300, HP 150, Leading Edge PC, Apple Macintosh, IBM PC, and IBM PC AT-- are reliable, and the companies have good reputations for product support. Thus, these six personal computers fulfill all the usage/maintenance criteria.

Future Work

The five personal computers that satisfy both the technical and usage/maintenance criteria will be evaluated on the basis of the financial criteria. I will use a cost-benefit analysis involving decision matrices that display quantitatively the various outcomes for each alternative course of action.

Conclusion

The investigation is expected to proceed on schedule (see the attached task schedule), and no problems are foreseen. Of the five remaining personal computers, those fulfilling the financial criteria will be recommended. If none of these five PC's is financially feasible, then we will need to discuss three alternatives:

(margin annotations)

This heading is a combination of generic and specific phrases.

This paragraph explains how the writer chose the PC's for the investigation.

The writer demonstrates her working method: to compare the available systems with her department's needs, using the process of elimination.

The writer clearly describes the work she is now doing.

The conclusion states that the work is proceeding as planned and suggests how the final phase of the investi-

gation will deter-
mine the ultimate
recommendation,
which will be in-
cluded in the com-
pletion report.

1. consider one of the other PC's that fulfills most of the criteria
2. reconsider the criteria and their importance
3. consider hiring someone to manually enter the records

The completion report with my results, conclusions, and recommendations will be submitted on February 20.

Updated Task Schedule

The writer updates
her task schedule.

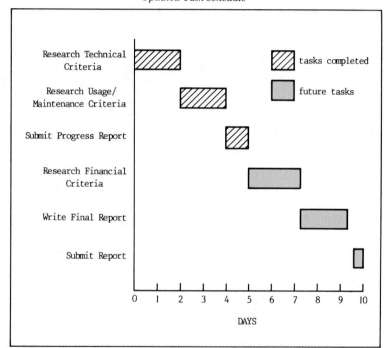

The type of report.

The subject of the report.

PROGRESS REPORT:

REMOVING MISCIBLE ORGANIC
POLLUTANTS FROM WATER
EFFLUENTS

Submitted by: Martin Charles
                 Biologist Class I
                 Marshall Chemicals, Inc.

        to: Dr. Helen Jenners
                 Chief, Biological Division
                 Marshall Chemicals, Inc.

May 11, 19--

The writer clearly defines the nature of this report and places this progress report within the sequence of reports.

The purpose of the study is defined.

The writer fills in the background of the problem.

Introduction

This progress report describes my findings after the first week of the investigation. The second progress report will follow in one week. The project is expected to be completed within two weeks, at which time a completion report will be submitted.

The purpose of this project is to devise a method for removing miscible organic pollutants from water effluents.

Discussion

The Problem

Currently, there are two methods of removing miscible organic pollutants from water effluents: column filtration and absorbent screens.

Column filtration, in which the effluent is pumped through an absorbent-packed column, is ineffective for rivers and streams, because of their great flowrates. The result is that a large percentage of the effluent remains unfiltered.

Absorbent screens, in which the effluent passes through an absorbent screen, are ineffective because the

screens create resistance, causing the effluent to flow
around, rather than through, them.

The writer repeats
the goal of the
project.

        The goal of this project is to devise a more effec-
tive method than column filtration or absorbent screens
for removing miscible organic pollutants from water effluents.

<u>Work</u> Completed

This introductory
paragraph forecasts
the discussion that
follows.

        To date, we have devised the general principle for
the new method--the use of floatable absorbent particles
--and selected an absorbent.

        The general principle behind the new method is to
introduce floatable absorbent particles into the bottom
of the stream or river. As the particles float to the surface, they come in contact with and absorb pollutants.
Particle floatability is limited, to allow for resub-
mergence in turbulent water conditions. At a calm section
of water downstream, the particles are surface-skimmed.

Notice how the
writer uses chronol-
ogy to emphasize the
coherence of the dis-
cussion.

        We then tested three absorbents--Absorbtex 115,
Filtrasorb 553, and activated carbon--to determine cost-
effectiveness. These data are listed below.

Absorbent	Cost/ Lb	Absorption/ Lb Absorbent	Lbs Absorbed/ $1
(1) Absorbtex 115	$9.50	20.9 lb CCl_4/lb	2.2 lb
(2) Filtrasorb 553	$7.65	18.6 lb CCl_4/lb	2.4 lb
(3) Activated Carbon	$2.89	10.9 lb CCl_4/lb	3.7 lb

These data show that although activated carbon is the least-effective absorbent, it is--by far--the most cost-effective of the three.

Future Work

Here the writer explains how the nature of the problem affected his research plan.

In our preliminary tests, each of the three absorbents presented different technical problems. Because of the substantial cost advantage of activated carbon, we decided to address the technical problems associated with that absorbent before investigating the other two absorbents.

In our experiments with activated carbon, we discovered that it loses buoyancy after short periods of contact with water. The necessary flotation must be induced by attaching the carbon to a buoyant material. Currently we are testing two materials--polyethylene and paraffin--and one other alternative: affixing the carbon to the outside of glass spheres.

In this paragraph, as well as in the next, the writer uses chronology effectively.

The next phase of study will involve determining the optimum size for the carbon particles. The smaller the size, the greater the surface area, of course, per gram of carbon. On the other hand, smallness increases the risk of water saturation, which causes the particles to sink and makes recovery impossible.

Finally, a method of introducing the carbon into the stream or river must be devised. If coated carbon is the absorbent, weighted milk jugs with water-soluble caps will probably be effective.

Conclusion

The writer attempts to forecast the future of the project.

We foresee no special problems in completing the next phases of the project. The second progress report should answer the questions of coatings and particle size. We expect to conclude the project successfully on time.

WRITER'S
CHECKLIST

Even though progress reports vary considerably in format and appearance, the basic strategy behind them remains the same. The purpose of this checklist is to help you make sure you have included the major elements of a progress report.

1. Does the introduction
   a. Identify the document as a progress report?
   b. Indicate the period the progress report covers?
   c. Place the progress report within the sequence of any other progress reports?
   d. State the objectives of the project?
   e. Outline the major phases of the project?

2. Does the summary
   a. Present the major accomplishments of the period covered by the report?
   b. Present any necessary comments on the current work?
   c. Direct the reader to crucial portions of the discussion section of the progress report?

3. Does the discussion
   a. Describe the problem that motivated the project?
   b. Describe all the work completed during the period covered by the report?
   c. Describe any problems that arose, and how they were confronted?
   d. Describe the work remaining to be done?

4. Does the conclusion
   a. Accurately evaluate the progress on the project to date?
   b. Forecast the problems and possibilities of the future work?

5. Do the appendixes include the supporting materials that substantiate the discussion?

EXERCISES

1. Write a progress report describing the work you are doing for the major assignment you proposed in Chapter 13.

2. The following progress report, titled "The Future of Municipal Sludge Composting," was written by Walter Prentice, an engineer working for the waste resources department of a medium-sized city. The reader is the head of the writer's department. In an essay, evaluate the report from the points of view of clarity, completeness, and writing style.

Introduction
      Sludge composting is a 21-day process by which wastewater
sludge is converted into organic fertilizer that is aesthetically
acceptable, pathogen-free, and easy to handle. Composted sludge can
be used to improve soil structure, increase the soil's water retention, and provide nutrients for plant growth.

Discussion
      Sludge composting is essentially a two-step process:

1. Aerated-Pile Composting
    Dump trucks deliver the dewatered raw sludge to the compost site.
    Approximately 10 tons of sludge is dumped on a 25 yd^3 bed of bulk-
    ing agent (usually woodchips). A front-end loader mixes the
    sludge into the bulking agent. The mixture is then placed on a
    compost pad and covered with a blanket of unscreened compost 1 ft
    thick. This layer is applied to insulate the sludge-bulking agent
    for ambient temperatures and for preventing the escape of odors
    from the pile. The air and odors are sucked out of the bulking
    agent base by an aeration system of pipes under the compost pad.
    After three weeks, the sludge in the aerated compost pile is es-
    sentially free of pathogens and stabilized.
2. Drying, Screening, and Curing the Composted Sludge
    After the aerated pile composting is completed, the pile is
    spread out and harrowed periodically until it is dry enough to
    screen. Screening is desirable, because it recovers 80 percent of
    the costly bulking agent for reuse with new sludge. The screened
    compost is stored for at least 30 days before being distributed
    for use. During the curing period, the compost continues to de-
    compose, ensuring an odor- and pathogen-free product.

Future Work
      The next step is to determine the cost of sludge composting.
Sludge composting using the aerated-compost-pile method is estimated to cost between $35 and $50 per dry ton ($35 for a 50-dry-ton per
day operation, $50 for a 10-dry-ton per day operation). These esti-
mates include all facilities, equipment, and labor necessary to compost at a site separate from the treatment plant. Not included are the
costs of sludge dewatering, transportation to and from the site, and
runoff treatment. These additional factors can raise the cost of
composting to $160 per dry ton.
      The breakdown of the capital costs for composting follows:

1. Site development--one acre of land is required for every three
    dry tons per day capacity. Half of the site should be surfaced.
    Asphalt paving costs about $60,000 per acre. In addition, the
    site requires electricity for the aeration blowers.
2. Equipment--front-end loader, trucks, tractors, screens, blow-
    ers, and pipes are required.
3. Labor--labor represents between a third and a half of the operat-
    ing costs. Labor is estimated to cost $6 per hour, with five weeks
    of paid sick or vacation time.

      However, a potential market exists for compost. Although the
high levels of certain heavy metals in the compost restrict its use in
some cases, compost can be used by businesses such as nurseries, golf
courses, landscaping, and surface mining. Transportation costs can
be high, however.

Conclusion

This information comes from published reports by federal agencies and journal articles. Since the government banned ocean sludge dumping in 1981, composting has become a viable method of waste treatment, and much has been written about it.

Before we can reach a final decision about whether composting would be economically justifiable for Corinth, we must add our numbers to the costs above. This process should take about two more weeks, provided that we can get all the information.

3. The following progress report was written by a college student who worked part-time in a tavern/restaurant in a major city. In an essay, evaluate the report from the points of view of clarity, completeness, and writing style.

HIRING ACCOUNTING HELP OR BUYING
AN ELECTRONIC REGISTER FOR THE CROW'S NEST:
A PROGRESS REPORT

Background

The arrival of the Saratoga for a 30-month renovation program has increased our business about 60 percent. As a result, the preparation of daily reports and inventory calculations has become a lengthy and cumbersome task. What used to take two hours a day now takes almost four. We have two alternatives: hire a part-time accountant (such as a local university student), or invest in an electronic cash register system. Following is a report on my first week's findings in the investigation of these two alternatives.

Work Completed

The going rate for an accounting student is about $4.50/hour. At two hours per night, seven nights per week, our annual costs would be about $3,500, including the applicable taxes. If the student were to take over all your bookkeeping tasks--about four hours per night-- the cost would be about $7,000 annually.

Both local colleges have told me on the phone that we would have no trouble locating one or more students who would be interested in such work. Break-in time would probably be short; they could learn our system in a few hours.

The analysis of the electronic cash registers is more complicated. So far, I have figured out a way to compare the various systems and begun to gather my information. Five criteria are important for our situation:

1. overall quality
2. cost
3. adaptability to our needs
4. dealer servicing
5. availability of buy-back option

To determine which machines are the most reputable, several magazine articles were checked. Five brands--NCR, TEC, Federal, CASIO, and TOWA--were on everyone's list of best machines.

I am now in the process of visiting the four local dealers in business machines. I am asking each of them the same questions--about quality, cost, versatility, frequency-of-repair records, buy-back options, etc.

Work Remaining
        Although I haven't completed my survey of the four local deal-
ers, one thing seems certain: a machine will be cheaper than hiring a
part-time accountant. The five machines range in price from $1,000 to
$2,000. Yearly maintenance contracts are available for at least some
of them. Also, buy-back options are available, so we won't be stuck
with a machine that is too big or obsolete when the Saratoga repairs
are complete.
        I expect to have the final report ready by next Tuesday.

REFERENCE   Adams, C. 1986. Progress report on the investigation of personal com-
             puters for the Sensory Evaluation Department at Cavendish Labora-
             tories, Inc. Unpublished manuscript.

CHAPTER FIFTEEN

COMPLETION REPORTS

A completion report is generally the culmination of a substantial research project. Two other reports often precede it. A proposal (see Chapter 13) argues that the writer or writers be allowed to begin and carry out a project. A progress report (see Chapter 14) describes the status of a project that is not yet completed; its purpose is to inform the sponsors of the project how the work is proceeding. The completion report, written when the work is finished, provides a permanent record of the entire project, including the circumstances that led to its beginning.

FUNCTIONS OF THE COMPLETION REPORT

A completion report has two basic functions. The first is immediate documentation. For the sponsors of the project, the report provides the necessary facts and figures, linked by narrative discussion, which enable them to understand how the project was carried out, what it found, and, most important, what those findings mean. All completion reports lead, at least, to a discussion of results. For example, a limited project might call for the writer to determine the operating characteristics of three competing models of a piece of lab equipment. The heart of that completion report will be the presentation of those results. Many completion reports call for the writer to go beyond the results and analyze the results and present conclusions. The writer of the report on the lab equipment might be authorized to inform the readers which of the three machines appears to be the most appropriate for his organization's needs. And finally, many completion reports go one step further and present recommendations: suggestions about how to proceed in light of the conclusions. The writer of the lab equipment report might have been asked to recommend which of the three machines—if any—should be purchased.

The second basic function of the completion report is to serve as a future reference. Three common situations send employees searching for old reports.

The first such situation is a personnel change: a new employee is likely to consult filed reports to determine the kinds of projects the organization has completed recently. Reports are not only the best source of this information—they are often the only source, because the employees who participated in the project might have left the organization.

Second, when the organization contemplates a major new project, it usually will want to determine how the new project would affect existing procedures or operations. The completion reports in the files will be the best source of this information, too. For example, if the owner of an office complex wants to computerize the temperature control of his buildings, he will bring in an expert to consult the reports on the electrical wiring and the heating, ventilating, and air-conditioning systems to determine whether computerization is technically feasible and economically justifiable.

Third, and perhaps most important, if a problem develops after the project has been completed, employees will turn first to the project's completion report to try to figure out what went wrong. An analysis of a breakdown in a production line requires the technical description of the production line—the completion report that was written when the line was implemented. In these three situations, completion reports are valuable long after the projects they describe have been completed.

TYPES OF COMPLETION REPORTS

Every completion report is unique, because the needs of every audience are unique. However, it is possible to classify completion reports into two broad categories: physical research reports and feasibility reports.

Terminology varies from organization to organization. In some organizations procedures manuals and policy descriptions are considered completion reports because they report on the completion of projects: projects to establish procedures or policies. Some people consider a proposal to be a kind of completion report because it completes a recognizable phase of a project. And because most external proposals—and many internal proposals—are never funded, they are really unintentional completion reports.

PHYSICAL
RESEARCH
REPORTS
A physical research report is written about a project that involved substantial empirical research, whether it was carried out in a lab or in the field. Empirical research lies at the heart of the scientific method: you begin with a hypothesis, conduct experiments to test it, record your results, and determine whether your hypothesis was correct as it stands or needs to be modified. The key to empirical research is that you conduct tests and generate original research.

For instance, you work for the Navy. Your supervisors are considering building the hulls of minesweeper ships from glass-reinforced plastic—GRP—instead of wood. GRP would seem to offer several advantages, but first its properties have to be established. How strong is it? How flexible? How well does it deal with temperature variations? The list of questions that have to be answered is long. Each question will be answered by physical research on GRP in the laboratory, on mock-ups of hulls, and even on computer simulations of the material.

A feasibility report documents a study that attempts to evaluate at least two alternative courses of action. For example, should our company hire a programmer to write a program we need, or should we have an outside company write it for us? Should we expand our product line to include a new item, or should we make changes in an existing product?

FEASIBILITY
REPORTS
A feasibility study seeks to answer questions of possibility. We would like to build a new rail line to link our warehouse and our retail outlet, but if we cannot raise the cash the project is not possible at this time. Even if we have the money, do we have the necessary permission from government authorities? If we do, are the soil conditions adequate for the rail line?

A feasibility study can also consider questions of economic wisdom. Even if we can raise the money to build the rail line, is it a wise thing to do? If we use up all our credit on this project, what other projects will have to be postponed or canceled? Is there a less expensive—or a less risky—way to achieve the same goals as a rail link?

Finally, a feasibility study can also consider how an action will be received by interested parties. For instance, if your company's workers have recently accepted a temporary wage freeze, they might view the rail link as an unnecessary expenditure. The truckers' union might see it as a threat to their job security. Some members of the general public might be interested parties. Any sort of large-scale construction might affect the environment. Even though your plan might be perfectly acceptable according to its environmental-impact statement—the study required by the government—some citizens might disagree with the statement or might still oppose the project on aesthetic grounds. Whether or not you agree with the objections, going ahead with the project might create adverse publicity.

Feasibility studies often involve physical research. The physical research on GRP might be the first phase in a feasibility study comparing the existing material, wood, with GRP. The physical research answers the technical question of whether GRP would be an effective substitute for wood in the hulls of minesweepers. The feasibility study answers the question of whether it is possible and wise to use GRP.

THE STRUCTURE
OF THE COMPLETION REPORT

Like proposals and progress reports, completion reports must be self-sufficient: that is, they have to make sense without the authors there to explain them. The difficulty for you as a writer is that you can never be sure when your report will be read—or by whom. All you can be sure of is that some of your readers will be managers who are *not* technically competent in your field and who need only an overview of the project, and that others will be technical personnel who *are* competent in your field and who need detailed information. There will be yet other readers, of course—such as technical personnel in related fields—but in most cases the divergent needs of managers and of technical personnel are all you need to consider.

To accommodate these two basic types of readers, completion reports today generally contain an executive summary that precedes the body (the full discussion). These two elements overlap but remain independent; each has its own beginning, middle, and end. Most readers will be interested in one of the two, but probably not in both. As a formal report, the typical completion report will contain other standard elements:

title page
abstract
table of contents
list of illustrations
executive summary
glossary
list of symbols
body
appendix

This chapter will concentrate on the body of a completion report; the other elements common to most formal reports are discussed in Chapter 11.

THE BODY OF THE COMPLETION REPORT

The body of a typical completion report contains the following five elements:

1. introduction
2. methods
3. results
4. conclusions
5. recommendations

Some writers like to draft these elements in the order in which they will be presented. These writers like to compose the introduction first because they want to be sure that they have a clear sense of direction before they draft the discussion and the findings. Other writers prefer to put off the introduction until they have completed the other elements of the body. Their reasoning is that in writing the discussion and the findings they will inevitably have to make some substantive changes; therefore, they would have to revise the introduction, if they wrote it first. In either case, careful brainstorming and outlining are necessary before you begin to write.

THE
INTRODUCTION

The first section of the detailed discussion is the introduction, which enables the readers to understand the technical discussion that follows. Usually, the introduction contains most or all of the following elements:

1. An explanation of the problem or opportunity that led to the project. What was not working, or not working well, in the organization? What improvements in the operation of the organization could be considered if more information were known? It might be useful or even necessary here to include a few paragraphs of background to orient the readers.

2. A statement of the purpose of the project. What exactly was the project intended to accomplish? What information was it intended to gather or create, or what action was it intended to facilitate?

3. A statement of the scope of the project. What aspects of the problem or opportunity were included in the project and what aspects excluded? For example, a report on new microcomputers might be limited to those that cost less than $4,000, or those that have at least 512K of memory. Another approach to defining the scope would be to explain briefly the major technical tasks you had to perform.

4. An explanation of the organization of the report. Readers understand better if they know where you are going and why. Explain your organizational pattern so that readers are not surprised or puzzled.

5. A review of the relevant literature. Sometimes the literature will be internal—reports and memos produced within the organization. Sometimes the literature will be external—published articles or even books that help your readers understand the context of your work.

Figure 15-1 shows an introduction to the body of a completion report. The subject of the report (MacBride 1986) is an investigation to determine whether a retail store can increase its sales and reduce its heating bills by instituting an energy-management system. The writer was a student working for the store in its Engineering and Planning Department. The report is titled "Recommendation for Minimizing the Steam Consumption and Regulating the Ambient Temperature at Bridgeport Department Store." Marginal notes have been added.

THE METHODS

In the methods section of the report, you describe the technical tasks or procedures you performed. If you are reporting on a physical research project carried out in the lab or the field, this section will closely resemble the discussion section of a traditional lab report. If several research methods were available to you, begin by describing why you chose the method(s) you did. Either list the equipment and materials you used before the description of the re-

FIGURE 15-1

Introduction to a
Completion Report

The background.

INTRODUCTION

Bridgeport Department Store, built in 1912, is an expensive building to heat because of its high ceilings, large windows, and lack of wall insulation. Every increase in heating costs hits us harder than it hits our competitors in more modern facilities. Therefore, we must make sure we are using the most modern and cost-effective methods for regulating our store temperature.

Currently, we have an energy-management system, but it regulates only electricity for lights and air-conditioning, not our steam used for heating.

The problem, discussed in monetary terms.

In 1986, Bridgeport experienced a 6 percent increase in steam consumption over 1985 figures. This rise, combined with a 7 percent utility rate increase, resulted in a 15 percent ($67,000) rise in related operating expenses.

In addition, the number of customer complaints relating to the uneven temperatures throughout the building rose by 135 percent over the 1985 figures.

The purpose of the investigation.

The purpose of the study is to determine whether it would be cost effective to expand the capabilities of our current energy-management system to include the monitoring and regulating of steam consumption.

search or mention them within the description itself. The preliminary listing is more common when some of the readers are going to duplicate the research. If they simply want to understand what you did, the listing is probably not necessary.

For a feasibility study, you should also begin by justifying your methods. Describe what you did: physical experiments, theoretical studies, site visits, interviews, library research, and so forth. Your goal is to show your readers that you have conducted your research in a thorough, professional manner. This will increase not only your readers' ability to understand the findings that follow but also your credibility.

FIGURE 15-1

*Introduction to a
Completion Report
(Continued)*

The scope of the
investigation.

The relevant litera-
ture.

This study was restricted to changes in software. We do not have the funds to finance hardware changes to our main computer system, which was installed earlier this year.

Recent articles by Cottrell and by Gorham on energy management indicate that an effective energy management system can reduce steam usage by 20-25 percent. Donovan and Rossiter, writing on the relationship between store climate and consumer buying behavior, suggest that the more comfortable the shoppers, the longer they stay and the more money they spend.

Figure 15-2 shows the brief methods section from the report on the energy-management system at the department store.

THE RESULTS   The results are the data you observed, discovered, or created. You should present the results objectively, so that the readers can "experience" the methods just as you did. Save the interpretation of the results—the conclusion—for later. If you intermix results and conclusions, your readers might be unable to follow your reasoning process. Consequently, they will not be able to tell whether your conclusions are justified by the evidence—the results.

METHODS

Before studying the possible enhancement to our current

energy-management system, we devised a set of technical, manage-

ment/maintenance, and financial criteria by which to evaluate

the enhancement.

Then we studied the literature on energy-management sys-

tems (see References, page 11). We phoned Cottrell and Gorham to

follow up on points they raised in their articles. We visited two

sites (Jarvis Electronics and the Foundation Society)--local

facilities that are mentioned in the literature.

Next we studied our internal records on energy usage and the

blueprints, plans, and installation and operating manuals for

our existing system. Then we surveyed our site to determine

exactly how difficult it would be to convert to an effective

energy-management system.

Just as the methods section answers the question, "What did you do?" the results section answers the question, "What did you see?"

The nature of the project will help you decide how to structure the results. For physical-research reports, you can often present the results as a brief series of paragraphs and graphic aids. If in the methods section you described a series of tests, in the results section you simply report the data in the same sequence you used for the methods. For feasibility studies, a comparison-and-contrast structure (see Chapter 7) is generally the most accessible. When you are evaluating a number of different alternatives, the whole-

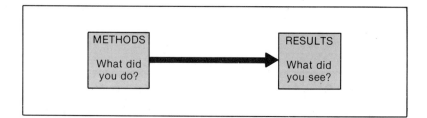

by-whole pattern might be most appropriate. When you are evaluating only a few alternatives, the part-by-part structure might
work best. Of course, you must always try to consider the needs of
your readers. How much they know about the subject, what they
plan to do with the report, what they anticipate your recommendation will be—these and many other factors will affect your decision on how to structure the discussion.

For instance, suppose that your company is considering installing a word-processing system. In the introduction you have already discussed the company's current system and its disadvantages. In the methods section you have described how you
established the criteria to apply to the available systems, as well as
the research procedures you carried out. In the results section, you
provide the details of each system.

Figure 15-3 is an excerpt from the results section of the report
on the energy-management system for the department store.

THE
CONCLUSIONS

The conclusions are the implications—the "meaning" of the
results. Drawing valid conclusions from results requires great
care. Suppose, for example, that you work for a company that
manufactures and sells clock radios. Your records tell you that in
1986, 2.3 percent of the clock radios your company produced were
returned as defective. An analysis of company records over the
previous five years yields these results:

YEAR	% RETURNED AS DEFECTIVE
1985	1.3
1984	1.6
1983	1.2
1982	1.4
1981	1.3

One obvious conclusion can be drawn: a 2.3 percent defective rate
is a lot higher than the rate for any of the last five years. And that
conclusion is certainly a cause for concern.

FIGURE 15-3

*Excerpt from the
Results Section in
the Body of a Com-
pletion Report*

Major Technical Components of an Efficient Energy-Management System

    There are two main technical components to an efficient energy-management system: a network of sensors and customized software.

Sensors

    Because climate is a function of humidity, entropy, and temperature levels, a network of sensors would have to be installed at strategic points throughout the building. These sensors would form the communication link to and from a particular area. Our energy-management system would continuously measure humidity, entropy, temperature, and steam-demand levels during a standard time interval. These actual climate measurements would be compared with predetermined target levels every few minutes. On the basis of these comparisons, the system would initiate the appropriate action (open or close the steam valves) to maintain the target temperature.

But do those results indicate that your company's clock radios are less well made than they used to be? Perhaps—but in order to reach a reasonable conclusion from these results, you must consider two other factors. First, you must account for consumer behavior trends. Perhaps consumers were more sensitive to quality in 1986 than they had been in previous years. A general increase in awareness—or a widely reported news item about clock radios—might account in part or in whole for the increase in consumer complaints. (Presumably, other manufacturers of similar products have experienced similar patterns of returns if general consumer

FIGURE 15-3

Excerpt from the Results Section in the Body of a Completion Report (Continued)

Software

A computer program is needed to process the signals from the sensors, determine whether adjustments are required, and transmit instructions. This program would compare the parameter readings to the predetermined target levels, initiate the necessary adjustments (open or close the steam valves) for the target climate to be maintained, and record these measurements as data. These data could then be used to analyze current steam efficiency and potential energy-management opportunities.

Tasks Involved in Converting Our System

As discussed in the system description, a sophisticated energy-management system is based on a set of sensors and customized software. This section discusses the results of our analysis of the tasks required to convert our system to a sophisticated, computerized energy-management system.

Sensors

One complication is that the current system employs steam valves that are operated by a pneumatic signal system. The system works on the principle of air pressure against a diaphragm; the expansion and contraction of the diaphragm opens and closes the

---

trends are at work.) A second factor to examine is your company's policy on defective clock radios. If a new, broader policy was instituted in 1986, the increase in the number of returns might imply nothing about the quality of the product. In fact, the clock radios sold in 1986 might even be better than the older models. In other words, beware of drawing hasty conclusions. Examine all the relevant information.

Just as the results section answers the question, "What did you see?" the conclusions section answers the question, "What does it mean?"

valve. However, a sophisticated energy-management system can recognize only digital signals. Therefore, pneumatic-to-digital converters would have to be purchased and installed so that our system could regulate steam flow throughout the building. These converters could be installed by our electricians. The cost of the units, installed, would be less than $3,500. See Appendix C, page 13, for descriptions and cost figures for the converters.

Software

The program required to monitor and regulate the steam would be quite similar to the current program used to monitor and regulate the air-conditioning units throughout the building. For this reason, developing the steam-regulating program would require only about four person-weeks of work by our programming staff. This would cost less than $2,000. In addition, several of the subprograms used by the air-conditioning system could also be used directly by the steam-heating program. This would reduce the amount of computer memory space needed. Therefore, the memory space currently available would be sufficient. See Appendix D, page 16, for the details of the programming needs.

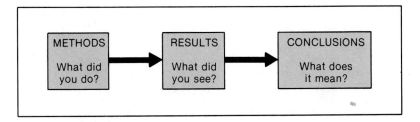

The conclusion of the energy-management system report is shown in Figure 15-4.

FIGURE 15-4

*Conclusion Section
from the Body of a
Completion Report*

CONCLUSION

Incorporating our current steam-delivery system into an
efficient energy-management system would cost less than $6,000
and could be accomplished in less than two weeks without our
having to employ any additional personnel. This action would be
the most cost-effective way to contain rising energy costs at the
Bridgeport Department Store.

**THE RECOM-
MENDATIONS**   Recommendations are statements of action. Just as the conclusions
section answers the question, "What does it mean?" the recommendations section answers the question, "What should we do
now?"

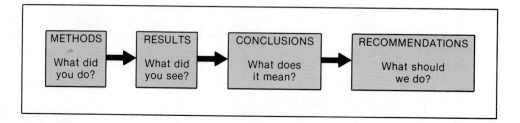

FIGURE 15-5

Recommendation Section from the Body of a Completion Report

RECOMMENDATION

We recommend that Bridgeport Department Store incorporate the steam-delivery system into its existing energy-management system as soon as practicable. The expenditure would pay for itself in less than one month. The Engineering and Planning Department stands ready to undertake this task.

The recommendations section is always placed at the end of the body; because of its importance, however, the recommendations section is often summarized—or inserted verbatim—after the executive summary.

If the conclusion of the report leads to more than one recommendation, use a numbered list. If the report leads to only one recommendation, use traditional paragraphs.

Of more importance than the form of the recommendations section are its content and tone. Remember that when you tell your readers what you think they ought to do next, you must be clear, complete, and polite. If the project you are describing has been unsuccessful, don't simply recommend that your readers "try some other alternatives." Be specific: What other alternatives do you recommend, and why?

And keep in mind that when you recommend a new course of action, you might well be rejecting a previous course of action. Thus you run the risk of offending whoever formulated the earlier course. Do not write that your new direction will "correct the mistakes" that have been made recently. Instead, write that your new direction "offers great promise for success." If the previous direction was not proving successful, your readers will probably already know that. A restrained, understated tone is not only more polite but also more persuasive: you appear to be interested only in the good of your company, not in personal rivalries.

Figure 15-5 shows the recommendations section of the energy-management system report.

This report delivers good news: a relatively small expenditure will save the company a lot of money. Often, however, feasibility reports such as this one yield mixed news or bad news; that is, none of the available options would be an unqualified success, or none of the options would work at all. Don't feel that a negative recommendation reflects negatively on you. If the problem being studied were easy to solve, it probably would have been solved before you came along. Give the best advice you can, even if that advice is to do nothing. The last thing you want to do is to recommend a course of action that will not live up to the organization's expectations.

SAMPLE COMPLETION REPORT

The following completion report was written by the student working in the sensory evaluation department at a pharmaceutical corporation. Her project was to determine whether personal computers would help her department better maintain its records. The proposal for this project is included in Chapter 13. The progress report is in Chapter 14. Marginal comments have been added.

CAVENDISH LABORATORIES, INC.

February 20, 1986

Ms. Laura Smith, Manager
Sensory Evaluation
Cavendish Laboratories, Inc.
Princeton, NJ 08540

Dear Ms. Smith:

The purpose of the report.

Attached is the report on the investigation of personal computers for the Sensory Evaluation Department, originally proposed February 10, 1986. A personal computer would greatly reduce the time spent manually entering data in the keywords-filing system.

The general conclusion.

For this investigation, I established technical, usage/maintenance, and financial criteria. Five PC's—AT&T PC 6300, HP 150, IBM PC, IBM PC AT, and Leading Edge PC—satisfied all criteria.

The methods.

The results.

Each of these PC's is an effective solution. At a cost ranging from $3,600 to $4,500, each would pay for itself in less than two years. It is not financially feasible to employ someone to manually record information.

The conclusion.

The recommendation.

I recommend that Sensory Evaluation purchase one of the five PC's. If you have any questions concerning this report, please feel free to contact me. I shall be happy to discuss them with you at your convenience.

A polite conclusion to the letter.

Sincerely,

Cynthia Adams

Cynthia Adams
Sensory Analyst

PERSONAL COMPUTERS FOR

THE SENSORY EVALUATION DEPARTMENT:

A RECOMMENDATION

Submitted to: Ms. Laura Smith

Manager, Sensory Evaluation

Cavendish Laboratories, Inc.

Submitted by: Cynthia Adams

Sensory Analyst, Sensory Evaluation

Cavendish Laboratories, Inc.

February 20, 1986

ABSTRACT

Title of the report.
Name and position
of the writer.

"An Investigation of Personal Computers for the Sensory Evaluation Department," Cynthia Adams, Sensory Analyst, Cavendish Laboratories, Inc.

Background of the
problem.

Within the last three months, Sensory repeated five tests that
had already been conducted, resulting in a loss of $1,200 to the de-
partment. This error occurred because the keywords-filing system,
which records the tests conducted, was not up to date. Personal computers were investigated to determine the feasibility of storing the
keywords-filing system in a PC. A leading computer journal indicated
that this would cut in half the time spent entering and retrieving
data, and thus assist us in keeping the keywords-filing system up to

Conclusion.

date. According to this analysis, five personal computers--AT&T PC
6300, HP 150, IBM PC, IBM PC AT, and Leading Edge PC--best fit Senso-

Technical descrip-
tion.

ry's needs. Each PC uses software that will allow sensory analysts to
enter keyword codes and keywords, and will allow classification of
tests by product. In addition, the analysts can determine how many
tests were conducted of a certain type or with a certain variable, and
each PC is equipped with a printer for obtaining hard copies of the

Recommendation.

data. Because the use of a personal computer would greatly reduce the
amount of time required to enter data into the keywords-filing system
and would reduce future errors when conducting tests, I recommend
that one of the PC's be purchased.

Prefatory pages are
numbered with
lower-case Roman
numerals, centered
horizontally.

i

CONTENTS

1

I. EXECUTIVE SUMMARY

The problem.

The keywords-filing system, which records all tests conducted in Sensory Evaluation, has not been kept up to date. No one in the department has had enough time to manually enter into the filing system the tests that have been conducted. As a result of this missing information, several tests were conducted that had been conducted six months ago, and this error cost the department $1,200 in lost time and products.

The methods.

Because of this problem, I proposed an investigation of 11 leading personal computers to determine the feasibility of storing the keywords-filing system in a PC. A PC would save time in entering and retrieving data, and thus would help keep the keywords-filing system up to date. The PC's were evaluated using technical, usage/maintenance, and financial criteria.

The conclusion.

On the basis of my analysis, five personal computers best fit our needs: AT&T PC 6300, HP 150, IBM PC, IBM PC AT, and Leading Edge PC. Each of these systems has adequate memory to store the keywords-filing system, uses software that handles word files, is IBM compatible, and comes with a printer. Each has a video display that is easy to read, has good documentation, and has a good reliability record. In addition, the companies manufacturing the PC's have good reputations for product support.

Costs and benefits.

These five personal computers cost from $3,600 for the Leading Edge PC to $4,500 for the IBM PC AT. A cost-benefit analysis indicates that between $2.74 and $3.36 in benefits is gained for every dollar expended on the systems over a five-year period, and that each system will pay for itself in under two years. (Hiring an employee to enter the data manually yielded a poor cost-benefit ratio).

The recommendation.

Because each of these personal computers would be an effective solution to the problem with the keywords-filing system in Sensory Evaluation, I recommend that one of them be purchased.

II. INTRODUCTION

Background.

    Each time a sensory evaluation test is conducted, it is assigned a unique code of seven characters. The test is also assigned several keywords that list the variables in the test, the type of test conducted, and the type of statistical analysis performed on the data. All this information is written on index cards. Sensory has about 300 cards that make up the keywords-filing system. This filing system is intended to keep a record of all tests conducted and enable sensory analysts to quickly find previous tests that have been conducted.

Problem.

    However, the keywords-filing system has not been kept up to date, because no one in the department has had enough time to manually record the information on the cards. Within the last three months, five tests were repeated that had been conducted six months ago. Because Sensory did not have this information, this error cost the department $1,200 in lost time and products. If the filing system had been kept up to date, this would not have occurred.

Purpose of the report.

Rationale for the methods.

    This report describes my investigation of 11 personal computers to determine the feasibility of storing the keywords-filing system in a PC. The 11 PC's chosen for the investigation represent major domestic and foreign manufacturers. These 11 machines (listed here in alphabetical order) are the ones mentioned most often in the numerous articles I have researched:

        Apple Macintosh

        Apple IIc

        AT&T PC 6300

        Coleco Adam

        HP 150

        IBM PC

        IBM PC AT

        IBM PCjr.

        Leading Edge PC

        NEC PC-8401A

        Tandy 200

3

III. <u>METHODS</u>

    To perform the analysis, I consulted computer journals that are not affiliated with any company. The PC's had been reviewed by independent consultants, and I considered reviews of them from at least two different journals, to avoid any bias the reviewer might have. I evaluated the PC's on the basis of technical, usage/maintenance, and financial criteria. I then performed cost-benefit analyses of the PC's that fulfilled the technical and usage/maintenance criteria, as well as of the option of hiring an employee. Then, using decision matrices, I compared the PC's and the employee.

Cross-reference to the next section of the report.

    The criteria and my calculations are explained in detail in the following section, Results.

IV. <u>RESULTS</u>

    The 11 PC's were evaluated according to three sets of criteria: technical, usage/maintenance, and financial. This section describes how well the 11 PC's fulfilled these criteria.

A. TECHNICAL CRITERIA

The results of each set of criteria are previewed in the introductory paragraphs of the discussion.

    In analyzing the personal computers, I emphasized four aspects of their technical characteristics: (1) file size, (2) software for word files, (3) IBM compatibility, and (4) the printer. Five PC's--AT&T PC 6300, HP 150, IBM PC, IBM PC AT, and Leading Edge PC--fulfilled these technical criteria.

1. <u>File Size</u>

    In order to store approximately 300 keywords, the PC would require a memory storage of at least 256K bytes RAM. For efficiency, a memory of this size is needed so that entire categories of data can be loaded into RAM at one time.

    The Tandy 200 and NEC PC-8401A, both new laptop computers, did not have nearly enough memory, at 24K and 32K, respectively. The Coleco Adam, a home computer with 80K, also did not have adequate memory. The Apple Macintosh, Apple IIc, and IBM PCjr each have 128K; although close to Sensory's requirement, they were

4

eliminated. Only five PC's remained: AT&T PC 6300, HP 150, IBM PC,
IBM PC AT, and Leading Edge PC.

2. Software for Word Files

    In order to store the keywords-filing system, the PC would
need to use software that handles word files, that is, software
designed as a filing system into which sensory analysts can enter
words and keyword codes. Several of the popular pieces of software of this type--Personal Card File, Easyfiler, Filecommand,
PC-File, and Quickfile III--can be used on each of the five PC's.

3. IBM Compatibility

    The criterion of IBM compatibility was chosen because the
mainframes that Sensory uses for the statistical analysis of data
are IBM computers. Although Sensory does not now link the PC to
the mainframe, a possible use for such a linkage might exist in
the future. Therefore, the possible PC's were limited to those
that were either IBM computers or IBM compatible. Of the 11 PC's,
five fulfilled this criterion: AT&T PC 6300, HP 150, IBM PC, IBM
PC AT, and Leading Edge PC.

4. Printer

    Any PC that Sensory would consider purchasing must be
equipped with a printer so that a sensory analyst could make a
hard copy of a file and then look for the corresponding reports in
the filing cabinet. Thus, another analyst could use the PC at the
same time. A dot-matrix printer is the cheapest type and is adequate for printing out a list of tests. The five PC's can use several different printers, so this criterion was easily satisfied.

B. USAGE/MAINTENANCE CRITERIA

    When considering usage/maintenance criteria, I began with
all 11 PC's and evaluated them, whether or not they had been eliminated by the technical criteria. The usage/maintenance criteria
are user friendliness, ease of use, accessibility of maintenance
and service personnel, and reliability. Six PC's--Apple Mac-

Preview of the
results.

intosh, AT&T PC 6300, HP 150, IBM PC, IBM PC AT, and Leading Edge
PC--fulfilled the usage/maintenance criteria. Except for the Apple Macintosh, these are the PC's that fulfilled the technical criteria.

1. User Friendliness

The "user friendliness" of the PC was an important criterion because the sensory analysts have limited computer experience. The PC would not be an effective solution if it were difficult to operate. Two aspects of user friendliness were identified
as important: video display and documentation.

a. Video Display

Of the 11 PC's evaluated, two--Coleco Adam and IBM PCjr--
are not equipped with a monitor but are connected to a television for viewing. Therefore, these personal computers were
eliminated because a television does not provide so clear an
image as a monitor. Several of the PC's use an LCD display instead of a CRT display. An LCD screen is slower and less legible
than a CRT display (Hartmann 1985), and as one reviewer com-
mented about the NEC PC-8401A, "This laptop computer has possibilities, but since it is impossible to read the display, you
cannot begin to use the PC" (Krause 1985). Of the remaining
personal computers, the AT&T PC 6300 was described as having
"excellent contrast" (Satchell 1985), the HP 150 has high resolution (Talmy 1984), and the Apple Macintosh also has very
high resolution, at 512 × 342 pixels (Webster 1984).

b. Documentation

Documentation, the manuals explaining the commands and
operating system, needs to be clear and understandable for
someone without previous computer experience. Possible rat-
ings for the PC's were poor, fair, good, or excellent, so I considered a rating of either "good" or "excellent" as satisfy-
ing this criterion.

Of the six remaining PC's, the AT&T PC 6300, HP 150, and

IBM PC AT have documentation rated "good" by the reviewers; the Apple Macintosh, IBM PC, and Leading Edge PC have documentation rated "excellent."

2. Ease of Use

Because the sensory analysts are quite busy with other responsibilities, using the PC should require as little effort and time as possible. The Coleco Adam is considered "excellent" in this category but has been eliminated because it requires a television set and has inadequate memory. The Apple Macintosh is also rated "excellent" because of its extensive use of icons but has been eliminated because of its inadequate memory.

In benchmark tests by the reviewers, the AT&T PC 6300 ran programs two to three times faster than the IBM PC (Troiano 1985), the IBM PCjr ran roughly twice as fast as the Apple IIc (Casta and Bernard 1984), and the Leading Edge PC operated up to 50 percent faster than the IBM PC (Mazur 1984). The HP 150 ran slightly slower than the IBM PC (Haas 1984). Thus, no one personal computer was superior to the others in amount of effort and time required.

3. Accessibility of Maintenance and Service Personnel

If problems arise with the PC, maintenance and service personnel should be available quickly. Six PCs--Apple Macintosh, AT&T PC 6300, HP 150, IBM PC, IBM PC AT, and Leading Edge PC--all have good serviceability records, are supplied with telephone numbers to call for technical help, and have service centers either in northern New Jersey or in New York City. Two of the manufacturers, AT&T and IBM, were noted as having good reputations for product support (Troiano 1985).

4. Reliability

Of the 11 PC's evaluated, one is clearly not reliable: the Coleco Adam. The tapes become damaged if stored near the central processing unit, backups are difficult, and the software can be erased from a tape (Gilder 1984). The other PC's do not have reliability problems, and are rated either "good" or "excellent,"

7

with the Apple Macintosh, AT&T PC 6300, IBM PC AT, and Leading
Edge PC rated "excellent." With the Leading Edge PC, the mean
time between failure (MTBF) of the disk drives is listed at 20,000
hours; the drives of the IBM PC are rated at 8,000 hours MTBF, and
the Leading Edge PC is recommended as a reasonable alternative to
the IBM PC (Mazur 1984).

C. FINANCIAL CRITERIA

According to my evaluation of the 11 personal computers on
technical and usage/maintenance criteria, 5 of them fully satisfy
both sets of criteria: AT&T PC 6300, HP 150, IBM PC, IBM PC AT, and
Leading Edge PC. These five PC's were evaluated on the basis of two
financial criteria: (1) the initial cost of the PC and printer
should not exceed $6,000, and (2) the annual maintenance costs of
the PC should not exceed $1,000. Costs for the personal computers
and for hiring an employee were calculated, and these costs were
then analyzed using decision matrices. According to these calculations, the five PC's that fulfill the technical and usage/mainte-
nance criteria also fulfill the financial criteria; the option of
hiring an employee does not.

Preview of the results.

1. Calculated Costs

The costs I calculated are the following: the initial cost
of the personal computers, the annual maintenance costs of the
personal computers, and the annual cost of an employee to enter
the data manually.

a. Initial Cost of Personal Computers

The initial cost for each of the PC's is the list price
stated in the journal articles. These figures are the most reliable and are used in the cost-benefit analysis. I realize
that these figures might be somewhat elevated, because the
price might have dropped since the article was written. It
would be possible to purchase a PC at a discounted price through
a supplier unaffiliated with the company, but this would involve risks that Sensory does not want to take (credibility of

8

Cross-reference to
Appendix A.

the supplier, guarantee of shipment, guarantee of equipment).
The initial costs for the PC's can be found in Appendix A.

b. Annual Maintenance Costs of Personal Computers

According to Brian Owen, a management consultant spe-
cializing in computer applications for business, the annual
maintenance costs for a PC can be estimated at 7 percent of the
total initial costs for the system (Owen 1985). For us this

Cross-reference to
Appendix A.

would include the PC and the printer. The annual maintenance
costs can be found in Appendix A.

c. Annual Cost of an Employee

A possible solution to the problem of the ineffective keywords-filing system would be to employ one of the sensory analysts for 2½ hours per week to manually record the reports in
the filing system. Because the sensory analysts are working at
capacity, this extra employment would have to be overtime and
would result in payment of time-and-a-half, or about $13.88 per
hour. When calculated for 2½ hours per week over a one-year period, this amounts to $1,803.75. Although this option entails
no payback period, the Personnel Department informs us that the
cost/benefit ratio is poor because of fringes; for every $1.00
expended, only $0.79 is gained in benefits. Therefore, this so-
lution is not financially feasible.

2. Analysis of Costs Using Decision Matrices

Explanation of the
use of decision ma-
trices.

Owen has developed a model for analyzing the costs and benefits of converting manual systems to computer systems. The total
costs are defined as the initial cost of the PC plus the annual
maintenance costs plus the cost in time spent converting from the
manual system to the computer system. The total benefits are defined as the time saved by using a computer system plus the objective benefits plus the subjective benefits. The cost-benefit
model can be applied to any number of years and will also determine the payback period for the system.

The cost in time spent converting from manual to computer
was estimated at 35 hours. At a salary of $9.25 per hour for a sen-

sory analyst, this amounted to $323.75. Approximately 130 hours per year and $1,202.50 are required to enter information manually into the keywords-filing system. A PC would cut this time in half, and the cost would be only $601.25 per year. The objective benefits are equal to the costs before converting from manual to computer, which is the $1,200 lost by the department within six months. For subjective benefits, Owens estimates $500 per year when using a personal computer.

In managerial decision-making, matrices can be used to eliminate potential solutions that would be unacceptable. Owen states that the decision-maker can set a minimum figure for benefits accrued per dollar expended. Any possible solution yielding fewer benefits than this minimum figure would then be eliminated. Likewise, a maximum payback period is set, and those solutions with payback periods exceeding the maximum would also be eliminated. Thus, the decision-maker is left with only those possible solutions that would be acceptable. In selecting a personal computer, our financial constraints were not benefits and payback period but rather the initial cost of the PC and the annual maintenance costs. The financial constraint of a purchase price not exceeding $6,000 eliminated one well-known model from the original sample: the IBM PC XT with 512K bytes RAM. Another model of the IBM PC--the AT, which has 256K--was not eliminated, and the constraint of annual maintenance costs did not eliminate any PC's.

Cross-reference to
Appendix B.

Appendix B shows both a cost-benefit analysis performed on the five PC's over a five-year period, and the payback periods for the PC's. According to this analysis, between $2.74 for the IBM PC AT (256K) and $3.36 for the Leading Edge PC is gained for every dollar expended on the systems. All the PC's will pay for themselves in under two years, with the lowest payback period at 1.51 years for the IBM PC AT.

V. CONCLUSION

Sensory Evaluation cannot continue to have information missing from the keywords-filing system, which can result in losses to the department. It is not financially feasible to employ someone to manu-

ally enter information in the filing system. However, it is financially feasible to purchase a PC. Five personal computers--AT&T PC
6300, HP 150, IBM PC, IBM PC AT (256K), and Leading Edge PC--are effective solutions to the problem with the keywords-filing system in Sensory Evaluation.

VI. RECOMMENDATION

     I recommend that the Sensory Evaluation Department purchase any
one of the five PC's--AT&T PC 6300, HP 150, IBM PC, IBM PC AT (256K),
and Leading Edge PC--that meet all our stated criteria.

VII. BIBLIOGRAPHY

The bibliography is lengthy because of the nature of the investigation: it relies on published analyses of the PC's.

Archer, R. 1984. The IBM PCjr. BYTE 9, no. 8:254-68.

Bullard, B. 1983. Comparing the IBM PC and the TI PC. BYTE 8, no.
    11:232-42.

Casella, P. 1985. NEC PC-8401A. InfoWorld 7, no. 11:42-43.

Casta, P., and J. Bernard. 1984. Face off: The Apple IIc vs. IBM's new
    PCjr. Computers & Electronics 22, no. 11:50-53.

Finger, A. 1985. IBM PC AT. BYTE 10, no. 5:270-77.

Gilder, J. 1984. The Coleco Adam. BYTE 9, no. 4:206-20.

Haas, M. 1984. The HP 150 Computer. BYTE 9, no. 12:262-75.

Hartmann, T. 1985. The Tandy 200: Tandy surpasses its Model 100 with
    an enhanced second-generation laptop computer. Popular Comput-
    ing 4, no. 9:82-85.

Krause, H. 1985. The Tandy 200 and the NEC PC-8401A. BYTE 10, no.
    13:306-14.

Markoff, J. 1984. The Apple IIc Personal Computer. BYTE 9, no. 5:
    276-84.

Mazur, J. 1984. The Apple IIc Personal Computer. BYTE 9, no. 10:
    312-20.

Neudecker, T. 1984. Macintosh. InfoWorld 6, no. 13:82-90.

Owen, B. 1985. Fifteen real-life productivity solutions. PC Magazine
    4, no. 24:111-135.

Satchell, S. 1985. AT&T 6300 Personal Computer. InfoWorld 7, no.
    1&2:49-54.

Schofield, J. 1984. IBM PC AT. Practical Computing 7, no. 12:72-74.

11

Talmy, S. 1984. The HP 150 Personal Computer. Creative Computing 10, no. 4:26-37.

Troiano, B. 1985. The AT&T PC 6300. BYTE 10, no. 13:294-302.

Vose, G.M., and R. Shuford. 1984. A closer look at the IBM PCjr. BYTE 9, no. 3:320-32.

Webster, B. 1984. The Macintosh. BYTE 9, no. 8:238-47.

Williams, G. 1984. The Apple Macintosh Computer. BYTE 9, no. 2:30-54.

Wilson, D. 1984. Adam: Lowered price, some cured ills make a better machine. InfoWorld 6, no. 53:39-40.

Wood, L. 1985. The Leading Edge PC: Genuine enhancements make this computer stand out from the pack. Popular Computing 4, no. 3:104-7.

APPENDIX A. COST DATA OF PERSONAL COMPUTERS

Personal Computer	Initial Cost[a]	Annual Maintenance Costs[b]
AT&T PC 6300	$4475	$313.25
Hewlett Packard PC	3995	279.65
IBM PC	4095	286.65
IBM PC AT (256K)	4495	314.65
Leading Edge PC	3620	253.40

[a]Includes PC and printer.

[b]Calculated 7% per year.

APPENDIX B. COST-BENEFIT ANALYSIS AND PAYBACK PERIOD OF THE PERSONAL COMPUTERS

Cost-benefit analysis over a five-year period

$$\frac{\text{Benefit}}{\text{Cost}} = \frac{n(\underline{E} + \underline{O} + \underline{S})}{\underline{I} + \underline{C} + n(\underline{A})}$$

$\underline{n}$ = number of years the system is to be in place
$\underline{E}$ = money saved by entering reports with computer
$\underline{O}$ = objective benefits (costs before converting to computer)
$\underline{S}$ = subjective benefits
$\underline{I}$ = initial cost of system
$\underline{C}$ = cost in time spent converting from manual to computer
$\underline{A}$ = annual maintenance costs

Payback Period

The payback period is the time at which the costs equal the benefits, and indicates how much time is required for the system to pay for itself.

$$\underline{n}(\underline{E} + \underline{O} + \underline{S}) = \underline{I} + \underline{C} + \underline{n}(\underline{A})$$

Personal Computer	Benefit/Cost	Payback Period (years)
AT&T PC 6300	2.75	1.50
Hewlett Packard PC	3.06	1.34
IBM PC	2.99	1.38
IBM PC AT (256K)	2.74	1.51
Leading Edge PC	3.36	1.21

WRITER'S CHECKLIST

The following checklist applies only to the body of the completion report. The checklist pertaining to the other report elements is included in Chapter 11.

1. Does the introduction
   a. Identify the problem that led to the project?
   b. Identify the purpose of the project?
   c. Identify the scope of the project?
   d. Explain the organization of the report?

2. Does the methods section
   a. Provide a complete description of your methods?
   b. List or mention the equipment or materials?

3. Are the results presented clearly and objectively, and without interpretation?

4. Are the conclusions
   a. Presented clearly?
   b. Drawn logically from the results?

5. Are the recommendations stated directly, diplomatically, and objectively?

EXERCISES

1. Write the completion report for the major assignment you proposed in Chapter 13.

2. Following is a report titled "A Procedure for the Aseptic Propagation of Yeast" written by William J. DeMedio, a chemist working at a brewery (DeMedio 1980). It was submitted to the brewmaster, William Miller. In an essay, evaluate the effectiveness of the report.

American Brewing Company

May 28, 19--

Mr. William Miller
Brewmaster
American Brewing Company
Center and Elm Streets
Philadelphia, PA 19123

Dear Mr. Miller:

This report covers the observations and results of the near beer
yeast propagation procedure, originally proposed April 19,
19--. The procedure was found to be highly effective in preventing yeast infection.

The procedure uses the logarithmic growth pattern of the yeast,
as well as certain sterilization procedures fitted to the brewery, to grow yeast aseptically from test tube slants to the eight
barrel pitching amount.

The propagation is very inexpensive: $840.00 for labor and equip-
ment. It is also a rapid procedure and can be started in a day's
notice. The procedure should save the brewery any troubles associated with infected near beer brews.

If you have any questions concerning this report, please feel
free to contact me. I shall be happy to discuss them with you, at
your convenience.

Sincerely,

William DeMedio
Chemist
Brewing Laboratory

ABSTRACT

　　The brewing laboratory was given the task of developing a method for the aseptic propagation of near beer yeast from test tube slants. Vague details of the propagation were sent from Switzerland, but it was the task of the brewing laboratory to develop a method that would fit the needs of the brewery. A highly successful and inexpensive method was developed, which involved specialized sterilization procedures and the utilization of the logarithmic growth pattern of the yeast. The total cost of the method (material and labor) was $840.00. If this is compared to the cost of an infected brew, which would be a direct consequence of an improper yeast propagation, one sees a considerable savings. In my opinion, the method of yeast propagation developed in the brewing laboratory is the safest and least expensive way to grow yeast.

William DeMedio
Chemist
Brewing Laboratory

i

CONTENTS

I. EXECUTIVE SUMMARY

If a near-beer yeast culture becomes infected, serious problems can result. First, an entire new brew would be ruined, resulting in a loss of the range $700-$1,000 for materials alone.
Second, the company would have to go through government clear-
ance procedures in order to dispose of the brew. Third, drain-
ing the beer would increase the sewage costs somewhat. Finally,
if infected near beer were canned and sold, the company would be
liable to severe fines.

An infected brew can be prevented by pitching it with
infection-free yeast.

The brewing laboratory was given the task of developing a
procedure for the aseptic propagation of near-beer yeast from
test-tube slants to an amount feasible for pitching a commercial brew. A successful and inexpensive method was developed,
utilizing the growth patterns of yeast. The procedure was shown
to be extremely effective in preventing costly infection.

As well as being safe, the method was also very inexpensive. Total labor costs amounted to $640.00, while equipment
costs were $200.00. (Please refer to Appendix 1, on page 5.)
Also, the equipment that was purchased is a permanent investment, since it can be used in repeated propagations.

The procedure takes eight days to produce an amount of
yeast feasible for pitching a 160-barrel brew. This is enough
beer to can 5,000 cases. The cost of propagation is a small
fraction (6%) of the total cost of production of near beer.

The propagation is based on the vague details sent from
Switzerland. (Please see Appendix 2, on page 5.) It was the task
of the brewing laboratory to develop a method that both meets
the requirements and fits the needs of the brewery.

II. THE PROPAGATION PROCEDURE

1. Background
    The method of propagation contained four major steps. The
    first step was the transfer of a loopful of each of the two
    yeast strains into individual 100 ml bottles of wort, each
    containing 5 percent sucrose (saccharose). After two days
    of fermentation time, each 100-ml sample of yeast culture
    was transferred into individual 1,000-ml wort samples,
    containing 5 percent sucrose. Again, these were allowed to
    ferment for two days. In the third step of the propagation,
    each 1,000-ml sample of yeast culture was transferred into
    individual 10,000-ml wort samples, containing 5 percent
    sucrose. Finally, in the fourth step of the propagation,
    the two strains of yeast were mixed in eight barrels of
    plain wort. This was allowed to ferment for two days to
    obtain the objective: eight barrels of aseptic pitching
    yeast. The procedure obviously involved taking advantage
    of the logarithmic growth pattern of yeast.

        In order to prevent infection, standard aseptic
    technique was used in all yeast transfers. (Please see H.
    J. Benson, Microbiological Applications, W. C. Brown and

Co., 1979, p. 78, for an example of standard aseptic technique.)

Three germicides were used in order to sterilize utensils: heat, iodophor, and anhydrous isopropyl alcohol. First, each utensil was rinsed in a 50 PPM iodophor solution. Then, each was rinsed clean with anhydrous isopropyl alcohol in sparing amounts. Finally, the utensils were placed in an oven at 100°C to maintain sterility. Presterilized absorbent cotton was used to plug all flasks in the propagation.

In order to maintain the pH at the bacteria-inhibiting 4.0-4.5 level, water corrective was employed (85% phosphoric acid) in the final steps of the propagation. A negligible amount was used.

The wort used was ordinary lager kettle wort. In order to differentiate between strains, one was called strain "A" and the other was called strain "B."

2. Step One
   a) Preparation: First, we sterilized three 250-ml centrifuge bottles. Then, we cleared sufficient bench space and lit two bunsen burners and a propane torch for a hood effect. We then got all materials ready: two 7-g packets of sucrose, 2 yeast slants in a test-tube stand, inoculating loops, towels. Towels were used to handle all hot material.
   b) Wort Obtaining: Two technicians were needed for this procedure. First, one technician took a 2½-gallon bucket, with a 12-foot rope attached, to the kettle. With another technician assisting, he then lowered the bucket into the hot (214°F) wort. The bucket was allowed to fill with wort, and then the technicians carefully raised it out of the kettle. This was the wort to be used as a medium for growing the yeast. The technicians then took this hot wort up to the laboratory, where they placed it on the stand. Since the wort was boiling, it was already sterile.
   c) Preparation of 100-ml Wort Media: Two technicians were needed for this procedure. One technician carefully placed 7 g of sucrose into each of two centrifuge bottles, marked "A" and "B." A third bottle, marked "C," was used as a control. After placing the sucrose into the sterile bottles, the technician placed a sterile ½" × 5" stainless steel funnel into the mouths of bottles "A" and "B." Another technician dipped hot wort fresh from the kettle and filled first the "A" bottle, and then the "B" bottle up to the 100-ml mark. The "B" bottle funnel was then placed in the mouth of the "C" bottle and this was filled to the 100-ml mark. A sterile thermometer was placed in the "C" bottle, and all three bottles were placed in the refrigerator. When the "C" bottle registered 70°F, all three were taken from the refrigerator.

2

d) Inoculation: One slant was marked "A" and the other
    "B" to differentiate them. A loopful of the yeast from
    the "A" slant was transferred into the bottle marked
    "A." The same was done with the slant marked "B." The
    bench area had been cleaned in advance with a 50 PPM
    iodophor solution. After inoculation, all of the
    bottles were placed in a dark place at 70°F. The
    uninoculated "C" bottle was used to determine
    whether any infection could have invaded the "A" or
    "B" bottles, with the idea that if the "C" bottle
    remained uninfected, so would the "A" and "B"
    bottles.

e) Fermentation: The bottles were allowed to ferment for
    two days. Bubbles appeared in the "A" bottle after 24
    hours and in the "B" bottle after 28 hours. The
    bottles were periodically swirled to disperse the
    yeast. Each bottle had a very clean smell. After two
    days, no growth was seen in the "C" bottle, nearly
    confirming the asepticness of bottles "A" and "B." A
    simple taste test showed no infection present in the
    "C" bottle.

3. Step Two
    a) Preparation: We cleared and prepared a bench area as in
        "Step One--Preparation," on page 2, except for
        obtaining the slants, test-tube stand, and inoculat-
        ing loops. Also, we used three 2,000-ml Erlenmeyer
        flasks and two 1" X 7" stainless steel funnels in this
        step.
    b) Wort Obtaining: The wort was obtained in exactly the
        same way as in "Step One--Wort Obtaining," on page 2.
    c) Preparation of 1,000-ml Wort Media: The method of
        preparation was the same as in "Step One--Preparation
        of 100-ml Wort Media," on page 3, except that 2,000-ml
        Erlenmeyer flasks, 1,000-ml wort samples, 70-g
        packets of sucrose, and 1" × 7" funnels were used.
    d) Inoculation: Each 100-ml yeast culture was trans-
        ferred into its respective flask, marked "A" and
        "B." This was done aseptically by simply pouring the
        contents of each bottle into the other flask.
    e) Microbial Yeast Examination: A wet mount slide was
        prepared from the residue in each bottle marked "A"
        and "B." The slide was observed under high power. A
        good number (3/4) of the yeast cells showed signs of
        budding. There was no sign of any infection microbes.
    f) Fermentation: The observations on maintenance of the
        flasks were exactly the same as in "Step One--
        Fermentation," on page 3. The only difference was
        size.

4. Step Three
    a) Preparation: The preparation for Step Three was the
        same as that mentioned in "Step Two--Preparation,"
        on page 3, except that 12,000-ml Florence flasks,

3

10,000-ml wort samples, and 700-g packets of sucrose were obtained. Also, 1½"×12" stainless steel funnels were used.

b) Wort Obtaining: The wort was obtained in the same way as mentioned in "Step One--Wort Obtaining," on page 2. Several trips to the kettle were necessary in order to get enough. It was found that one 2½ gallon bucketful of wort is approximately 10,000-ml of wort.

c) Preparation of 10,000-ml Wort Media: The preparation of media for step three is the same as in "Step One--Preparation of 100 ml Wort Media," on page 2, except 12,000-ml Florence flasks, 700-g sucrose packets, and 1½" × 12" funnels were used.

d) Inoculation: We transferred the 1,000-ml yeast cultures into their respective flasks as in the method described in "Step Two--Inoculation," on page 3.

e) Microbial Yeast Examination: The same wet mount procedure was performed here as in "Step Two--Microbial Yeast Examination," on page 3. Again, we found much yeast activity and little infection.

f) Fermentation: The same procedure and observations were made here as in "Step Two--Fermentation," on page 3.

5. Step Four

a) Preparation: An eight-barrel yeast tank was filled with wort by pumping it directly from the kettle. The lines, pump, and yeast tank were sterilized with 50 PPM iodophor in advance. The wort was adjusted to pH 4.3 by adding 4 ounces of 85 percent H_3PO_4 in a slurry. (Please refer to Appendix 3, on page 5, to see the experiment used to determine this figure.)

b) Inoculation: The two 10,000-ml yeast cultures were taken to the yeast tank and poured directly into it together. The tank was then covered. No sucrose was added to the wort.

c) Microbial Yeast Examination: We periodically observed wet mounts of the wort-yeast mixture in the tank. No signs of infection were observed, and most yeast cells showed activity.

d) Fermentation: The inoculated wort in the yeast tank was allowed to ferment for two days at 70°F. At the end of two days, we had reached our objective: eight barrels of aseptic near beer yeast. This was then given to the fermenting-room men, who pitched a 160-barrel brew with it.

III. CONCLUSION

According to our results, we can conclude that the procedure we have developed is very effective in preventing the infection of near beer yeast. This procedure is also very inexpensive, involving little cost in labor and materials. The propagation takes only eight days to perform and can be started with a day's notice. Overall, we conclude that the procedure is safe, effective, inexpensive, and fits the needs of the brewery well.

APPENDIX 1. Costs of Equipment

ITEM	PRICE
Case of four 12,000-ml Florence Flasks	$ 50.00
6 Stainless-Steel Funnels	20.00
1 Carton Wax Pencils	3.00
Case of four 2,000-ml Erlenmeyer Flasks	25.00
Platinum Inoculating Loops	102.00
TOTAL	$200.00

APPENDIX 2. Original Propagation Instructions

BIRELL--Yeast Propagation

- The two yeast strains A and B must be propagated in separate flasks up to the 10-liter stage under strict sterile conditions.

- The following procedure is recommended:

- Inoculate from the agar slant tube into 100 ml of hopped wort, containing 5 percent sucrose.

- After two days at 20°C, pour the 100-ml culture into flasks containing 1 liter of wort with 5 percent sucrose.

- Pour the 1-liter culture, after two days at 20°C, into flasks containing 10 liters of wort with 5 percent sucrose.

- Pitch 5 ml of wort (as sterile as possible) with the 10-liter cultures of strains A and B.

Dr. K. Hammer

APPENDIX 3. Data from Experiment to Change the pH of Lager Wort from 5.10 to 4.30
Water corrective [85% H_3PO_4 (aq)] was added in varying amounts to one gallon of lager wort, and the pH was measured. The purpose was to find out how much corrective is required to lower the pH of lager wort from 5.10 to 4.3-4.5 range.

Trial	Amt. 85% H_3PO_4 Added	pH of Wort	Comments
1	0.00	5.10	Start
2	0.50	4.47	Considerable change
3	1.0	4.30	OK

CONCLUSION
We must add 1.00-ml corrective to one gallon lager wort to change the pH from 5.10 to 4.3-4.5, or we must add four ounces to eight barrels of lager wort.

6

REFERENCES   Adams, C. 1986. An investigation of personal computers for the Sensory Evaluation Department. Unpublished manuscript.

DeMedio, W. 1980. A procedure for the aseptic propagation of near beer yeast. Unpublished manuscript.

MacBride, M. 1986. Recommendation for minimizing the steam consumption and regulating the ambient temperature at Bridgeport Department Store. Unpublished manuscript.

PART
FOUR

OTHER
TECHNICAL
WRITING
APPLICATIONS

CHAPTER
SIXTEEN

TECHNICAL ARTICLES

Many popular magazines, from *Reader's Digest* to *Popular Mechanics* and *Personal Computing*, print journalistic articles about technical subjects. This chapter discusses technical articles, which in contrast might be defined as essays that are written by specialists—for other specialists—and that appear in professional journals. Technical articles are written principally to inform their readers, not to entertain them, and most of the readers of these articles read them for the information they convey.

TYPES OF TECHNICAL ARTICLES

Technical articles fall into three categories: (1) research articles, (2) review articles, and (3) conference papers. A research article reports on a research project. In most cases, the research is carried out by the writer in the laboratory or the field. A typical research article might describe the effect of a new chemical on a strain of bacterium, or the rate of contamination of a freshwater aquifer. A review article, on the other hand, is an analysis of the published research on a particular topic. It might be, for example, a study of the advances in chemotherapy research over the last two years. The purpose of a review article is not merely to provide a bibliography of the important research, but also to classify and evaluate

the work and perhaps to suggest fruitful directions for future research. The third kind of technical article—the conference paper—is the text of an oral presentation that the author gave at a
professional conference. Because conferences provide a useful fo-
rum for communicating tentative or partially completed research
findings, a paper printed in the "proceedings"—the published collection of the papers given at the meeting—is generally less authoritative and prestigious than an article published in a scholarly
journal. However, proceedings do give the reader a good idea of
the research currently being done in the field.

WHO WRITES TECHNICAL ARTICLES, AND WHY?

Technical articles are written by the people who actually perform
the research or, in the case of review articles, by specialists who
can evaluate the research of others. Often, the only characteristic
these authors share is that they themselves generated or gathered
the information they communicate: a physicist at a research organization writes up her experiment in laser technology, just as a
hospital administrator writes up his research on the impact of
health maintenance organizations on the private hospital industry.

Technical articles are the basic means by which professionals
communicate with each other. Most professional organizations
hold national conferences (and some hold regional and international ones, too) at which members present papers and discuss
their collective interests. But meetings occur infrequently, and not
all interested parties can attend; therefore, conferences are not the
principal means by which professionals stay current with the day-
to-day changes in their fields. Only technical journals can fulfill
that function, for they provide a convenient and inexpensive way
to transmit recent, authoritative information. A technical article
can be in a reader's hand less than two months after the writer sent
it off to the journal; most books, which deal with broader subjects
and contain more information than articles can, require at least
two years and therefore are not useful for communicating new
findings. And because it has been accepted for publication by a
group of referees—an editorial board of specialists who evaluate
its quality—a technical article is likely to be authoritative.

Because technical articles are a vital communications tool,
their authors are highly valued and frequently rewarded by employers and professional organizations. In some professions—no-

tably university teaching and research—regular and substantial publication is virtually a job requirement. In many other professions, occasional articles enhance the reputations and hasten the promotion and advancement of their authors.

TECHNICAL ARTICLES AND THE STUDENT

As an undergraduate you are unlikely to write a technical article that would make an important contribution to the scholarship in your field, although you might well write a journalistic article for a local publication about some subject you're studying in an advanced course or on a project one of your professors is working on. However, within a few years—probably far fewer than you think—you may find yourself applying for a job that requires publication in a professional journal. Or you may realize while working on a project that you have perfected a new technique for carrying out a particular task and that your technique warrants publication.

In the meantime most of your contact with technical articles will be as a reader. Nonetheless, when you read technical articles, don't limit your attention to the information that they convey. Notice the conventions and strategies that they employ, keeping in mind that at some point your résumé might benefit from an article or two.

When the occasion for writing a technical article arises, you will want to know how to approach your task. At such a point, you will probably not have a great deal of time to investigate your options, so it is important for you to understand from the start the fundamentals of publication. Even for a reader of technical articles, some familiarity with publication procedures can be helpful: recognizing the editorial positions of the journals in your field and understanding how an author's article has come to be published may shed some light on apparent mysteries; moreover, knowing the standard structure of technical articles will help you skim possible sources for research.

CHOOSING A JOURNAL

Having determined the topic of a possible article and perhaps having written a rough outline, a writer must think of a journal to which to submit the article. You might ask, "Why should anyone worry about placing an article that hasn't been written yet—

wouldn't it be more logical to write it first and then find an appropriate journal?" The answer would be yes if all journals operated in the same way. But they don't.

Most professional journals are run by volunteers. Although two or three persons at a journal might be salaried employees, the bulk of the work usually is done by eminent researchers and scholars whose principal goal is to strengthen the work in their fields. Consequently, these editors and their assistants are extremely serious about their work; they draw up and publish careful and comprehensive statements of purpose and editorial policies. They see themselves as addressing a particular audience, in a particular way, for a particular reason.

Editors expect a prospective author to take the journals they publish seriously and to follow their editorial policies. An article that does not follow stated editorial policies suggests that its author is careless, indifferent, or a bit smug. If, for instance, the journal uses one style of documentation and an article conforms to a different one, the author should not be surprised if the article is rejected without comment or explanation.

Sheer numbers contribute to this situation. Publication is so highly valued these days that few editors of reputable journals ever have a shortage of good articles from which to choose. The acceptance rate might be as low as 3 or 4 percent; rarely is it above 40 percent. Understandably, editors are unwilling to take the time or use their tiny staffs to make an article conform to their preferred style when they already have a dozen equally good articles that don't need stylistic revisions.

For these reasons, then, an author greatly enhances his or her chances of success by choosing a journal *before* starting to write. Only by doing so can an author tailor an article to its first audience—the editorial board of a particular journal.

WHAT TO
LOOK FOR

If you plan to write a professional article, start in the library. You already know the leading journals in your field. Get a few recent issues of each journal and find a quiet place to study them.

First, check the masthead of each journal. The masthead, which is generally located on one of the first few pages, is the statement of ownership and editorial policy. The editorial board is often listed on the masthead as well. The editor is likely to have a column—also near the front of the issue—that sometimes discusses the kinds of articles the journal is (or isn't) looking for. Study the editor's comments and the journal's editorial policy carefully; these sources are likely to provide crucial information that will save you considerable time, trouble, and expense. For instance, every once in a while a journal stops accepting submissions for a specified period (sometimes up to a year or two), because of a backlog of good articles. Of course, you would not write your article for that journal. Or you might learn that a particular journal does not accept unsolicited manuscripts: that is, that it commissions all its articles from well-known authorities or that it publishes only staff-written articles. Again, this is not the journal for you. Or maybe a journal is planning a special issue devoted entirely to one subject. Some journals print *only* special issues, and they announce the subjects in advance. If you choose one of these journals, make sure your idea fits the subject before you write the article.

Not all the information in the editorial policy or the editor's column is negative, of course. Most journals *do* accept unsolicited articles and are happy to explain their requirements: the explanation saves everyone a lot of frustration. Often, they will provide guidelines for sending in manuscripts (how to mail them, how many copies to send, etc.). Perhaps the most pertinent piece of information journals convey is the name of their preferred style sheet: the *Council of Biology Editors Style Manual, The Chicago Manual of Style*, the U.S. Government Printing Office *Style Manual*, or any other style guide. If the journal specifies a style manual, find it and follow it. Editors notice these things immediately. Figure 16-1 provides an example of an editorial policy statement.

After you have studied a journal's editorial policies, turn to what it actually publishes: the articles themselves. Read a number of articles and try to determine as much as you can about the following matters:

1. *Article length.* Sometimes, journals only print articles that fall within a certain range, such as 4,000 to 6,000 words. Even if a range is not stated explicitly in the journal, the editors might have one in mind as they read your article. If your article is going to be either much shorter or longer than the average length of the articles the journal publishes, you might not want to write it for that journal.

2. *Level of technicality.* Are the articles moderately technical or extremely technical? Do they include formulas, equations, figures?

3. *Prose style.* How long are the paragraphs? Do the authors use the passive voice ("The mixture was added . . .") or active voice ("I added the mixture . . .")? Is the writing formal or somewhat informal?

4. *Formal requirements.* Are the titles purely informative or are they "catchy"? Are subtitles included? Abstracts? Biographical sketches? Are the articles written with American or British spelling and punctuation?

Journal of Gerontology

General Information and Instructions to Authors

THE JOURNAL OF GERONTOLOGY is a bimonthly journal of the Gerontological Society of America. It publishes manuscripts dealing with or bearing on the problems of aging from the fields of biological, medical, psychological, and social sciences. The Journal publishes five types of contributed papers: 1) articles reporting original research; 2) brief reports and rapid communications; 3) letters to the editor; 4) invited reviews; and 5) theoretical/methodological papers. All manuscripts (1–5 above) should be addressed to Martha Storandt, PhD, Editor-in-Chief, Journal of Gerontology, Department of Psychology, Box 1125, Washington University, St. Louis, MO 63130.

1. Article Preparation

a. *Blind review.* Blind review is available upon request. Manuscripts should be prepared to conceal the identity of the author(s). The cover page and footnotes that identify author(s) should be omitted from two copies of the manuscript. Manuscripts not prepared in this manner will not receive anonymous review.

b. *Preparing the manuscript.* Papers must be typed *double spaced including references and tables* on 8½" × 11" white paper using 1" margins. Pages should be numbered consecutively, beginning with the abstract and including all pages of the submission. A total of *four* (4) copies must be submitted. Conciseness of expression is imperative.

c. *Title page.* A title page should include the title of the manuscript and the author's full name. A short running page headline not to exceed 40 letters and spaces should be placed at the foot of the title page.

d. *Abstract and key words*

1. A one paragraph abstract of not more than 150 words should be typed, *double spaced,* on a separate page. It should state the purpose of the study, basic procedures (study participants or experimental animals and observational and analytical methods), main findings and conclusions.

2. Key words. Below the abstract, provide 3 to 10 key words or short phrases that do not appear in the title and that will assist in cross-indexing the article. Add species of animal (e.g., rat, human) studied if this information does not appear in the title.

e. *Text.* The text of observational and experimental articles is usually (but not necessarily) divided into sections with the headings: Introduction, Method, Results, and Discussion. Articles may need subheadings within some sections to clarify their content. The Discussion may include conclusions deriving from the study and supported by the data.

f. *Text references.* Show all references in the text by citing in parentheses the author's surname and the year of publication. Example: ". . . a recent study (Jones, 1975) has shown . . . ." If a reference has two authors the citation includes the surnames of both authors each time the citation appears in the text. If a reference has more than two authors the citation includes only the surname of the senior author and the abbreviation "et al." If the reference list includes more than one publication in the same year by a given author or multiple authors, citations are distinguished by adding, a, b, etc. to the year of publication. Example: ". . . a recent study (Jones, 1975b) has shown . . . ." Several references cited at the same point in the text are separated by semicolons and enclosed in one pair of parentheses. The order is alphabetical by author's surname or, if all by the same author, in sequence by year of publication. Examples: "Recent studies (Brown, 1975; Jones, 1974) have shown . . . ." "Recent studies (Jones, 1975a, 1975b)."

g. *Reference list.* Type *double spaced.* Arrange entries alphabetically by author's surname; do *not* number. The reference list includes only references cited in the text and in most cases should not exceed 25 entries nor include references to private communications or unpublished work. Use the following form (*do not abbreviate titles of journals*):

Journals: Howard, D. V., McAndrews, M. P., & Lasaga, M. I. (1981). Semantic priming of lexical decisions in young and old adults. *Journal of Gerontology, 36,* 707–714.

Books: Rockstein, M. (Ed.). (1974). *Theoretical aspects of aging.* Academic Press, New York.

Chapters: Neugarten, B. L., & Hagestad, G. O. (1976). Age and the life course. In R. H. Binstock & E. Shanas (Eds.), *Handbook of aging and the social sciences.* Van Nostrand Reinhold, New York.

h. *Footnotes.* Footnotes related to the title, author's address, source of research support, and other acknowledgements should be numbered consecutively on a separate page placed after the references. Reference footnotes should *not* be used.

i. *Tables.* Type each table on a separate sheet, *remembering to double space.* Number tables consecutively and supply a brief title for each. Footnotes to a table are typed immediately below the table. The reference marks are superscript small letter (^{a,b,c} . . .) with the footnotes arranged alphabetically by their superscripts. The asterisk and double asterisk are used only for the probability level of tests of significance.

j. *Illustrations.* Each plate must be arranged by the author. For a graph or drawing, one glossy print must be included. If *photomicrographs* are submitted, *four glossy prints* should be supplied for review purposes. Color plates are at the expense of the author. *Sharp, glossy prints are required for high quality reproduction.* All lettering should be done with a lettering guide large enough to give 2-mm letters in the final reduction (*typewritten/handwritten lettering will not be acceptable*). Maximum plate size is 5¼" × 8¼". Final single column plate width is 2¾".

k. *Captions for illustrations.* Type captions *double spaced* on a separate page with numbers corresponding to the illustrations. Explain symbols, arrows, numbers, or letters used in illustrations. Explain internal scale and identify staining method in photomicrographs.

l. Other than as specified above, manuscripts should be prepared according to the Publications Manual of the American Psychological Association (3rd ed.), obtainable from APA, 1200 17th St. NW, Washington, DC 20036.

2. Brief Reports

Manuscripts (including references) must be no longer than six (6) *double spaced* typewritten pages, including no more than one (1) table, graph, or illustration. Four (4) copies should be submitted.

3. Rapid Communications

If findings are considered to be of immediate, significant importance, the author may request consideration as a rapid communication. *Full cooperation* in manuscript preparation as described in Section 1 is necessary to expedite such a request.

4. Letters to the Editor

Letters must be typewritten, *double spaced.* Submit four (4) copies. Letters should be short, approximately 500 to 750 words. If appropriate, a copy will be sent to the author of the original article to provide an opportunity for rebuttal. Letters and rebuttals will be reviewed and are subject to editing. Both letter and rebuttal will be published in the same issue.

5. Review Articles

These are solicited only by the Journal.

6. Theoretical/Methodological Papers

Theoretical papers must include an integration and critical analysis of existing positions in a specific area as well as proposed resolution(s) of controversial positions. Methodological papers should be supported with data if possible.

Miscellany:

a. *Reprints.* A reprint price list and order blank will be sent to the author with page proofs.

b. *Page proofs.* Substantive author's changes will be charged to the author.

c. *Other submissions of manuscripts.* Submission of a paper to the Journal implies that it has not been published elsewhere or is not under consideration elsewhere, and, if accepted for this journal, is not to be published elsewhere (except as Society Proceedings) without permission.

d. *Review.* Acceptance or rejection of manuscripts will be based on recommendations of the Section Editors.

FIGURE 16-1

A Typical Editorial Policy Statement

THE QUERY LETTER Once you are fairly certain of the journal you want to shape your article for, it's a good idea to find out what the editor thinks. Some editors will talk to you on the phone, but most prefer that you write a query letter.

A query letter is brief—usually less than a page. Your purpose is to answer briefly the following questions:

1. What is the subject of the article?
2. Why is the subject important?
3. What is your approach to the subject (that is, will the article be the report of a laboratory procedure, a rebuttal of another article, etc.)?
4. How long will the article be?
5. What are your credentials?
6. What is your phone number?

Figure 16-2 shows an example of a query letter.

THE STRUCTURE OF THE ARTICLE

Most of this text has been devoted to an explanation of multiple-audience documents—reports addressed both to people who know your subject and want to read all the details and to people who want to know only the basics. Although some readers of a technical article will skip one or several of the sections, for the most part you will be writing to fellow specialists who will read the whole article carefully. Because technical articles are intended for a national or international audience, a uniform structure has been developed to assist both writers and readers.

This structure is essentially the same as that of the *body* of the average report, except that an abstract generally precedes the article itself, with no other material intervening. The components of this structure are as follows:

1. title of article and name of author
2. abstract
3. introduction
4. materials and methods
5. results
6. discussion
7. references

FIGURE 16-2

Query Letter for an Article

Century Plastics

ALBION, MISSISSIPPI 34213

March 21, 19--

Dr. William Framingham
Editor
Plastics Technology
1913 Vine Street
Fowler, MS 34013

Dear Dr. Framingham:

I would like to know if you would be interested in considering an article for Plastics Technology.

My work at Century Plastics over the last two years has concentrated on laminate foamboard premixes. As you know, one of the major problems with these premixes is the loss of the blowing agent. In my research, I devised a new compound that effectively retards these losses.

We are currently in the process of patenting the compound, and I would now like to share my findings. I think this compound will reduce fabrication costs by 15%, as well as improve the quality of the premix.

The article, which will be a standard report of a laboratory procedure, will run about 2,500 words. Currently I am a Senior Chemist at Century Plastics and the author of fourteen articles in the leading plastics journals (including Plastics Technology, Winter 1972, Fall 1976, and Summer 1985).

If you think this article would interest your readers, please contact me at the above address or at (316) 437-1902.

Sincerely yours,

Anson Hawkins, Ph.D.
Senior Chemist

The following discussion describes each of these components and includes representative sections of a brief technical article from the journal *Stain Technology* as illustration.

The sequence of composition is not the same as the sequence of presentation in a technical article. Many writers like to start with the materials-and-methods section, because that is relatively simple to write, and then proceed to the end of the article. Then they go back and write the introduction, for that section has to lead smoothly into the following sections. Once the article itself is drafted, writers draft the abstract and, finally, the title.

> ## USE OF A FLUORESCENT STAIN FOR VISUALIZATION OF NUCLEAR MATERIAL IN LIVING OOCYTES AND EARLY EMBRYOS
>
> ELIZABETH S. CRITSER AND NEAL L. FIRST, *Department of Meat and Animal Science, University of Wisconsin, Madison, Wisconsin 53706*

FIGURE 16-3

Title and Author Listing of a Technical Article

TITLE AND AUTHOR
Technical articles tend to have long (and clumsy) titles, because of the need for specificity. Abstracting and indexing journals classify articles according to the key words in the title; therefore, you should be sure to include in your title those terms that will enable the readers to find your article.

Following the title comes your name (and the name of any co-authors). In addition, you should include your institutional affiliation, as shown in Figure 16-3.

Notice that the first words of the title—"Use of"—help the readers understand what kind of article it is: a report on a laboratory procedure. Notice, too, in this title that most of the nouns are modified by adjectives: "fluorescent stain," "nuclear material," "living oocytes," and "early embryos." The writers are extremely precise. This helps potential readers determine whether to study the article.

ABSTRACT
Technical articles contain abstracts for the same reasons that reports do: to enable readers to decide whether to read the whole discussion. Most journals today prefer informative abstracts to descriptive ones (see Chapter 11 for a discussion of these two basic types). Figure 16-4 shows the abstract for the article on stains.

> ABSTRACT. Hoechst dyes 33342 and 33258 were used to visualize pronuclei and nuclei of early preimplantation embryos. Murine one-cell zygotes exposed to dye stained rapidly over a range of concentrations (0, 0.02, 0.04, 0.1 or 0.2 μg/50 μl of media). Development to morula and blastocyst *in vitro* was reduced (39/70, 56%; $p < 0.05$) compared to controls (44/57, 77%) but not completely blocked. Porcine and bovine zygotes and embryos could also be stained but required incubation times up to 4 hr. Porcine embryos exposed to Hoechst 33342 had limited ($p < 0.01$) *in vitro* development (29/74, 39%) compared to unstained controls (49/64, 76%). Hoechst dyes stain embryos from different species but suitably adjusted incubation times are required. Limited preimplantation development *in vitro* may be expected following staining and exposure to ultraviolet light.

FIGURE 16-4

Abstract of a Technical Article

This is an excellent abstract. The first sentence provides an overview of the study. The next two sentences summarize the murine (mouse) experiments: the cells stained effectively, but growth of the cells was somewhat retarded by the process. The next two sentences summarize the pig and cattle experiments: these cells, too, stained well, but the growth of the pig cells was severely retarded. The last two sentences summarize the conclusions.

INTRODUCTION   The purpose of an introduction to a technical article is to define the problem that led to the investigation. To do so, the introduction generally must (1) fill in the background and (2) isolate the particular deficiencies that currently exist. Often, the introduction requires a brief review of the literature, that is, an annotated summary of the major research, either on the background of the problem or on the problem itself. This literature review provides a context for the discussion that is to follow. Perhaps more important, the literature review gives credence to the article by demonstrating that the author is thoroughly familiar with the pertinent research.

Some writers like to end the conclusion with a brief summary—usually just one sentence—of their main conclusion. This summary serves two functions: to enable some readers to skip parts of the discussion they don't need and to help other readers understand the discussion that follows:

Figure 16-5 shows the introduction to the article on stains.

The first paragraph of this introduction describes the background of the problem: genetic researchers need to be able to see the individual components of cells, but in some species (such as cattle and pigs) the cell structures make routine techniques ineffective. The second paragraph begins with references to the pertinent literature about the dye that is being studied in this experiment: it has been used in a number of applications. The remaining sentences—beginning with "We report here"—indicate the purpose and scope of the present study and explain the organization of the article.

MATERIALS   "Materials and methods" is a standard phrase used to describe the
AND METHODS   actual procedure the researcher followed. Basically, this section of the article has the same structure as a common recipe: these are the things you will need, and this is what you will do. Often, the materials and methods section is clearly divided into its two halves, and the steps of the procedure are numbered and expressed in the imperative mood (for example, "Mix the . . .").

The development of efficient nuclear and gene transfer techniques requires visualization of pronuclei in one-cell zygotes or nuclei in blastomeres of multicelled embryos. In some mammalian species (human, mouse) visualization can be accomplished by routine light microscopy and enhanced by phase or Nomarski optics. However, in most domestic species (*e.g.*, cattle, pigs) the dense cytoplasm of early zygotes precludes distinct visualization of nuclear material with routine techniques for light microscopy.

The fluorescent bisbenzimidazole Hoechst dyes, which have been widely used to study DNA synthesis (Latt and Stetten 1976), allow visualization of individual chromosomes in a variety of cell types (Hilwig and Gropp 1972, Bloom and Hsu 1975, Allen and Latt 1976, Arndt-Jovin and Jovin 1977). Hoechst 33342 and 33258 are nontoxic vital stains specific for DNA. We report here our investigation of the use of these dyes to stain chromosomes of mammalian preimplantation embryos, thus visualizing the nuclei of living cells. Our initial studies used murine embryos in which pronuclei are visible by light microscopy, thus allowing positive identification of the structures stained. Subsequent studies investigated the use of Hoechst dyes to visualize nuclei and pronuclei in embryos of domestic species in which the nuclei are normally obscured by the dense cytoplasm.

FIGURE 16-5

Introduction to a Technical Article

Figure 16-6 shows an excerpt from the materials-and-methods section of the article.

Notice that brand names of the equipment and materials used are specified so that readers can duplicate the experiment if they wish. Also note that the authors clearly explain the purpose of Experiment 1 before they describe how they performed it.

RESULTS   The results of the procedure are what happened. Keep in mind that the word *results* refers only to the observable or measurable effects of the methods; it does not attempt to explain or interpret those effects.

DISCUSSION   The discussion section answers two basic questions: (1) Why did the results happen? (2) What are the implications of the results? The two halves of the typical discussion section of an article are thus similar to the conclusion and recommendation sections of a report.

MATERIALS AND METHODS

Bisbenzimidazole dyes Hoechst 33258 and 33342 were obtained from Sigma Chemical Company. Stock solutions (10 μg/ml, 20 μg/ml and 60 μg/ml) were prepared in Dulbecco's phosphate buffered saline (pH 7.4) and stored at 4 C. Fresh reagents were prepared at least monthly, or more frequently if staining efficiency decreased.

The Nikon inverted microscope DIAPHOT-TMD equipped with epifluorescence attachment TMD-EF and ultraviolet filter was used. Excitation was induced at 365 nm and the emission is viewed through a 420 nm barrier filter. Exposure times were brief but sufficient to allow identification and recording of nuclear structures.

Experiment 1. The first experiment was designed to verify that Hoechst 33258 and 33342 would stain murine pronuclei. One-cell zygotes were obtained from the oviducts of mature female mice which had been bred the previous evening. Cumulus-enclosed one-cell zygotes were treated for 10 min with hyaluronidase (1 mg/ml) prepared in Dulbecco's phosphate buffered saline (DPBS) to remove cumulus cells and subsequently washed three times in DPBS supplemented with bovine serum albumin (BSA) (3 mg/ml). One-cell embryos were transferred to drops of Whitten's medium supplemented with 3 mg/ml BSA. Sixty-six embryos were randomly assigned to one of five concentrations of Hoechst 33342: 0, 0.02, 0.04, 0.1 or 0.2 μg dye/50 μl DPBS. Embryos were incubated for 15 min at 37 C in a high humidity incubator with 5% CO_2–95% air. After incubation, embryos were viewed by phase contrast and fluorescence microscopy to evaluate staining of pronuclei and polar bodies. In a subsequent study, 45 one-cell embryos were randomly assigned to one of four concentrations of Hoechst 33258: 0, 0.05, 0.1 or 0.2 μg dye/50 μl of medium. Embryos were incubated and evaluated as previously described.

FIGURE 16-6

Materials-and-Methods Section of a Technical Article

Figure 16-7 shows an excerpt from a combined results-and-discussion section. Such a combination is common in brief technical articles.

The paragraph explains that Experiment 1 verified one of the authors' hypotheses: that the dye being studied is effective in staining mouse cells. The paragraphs that follow discuss the results and conclusions of the other experiments.

REFERENCES   There are different styles for listing references, but each method has the same objective: to enable readers to locate the cited source,

---

RESULTS AND DISCUSSION

The Nikon inverted microscope DIAPHOT-TMD equipped with TMD-EF system permits visualization with low-level bright field and fluorescence optics simultaneously. The ability of Hoechst dyes 33342 and 33258 to stain chromatin of one-cell murine zygotes was verified in experiment 1. For both dyes, throughout the range of doses tested, distinct fluorescence was evident in the same locations that pronuclei were shown to be present by ordinary light microscopy. Both polar bodies did not always fluoresce.

FIGURE 16-7

Excerpt from a Results-and-Discussion Section of a Technical Article

should they wish to read it. Journals either specify that writers use a particular style sheet (which in turn defines a style for references) or provide their own guidelines in the discussion of editorial policies near the front of the issue. Following is the editorial statement on references from *Stain Technology*, the journal that printed the staining article.

References: Alphabetical listings (not numbered) at the end of the paper; text citations given by dates. Complete names, titles, and first and last pages are required; for example: Lyons, W. R. and Johnson, Ruth E. 1953. Embedding stained mammary glands in plastic. Stain Technol. 28: 201–204. References to books should include date, edition, publisher, and publisher's city.

Figure 16-8 shows the references section of the article.

This format highlights the names of the authors and the dates of publication of their works. Notice that journal titles are abbreviated. The italicized number that follows each title is the volume number; after the colon are the page numbers of the article. References to books are not common in technical articles, because most research is printed in journals.

The article on stains includes another section that has not been mentioned yet. An acknowledgment section (see Figure 16-9), which generally precedes the references, gives you the opportunity to express your appreciation to persons or institutions that have assisted you in your research.

The sample article that appears in this chapter demonstrates the usefulness of the basic article structure. You might not have understood the details of the discussion, but the logical flow of the argument should be clear: the authors were investigating whether two dyes would effectively indicate viability of hamster ova and

REFERENCES

Allen, J. W. and Latt, S. A. 1976. *In vivo* Brd U-33258 Hoechst analysis of DNA replication kinetics and sister chromatid exchange formation in mouse somatic and meiotic cells. Chromosoma (Berl) *58:* 325–340.

Arndt-Jovin, D. J. and Jovin, T. M. 1977. Analysis and sorting of living cells according to deoxyribonucleic acid content. J. Histochem. Cytochem. *25:* 585–589.

Ball, G. D., Leibfried, M. L., Lenz, R. W., Ax, R. L., Bavister, B. D. and First, N. L. 1983. Factors affecting successful *in vitro* fertilization of bovine follicular oocytes. Biol. Reprod. *28:* 717–725.

Bloom, S. E. and Hsu, T. C. 1975. Differential fluorescence of sister chromatids in chicken embryos exposed to 5-bromodeoxyuridine. Chromosoma (Berl.) *52:* 261–267.

Hilwig, I. and Gropp, A. 1972. Staining of constitutive heterochromatin in mammalian chromosomes with a new fluorochrome. Exp. Cell Res, *75:* 122–126.

Latt, S. A. and Stetten, G. 1976. Spectral studies on 33258 Hoechst and related bisbenzimidazole dyes useful for fluorescent detection of deoxyribonucleic acid synthesis. J. Histochem. Cytochem. *24:* 24–33.

FIGURE 16-8

References Section of a Technical Article

embryos. Both stains are effective; one is somewhat more effective than the other. The basic structure of the article makes it easier for the authors to write it—and easier for the readers to understand it.

Meat and Animal Science Paper No. 901. Supported by the College of Agricultural and Life Sciences, University of Wisconsin, Madison, and by a grant from W. R. Grace Company, NY.

1

FIGURE 16-9

Acknowledgment Section of a Technical Article

PREPARING AND MAILING THE MANUSCRIPT

Preparing and mailing the manuscript in a professional manner will not guarantee that an article is accepted for publication, of course. However, an unprofessional approach will almost certainly ensure that an article is *not* accepted.

TYPING THE ARTICLE   The article must be typed double-spaced. Whether or not *you* type it, keep in mind that editors are like any other readers: they don't like sloppy manuscripts. And they particularly don't like articles that violate the preferred stylistic guidelines. One reason is that it costs the journal money to revise a manuscript to make it conform stylistically. Perhaps a more important reason is that it is rude.

One new area of concern is word processing. Most journals address the question of whether they accept word-processed manuscripts at all and, if they do, what kind of printing they require.

The reason that some journals don't accept a word-processed article is that they want to discourage simultaneous submissions. That is, they don't want the writer to submit the same article to more than one journal at a time. Because most reviewers work for free, they understandably feel victimized when they spend long hours reviewing a manuscript only to have the author withdraw it and publish it in another journal. However, now that the letter-quality printers are commonplace, it is almost impossible to distinguish typed copy from word-processed copy.

Many journals will not accept dot-matrix printing because it is more difficult to read than letter-quality printing. Mathematical symbols can be especially difficult to decipher.

Increasingly common, however, are journals that accept electronic submissions. If the author can create the manuscript using the same system that the journal uses, the article can be transferred electronically through the phone lines. This saves money for both the author and the journal. The author saves postage and the journal saves retyping expenses. Most important, however, no errors are introduced into the article during retyping.

If the journal has stated its preferences, follow them to the letter. Most journals have clear guidelines about margins, pagination, spacing, number of copies, and so on. Look, for example, at the discussion from *Stain Technology*.

The second paragraph is a polite way of saying that the journal will correct minor stylistic problems, but the writer will have to retype the whole article if the deviations are great. What this statement does not say—but what any editor will admit—is that any manuscript that blatantly violates the journal's specifications will probably be rejected outright.

Preparation of Manuscript: Use letter-size paper, not larger than 8½ × 11 inches. Type *all* material in double-spaced lines (including footnotes, references, legends, *etc.*) on only one side of the paper. Allow a margin of at least 1 inch at the top, bottom, and *both sides* of the page. On the first 4 pages, place and mark items as follows: page 1, short title (running head), the name and address of the person to whom proof is to be sent, and key words to assist in indexing; page 2, full title only; page 3, names and addresses, including institutional affiliations, of author or authors as they are to appear in the journal; and page 4, abstract only. Use page 5 and succeeding pages exclusively for the textual material of the article, without footnotes or other material which is to be set in small type. The latter material consists of: references, footnotes, tables, and legends. Combine these on pages separate from the text proper as may be convenient; *e.g.*, a separate page for each legend or each footnote is not required. Legends should not be attached to photomicrographs or other illustrative material. Charts should have a typed legend, not one that is an integral part of the chart. Usually, a page for each table is advantageous for both author and typesetter. Submit manuscripts in duplicate.

By conforming with these specifications, authors can greatly facilitate both editing and typesetting. Manuscripts which do not conform may be either returned to the authors for retyping, or cut and rearranged when the deviations are minor.

If the journal has provided no typing guidelines, double-space everything and type on only one side of 8½- by 11-inch nonerasable bond paper. Use a good electric typewriter or a word processor with a printer the journal accepts. Also use a fresh ribbon. Send a photocopy with the original, and *make sure you keep a photocopy*. The percentage of manuscripts lost in the mail or in the journal's office is probably quite small, but when it happens to people who didn't keep a copy, they start to ask fundamental questions about life.

WRITING A COVER LETTER

An article should not arrive alone in an envelope. Enclose a cover letter. This letter should be brief, for your purpose is not to "sell" the article: any boasting would probably work against you. The abstract will tell your readers everything they want to know.

If the editor has responded to a query letter, you should work this fact into the first paragraph. If you didn't inquire first, simply state that you are enclosing an article that you would like the editor to consider. In the next paragraph, you might wish to define briefly the subject or approach of your article and give its title if you haven't done so already. Then conclude the letter politely. Figure 16-10 shows a sample cover letter.

MAILING THE ARTICLE

Follow the journal's mailing guidelines—about the number of copies to include, for example. Some journals request that you enclose a self-addressed, stamped envelope (SASE); others ask for loose postage. You might wish to state, in your cover letter, that you are enclosing an SASE or postage.

To keep the article from getting mangled, it's a good idea to enclose it between pieces of cardboard. Be especially careful if you are enclosing photographs or other artwork. If you are sending a total of about 100 pages or more, use a padded envelope, which can be obtained from the post office.

Some journals state that they acknowledge received manuscripts. If the journal you are writing to doesn't, enclose a self-addressed postcard and ask the editor to mail it when he or she receives your article.

RECEIVING A RESPONSE FROM THE JOURNAL

When your article arrives at a journal, it is read by the editor. If he or she believes that the subject matter is within the scope of the journal and that the article appears authoritative and is reasonably consistent with the specified stylistic guidelines, it will be sent out to two or three reviewers who are specialists in your subject.

FIGURE 16-10

Cover Letter for a
Technical Article

Century Plastics
ALBION, MISSISSIPPI 34213

April 3, 19--

Dr. William Framingham
Editor
Plastics Technology
1913 Vine Street
Fowler, MS 34013

Dear Dr. Framingham:

Thank you for replying promptly to my inquiry about the article on laminate foamboard premixes. As you requested, I am submitting the article for your consideration.

The article provides a step-by-step procedure for using Century Plastics' new compound in reducing blowing agent losses.

I look forward to learning your decision.

Sincerely yours,

Anson Hawkins

Anson Hawkins, Ph.D.
Senior Chemist
(316) 437-1902

WHAT THE
REVIEWERS
LOOK FOR

There is little mystery about what reviewers look for when they read an article. They want to determine, basically, if the subject of the article is worthy of publication in the journal, if the article is technically sound, and if it is reasonably well written. These three questions might be expanded as follows.

First, does the article fall within the scope of the journal? If it doesn't, the article will be rejected, regardless of its quality. If the article does fall within the journal's scope, the reviewers will want to see a convincing case that the subject is important enough to warrant publication.

Second, is the article sound? Is the problem defined clearly, and is the literature review comprehensive? Would the readers be able to reproduce your work on the basis of the methods and materials section? Are the results clearly expressed, and do they flow logically from the methods? Is the discussion clear and coherent? Have you drawn the proper conclusions from the results, and have you clearly suggested the implications of your research?

Third, is the article well written? Does it conform to the journal's stylistic requirements? Is it free of errors in grammar, punctuation, and spelling?

THE EDITOR'S RESPONSE

While the reviewers are studying your article, you wait—anywhere from three weeks to six months or a year. When you receive an acknowledgment from the editor, he or she will usually tell you how long the review process should take. If you haven't received an acknowledgment within three or four weeks, call or write the editor, asking if your manuscript arrived.

Finally, you will receive a response from the editor. Most likely, the response will take the form of an acceptance, a request to revise the article and resubmit it, or a rejection.

1. *Acceptance.* Very few articles—usually fewer than 5 percent of those submitted—are accepted outright. If you are one of the select few, congratulations. The acceptance letter might tell you the issue the article will appear in. Some journals edit the author's manuscripts; you might be asked to review the edited manuscript and to proofread the galleys—the first stage of typeset proof, which is later corrected and cut into pages. Other journals don't edit the manuscripts or ask the authors to proofread.

2. *Request to revise the article.* The majority of articles that eventually get published were revised according to the reviewers' recommendations. Sometimes reviewers call for major revisions: a total rethinking of the article. Sometimes the repairs are minor. Some journals call this response a provisional acceptance: they will accept your article if you make the specified revisions. Other journals do not commit themselves, but merely invite you to send it in again. If the reviewers' comments seem reasonable to you, revise the article and send it in. If they don't, you might want to consider trying another journal. Keep in mind, however, that another journal might require other revisions.

3. *Rejection.* The majority of articles submitted to most journals are simply rejected. A rejection means that the journal will not print it and does not want to see another version of it. All rejections hurt. The worst are the curt letters that offer no explanation: "We do not feel your article is appropriate for our journal." These rejections are frustrating, because you wonder if anyone read the article that you put so much time and effort into. Some

journals offer explanations, either by including the reviewers' comments or by excerpting them. If the criticism is valid, you don't feel too bad. If it isn't, you feel miserable.

Most experienced writers would advise that you develop a thick skin—but not too thick a skin. Many important articles are rejected by several journals before they are accepted. On the other hand, you should try to learn from the rejections. If your article has been rejected, study the reviewers' comments (if you received any) and try to react objectively. Consider it free advice, not a personal insult. If you haven't received any comments, take the article apart piece by piece, and see if you can discover any flaws. If you do, correct them and submit the article to another journal.

The first article is the hardest to write, and the one most likely to be rejected. After you publish two or three, you will develop a system that works well for you and learn to accept rejection as an inevitable part of publication. Knowing that most articles never get printed, you can take understandable pride in your successes.

WRITER'S
CHECKLIST

1. Does the article fall within the scope of the journal?

2. Is the subject of the article sufficiently important to justify publication in the journal?

3. Does the abstract effectively summarize the problem, methods, results, and conclusions?

4. Is the article technically sound?
   a. Is the problem clearly defined?
   b. Is the literature review comprehensive?
   c. Is the methods and materials section complete and logically arranged? Would it be sufficient to enable researchers to reproduce the technique?
   d. Are the results clear? Do they seem to follow logically from the methods?
   e. Is the discussion clear and coherent? Has the writer drawn the proper conclusions from the results and clearly suggested the implications of the research?

5. Is the article reasonably well written?
   a. Does it conform (at least basically) to the journal's stylistic requirements?
   b. Is the writing clear and unambiguous? Is it free of basic errors of grammar, punctuation, and spelling? In short, can it be understood easily?

EXERCISES

1. Choose a major journal in your field and write an essay analyzing the degree to which its contents correspond to its stated editorial policy. First, using the editorial policy statement, identify what the editor says the journal publishes. Consider such questions as range of subjects covered, type of article, writing style (paragraph and sentence length, technicality, etc.), and format (type of abstract, documentation style, etc.). Then, survey two or three articles in the issue. How closely do the articles reflect the editorial policy statement?

2. Write an essay evaluating an article in your field.

REFERENCE   Critser, E.S., and N.L. First. 1986. Use of a fluorescent stain for visualization of nuclear material in living oocytes and early embryos. *Stain Technology* 61, 1:1–5.

CHAPTER
SEVENTEEN

ORAL PRESENTATIONS

TYPES OF ORAL PRESENTATIONS

Although you will occasionally be called on for impromptu oral reports on some aspect of your work, most often you will have the opportunity to prepare what you have to say. This chapter covers two basic kinds of oral presentations that are prepared in advance: the extemporaneous presentation and the presentation read verbatim from manuscript. In an extemporaneous presentation, you might refer to notes or an outline, but you actually make up the sentences as you go along. Regardless of how much you have planned and rehearsed the presentation, you create it as you speak. An oral presentation read from manuscript is a talk that is completely written in advance; you simply read your text.

This chapter will not discuss the memorized oral presentation. Memorized presentations are not appropriate for most technical subjects, because of the difficulty involved in trying to remember technical data. In addition, few people other than trained actors are able to memorize oral presentations of more than a few minutes in length.

The extemporaneous presentation is the preferred type for all but the most formal occasions. At its best, it combines the virtues of clarity and spontaneity. If you have planned and rehearsed your presentation sufficiently, the information will be accurate, complete, logically arranged, and easy to follow. And if you can think well on your feet without grasping for words, the presentation will have a naturalness that will help your audience concentrate on what you are saying, just as if you were speaking to each person individually.

The presentation read from manuscript sacrifices naturalness for increased clarity. Most people sound stilted when they read from a text. But, obviously, having a complete text increases the chances that you will say precisely what you intend to say.

THE ROLE OF ORAL PRESENTATIONS

In certain respects, oral presentations are inefficient. For the speaker, preparing and rehearsing the presentation generally take more time than writing a report would. For the audience, physical conditions during the presentation—such as noises, poor lighting or acoustics, or an uncomfortable temperature—can interfere with the reception of information.

Yet oral presentations have one big advantage over written presentations: they permit a dialogue between the speaker and the audience. The listeners can offer alternative explanations and viewpoints, or simply ask questions that help the speaker clarify the information.

Oral presentations are therefore an increasingly popular means of technical communication. You can expect to give oral presentations to three different types of listeners: clients and customers, fellow employees in your organization, and fellow professionals in your field.

In representing your product or service to clients and customers, you will find oral presentations to be a valuable sales technique. Whether you are trying to interest them in a silicon chip or a bulldozer, you will have to make a presentation of the features of your product or service and the advantages it has over the competition's. Then, after the sale is made, you will likely give oral presentations to the client's employees, providing detailed operating instructions and outlining maintenance procedures. In both cases, extemporaneous presentations would probably be the more effective.

This same kind of informational oral presentation is necessary within an organization, too. If you are the resident expert on a certain piece of equipment or a procedure, you will be expected to instruct your fellow workers, both technical and nontechnical. After you return from an important conference or work on an out-of-town project, your supervisors will, whether or not you submit a written trip report, want an opportunity to ask questions: you must anticipate these questions and prepare answers. Another occasion for an oral presentation to your fellow workers comes up after you submit an informal proposal. If you have an idea for improving operations at your organization, you probably will first write a brief memo outlining your idea to your supervisor. If he or she agrees with you that the idea is promising, you might then be asked to present it orally to a small group of managers. This face-to-face meeting will give them an opportunity to question you in detail. In an hour or less, they can determine whether it is prudent to devote resources to studying the idea.

Oral presentations to fellow professionals generally are given at technical conferences and are often read from manuscript. You might speak on a research project with which you have been personally involved. Your subject might be a team project carried out at your organization. Or you may be invited to speak to professionals in other fields. If you are an economist, for example, you might be invited to speak to an association of realtors on the subject of interest rates.

The situations that demand oral presentations, therefore, are numerous, and they increase as you assume greater responsibility within an organization. For this reason, many managers and executives seek instruction from professionals in different kinds of oral presentations. You might not have had much experience in public speaking, and perhaps your few attempts have been difficult. But oral presentations on technical subjects are essentially the spoken form of technical writing. Just as there are a few writers who can produce effective reports without outlines or rough drafts, you might know a natural speaker who can talk to groups "off the cuff" effortlessly. For most persons, however, an oral presentation requires deliberate and careful preparation.

PREPARING TO GIVE AN ORAL PRESENTATION

Preparing to give an oral presentation requires four steps:

1. assessing the speaking situation
2. preparing an outline or note cards
3. preparing the graphic aids
4. rehearsing the presentation

ASSESSING THE
SPEAKING
SITUATION

The first step in preparing an oral presentation is to assess the speaking situation: audience and purpose are as important to oral presentations as they are to written reports.

Analyzing the audience requires careful thought. Who are the people who make up the audience? How much do they know about your subject? You must answer these questions in order to determine the level of technical vocabulary and concepts appropriate to the audience. Speaking over an audience's level of expertise is ineffective; oversimplifying and thereby appearing condescending might be just as bad. If the audience is relatively homogeneous—for example, a group of landscape architects or physical therapists—it is not difficult to choose the correct level of technicality. If, however, the group contains individuals of widely differing backgrounds, you should try in your presentation to accommodate everyone's needs by providing brief parenthetical definitions of those technical words and concepts that some of your listeners are not likely to know. Ask yourself the same kinds of questions you would ask about a group of readers: Why is the audience there? What do they want to accomplish as a result of having heard your presentation? Are they likely to be hostile, enthusiastic, or neutral toward your topic? A presentation on the virtues of free trade, for instance, will be received one way if the audience is a group of conservative economists and another way if it is a group of American steelworkers.

Equally important, analyze your purpose in making the presentation. Are you attempting to inform your audience, or to both inform and persuade them? If you are explaining how windmills can be used to generate power, you have one type of presentation. If you are explaining how *your* windmills are an economical means of generating power, you have another type of presentation. A successful oral presentation requires that you begin with a clear understanding of your purpose.

Finally, determine how the audience and purpose affect the strategy—the content and the form—of your presentation. You might have to emphasize some aspects of your subject or eliminate some altogether. You might have to arrange the elements of the presentation in a particular order to accommodate the audience's needs. You might have to replace technical vocabulary with common terms and add parenthetical definitions.

As you are planning, don't forget the time limitations established for your speech. At most professional conferences, the organizers clearly state a maximum time, such as 20 or 30 minutes, for each speaker. If you know that the question-and-answer period is part of your allotted time, plan accordingly. Even at an informal presentation, you probably will have to work within an unstated time limit—a limit that you must figure out from your understanding of the speaking situation. Claiming more than your share of an audience's time is rude and egotistical, and eventually the audience will simply stop paying attention. Most speakers need almost a minute to deliver a double-spaced page of text effectively.

PREPARING AN OUTLINE OR NOTE CARDS

After assessing the audience, purpose, and strategy of the presentation, you should prepare an outline or a set of note cards. In preparing these materials, keep in mind the difference between an oral presentation and a written one. An oral presentation is, in general, simpler than a written version of the same material, for the listeners cannot stop and reread a paragraph they do not understand. Statistics and equations, for example, generally should be kept to a minimum.

The oral presentation must be structured logically. The organizational patterns presented in Chapter 3 are as useful for structuring an oral presentation as they are for structuring a written document.

The introduction to the oral presentation is very important, for your opening sentences must gain and keep the audience's attention. An effective introduction might define the problem that led to the project, offer an interesting fact that the audience is unlikely to know, or present a brief quotation from an authoritative figure in the field or a famous person not generally associated with the field (for example, Abraham Lincoln on the value of American coal mines). All these techniques should lead into a clear statement of the purpose, scope, and organization of the presentation. If none of these techniques offers an appropriate introduction, a speaker can begin directly by defining the purpose, scope, and organization. A forecast of the major points is useful for long or complicated presentations. Don't be fancy. Use the words *scope* and *purpose*. And don't try to enliven the presentation by adding a joke. Humor is usually inappropriate in technical presentations.

The conclusion, too, is crucial in an oral presentation, for it summarizes the major points of the talk and clarifies their relationships with one another. Without a summary, many otherwise effective oral presentations would sound like a jumble of unrelated facts and theories.

With these points in mind, write an outline or note cards just as you would for a written report. Your own command of the facts will determine the degree of specificity necessary. Many people prefer a sentence outline (see Chapter 3) because of its specificity; others feel more comfortable with a topic outline, especially for

FIGURE 17-1

*Sentence/Topic
Outline for an Oral
Presentation*

OUTLINE

DESCRIBING, TO A GROUP OF CIVIL ENGINEERS, A NEW METHOD OF TREAT-
MENT AND DISPOSAL OF INDUSTRIAL WASTE.

I.   Introduction
      A.   The recent Resource Conservation Recovery Act places
           stringent restrictions on plant engineers.
      B.   With neutralization, precipitation, and filtration
           no longer available, plant engineers will have to
           turn to more sophisticated treatment and disposal
           techniques.

II.   The Principle Behind the New Techniques
      A.   Waste has to be converted into a cementitious load-
           supporting material with a low permeability coeffi-
           cient.
      B.   Conversion Dynamics, Inc., has devised a new tech-
           nique to accomplish this.
      C.   The technique is to combine pozzolan stabilization
           technology with the traditional treatment and dis-
           posal techniques.

III. The Applications of the New Technique
      A.   For new low-volume generators, there are two options.
           1.   Discussion of the San Diego plant.
           2.   Discussion of the Boston plant.
      B.   For existing low-volume generators, Conversion Dy-
           namics offers a range of portable disposal facili-
           ties.
           1.   Discussion of Montreal plant.
           2.   Discussion of Albany plant.
      C.   For new high-volume generators, Conversion Dynamics
           designs, constructs, and operates complete waste-
           disposal management facilities.
           1.   The Chicago plant now processes up to 1.5 million
                tons per year.
           2.   The Atlanta plant now processes up to 1.75 mil-
                lion tons per year.

note cards. Figure 17-1 shows a combined sentence/topic outline, and Figure 17-2 shows a topic outline. (The outline shown in Figure 17-2 could of course be written on note cards.) The speaker is a specialist in waste-treatment facilities. The audience is a group of civil engineers who are interested in gaining a general understanding of new developments in industrial waste disposal. The speaker's purpose is to provide this information and to suggest that his company is a leader in the field.

Notice the differences in both content and form between these two versions of the same outline. Whereas some speakers prefer to have full sentences before them, others find that full sentences in-

D. For existing high-volume generators, Conversion Dy-
    namics offers add-on facilities.
    1. The Roanoke plant already complies with the new
        RCRA requirements.
    2. The Houston plant will be in compliance within
        six months.

IV. Conclusion
     The Resource Conservation Recovery Act will necessi-
     tate substantial capital expenditures over the next
     decade.

terfere with the spontaneity of the presentation. (For presentations read from manuscript, of course, the outline is only a preliminary stage in the process of writing the text.)

PREPARING THE GRAPHIC AIDS   Graphic aids fulfill the same purpose in an oral presentation that
they do in a written one: they clarify or highlight important ideas
or facts. Statistical data, in particular, lend themselves to graphi-
cal presentation, as do representations of equipment or processes.
The same guidelines that apply to graphic aids in text apply here:
they should be clear and self-explanatory. The audience should
know immediately what each graphic aid is showing. In addition,

FIGURE 17-2

Topic Outline for an
Oral Presentation

```
                              OUTLINE

DESCRIBING, TO A GROUP OF CIVIL ENGINEERS,
A NEW METHOD OF TREATMENT AND DISPOSAL OF INDUSTRIAL WASTE.

      Introduction:
            -Implications of the RCRA

      Principle Behind New Technique
            -reduce permeability of waste
            -pozzolan stabilization technology and traditional
             techniques

      Applications of New Technique
            -for new low-volume generators (San Diego, Boston)
            -for existing low-volume generators (Montreal, Al-
             bany)
            -for new high-volume generators (Chicago, Atlanta)
            -for existing high-volume generators (Roanoke,
             Houston)

      Conclusion
```

the material conveyed should be simple. In making up a graphic aid for an oral presentation, be careful not to overload it with more information than the audience can absorb. Each graphic aid should illustrate a single idea. Remember that your listeners have not seen the graphic aid before, and that they, unlike readers, do not have the opportunity to linger over it.

In choosing a medium for the graphic aid, consider the room in which you will give the presentation. The people seated in the last row and near the sides of the room must be able to see each graphic aid clearly and easily (a flip chart, for instance, would be ineffective in an auditorium). If you make a transparency from a page of text, be sure to enlarge the picture or words; what is legible on a printed page is usually too small to see on a screen.

Although there are no firm guidelines on how many graphic aids to create, a good rule of thumb is to have a different graphic aid for every 30 seconds of the presentation. Changing from one to another helps you keep the presentation visually interesting, and it helps you signal transitions to your audience. It is far better to have a series of simple graphics than to have one complicated one that stays on the screen for 10 minutes.

After you have created your graphic aids, double check them for accuracy and correctness. Spelling errors are particularly embarrassing when the letters are six inches tall.

Following is a list of the basic media for graphic aids, with the major features cited.

1. *Slide projector:* projects previously prepared slides onto a screen.

ADVANTAGES:

It has a very professional appearance.

It is versatile—can handle photographs or artwork, color or black-and-white.

With a second projector, the pause between slides can be eliminated.

During the presentation, the speaker can easily advance and reverse the slides.

DISADVANTAGES:

Slides are expensive.

The room has to be kept relatively dark during the slide presentation.

2. *Overhead projector:* projects transparencies onto a screen.

ADVANTAGES:

Transparencies are inexpensive and easy to draw.

Speakers can draw transparencies "live."

Overlays can be created by placing one transparency over another.

Lights can remain on during the presentation.

The speaker can face the audience.

DISADVANTAGES:

It is not as professional-looking as slides.

Each transparency must be loaded separately by hand.

3. *Opaque projector:* projects a piece of paper onto a screen.

ADVANTAGES:

It can project single sheets or pages in a bound volume.

It requires no expense or advance preparation.

DISADVANTAGES:

The room has to be kept dark during the presentation.

It cannot magnify sufficiently for a large auditorium.

Each page must be loaded separately by hand.

The projector is noisy.

4. *Poster:* a graphic drawn on oak tag or other paper product.

ADVANTAGES:

It is inexpensive.

It requires no equipment.

Posters can be drawn or modified "live."

DISADVANTAGES:

It is ineffective in large rooms.

5. *Flip chart:* a series of posters, bound together at the top like a loose-leaf binder; generally placed on an easel.

ADVANTAGES:

It is relatively inexpensive.

It requires no equipment.

The speaker can easily flip back or forward.

Posters can be drawn or modified "live."

DISADVANTAGES:

It is ineffective in large rooms.

6. *Felt board:* a hard, flat surface covered with felt, onto which paper can be attached using doubled-over adhesive tape.

ADVANTAGES:

It is relatively inexpensive.

It is particularly effective if speaker wishes to rearrange the items on the board during the presentation.

It is versatile—can handle paper, photographs, cutouts.

DISADVANTAGES:

It has an informal appearance.

7. *Chalkboard:*

ADVANTAGES:

It is almost universally available.

The speaker has complete control—he or she can add, delete, or modify the graphic easily.

DISADVANTAGES:

Complicated or extensive graphics are difficult to create.

It is ineffective in large rooms.

It has a very informal appearance.

8. *Objects:* such as models or samples of material that can be held up or passed around through the audience.

ADVANTAGES:

They are very interesting for the audience.

They provide a very good look at the object.

DISADVANTAGES:

Audience members might not be listening while they are looking at the object.

The object might not survive intact.

9. *Handouts:* photocopies of written material given to each audience member.

ADVANTAGES:

Much material can be fit on paper.

Audience members can write on their copies and keep them.

DISADVANTAGE:

Audience members might read the handout rather than listen to the speaker.

A word of advice: before you design and create any graphic aids, make sure the room in which you will be giving the presentation has the equipment you need. Don't walk into the room carrying a stack of transparencies only to learn that there is no overhead projector. Even if you have arranged beforehand to have the necessary equipment delivered, check to make sure it is there; if possible, bring it with you.

REHEARSING
THE PRESEN-
TATION

Even the most gifted speakers have to rehearse their presentations. It is a good idea to set aside enough time to be able to rehearse your speech several times. For the first rehearsal of an extemporaneous presentation, simply sit at your desk with your outline or note cards before you. Don't worry about posture or voice projection. Just try to compose your presentation out loud. Many times speakers cut corners when they construct their outlines. In this first rehearsal, your goal is to see if the speech makes sense—if you can explain all the points you have listed and can forge effective transitions from point to point. If you have any trouble, stop and try to figure out the problem. If you need more information, get it. If you need a better transition, create one. You might well find that you have to revise your outline or notes. This is very common and no cause for alarm. Pick up where you left off and continue through the presentation, stopping again where necessary to revise the outline. When you have finished your first rehearsal, put the outline away and do something else.

Come back again when you are rested. Try the presentation once more. This time, it should flow more easily. Make any necessary revisions in the outline or notes. Once you have complete control over the structure and flow, check to see if you are within the time limits for the presentation.

After a satisfactory rehearsal, try the presentation again, under more realistic circumstances. If it is possible to rehearse in front of people, take advantage of the opportunity. The listeners might be able to offer constructive advice about parts of the presentation they didn't understand or about your speaking style. If people aren't available, use a tape recorder and then evaluate your own delivery. If you can visit the site of the presentation to get the feel of the room and rehearse there, you will find giving the actual speech a little easier.

Once you have rehearsed your presentation a few times and are satisfied with it, stop. Don't attempt to memorize it—if you do, you will surely panic the first time you forget the next phrase. During the presentation, you must be thinking of your subject, not about the words you used during the rehearsals.

Rehearse a written-out presentation in front of people, too, if possible, or use a tape recorder.

GIVING THE ORAL PRESENTATION

Most professional actors freely admit to being nervous before a performance, and so it is no wonder that most speakers are nervous. You might well fear that you will forget everything you have to say or that nobody will be able to hear you. These fears are common. But signs of nervousness are much less apparent to the audience than to the speaker. And after a few minutes most speakers relax and concentrate effectively on the subject.

All this sage advice, however, is unlikely to make you feel much better as you wait to give your presentation. When the moment comes, don't jump up to the lectern and start speaking quickly. Walk up slowly and arrange your text, outline, or note cards before you. If water is available, take a sip. Look out at the audience for a few seconds before you begin. It is polite to begin formal presentations with "Good morning" (or "Good afternoon," "Good evening") and to refer to the officers and dignitaries present. If your name has not been mentioned by the introductory speaker, identify yourself. In less formal contexts, just begin your presentation.

Your goal, of course, is to have the audience listen to you and have confidence in what you are saying. Try to project the same image that you would in a job interview: restrained self-confidence. Show your listeners that you are interested in your topic and that you know what you are talking about. As you give the presentation, this sense of control is conveyed chiefly through your voice and your body.

YOUR VOICE   Inexperienced speakers encounter problems with five aspects of vocalizing: volume, speed, pitch, clarity of pronunciation, and verbal fillers.

1. *Volume.* The acoustics of rooms vary greatly, so it is impossible to be sure how well your voice will carry in a room until you have heard someone speaking there. In some well-constructed auditoriums, speakers can use a conversational tone. Other rooms require a real effort at voice projection. An annoying echo plagues some rooms. These special circumstances aside, in general more people speak too softly than too loudly. After your first few sentences, ask if the people in the back of the room can hear you. When people speak into microphones, they tend to bend down toward the microphone and end up speaking too loudly. Glance at your audience to see if you are having volume problems.

2. *Speed.* Nervousness makes people speak more quickly. Even if you think you're speaking at the right rate, you might be going a little too fast for some of your audience. Remember, you know where you're going—you know the material well. Your listeners, however, are trying to understand new information. For particularly difficult points, slow down your delivery, for emphasis. After finishing a major point, pause before beginning the next point.

3. *Pitch.* In an effort to achieve voice control, many speakers end up flattening their pitch. The resulting monotone is boring—

and, for some listeners, actually distracting. Try to let the pitch of your voice go up or down as it would in a normal conversation. In fact, experienced speakers often exaggerate pitch variations.

4. *Clarity of pronunciation.* The nervousness that goes along with an oral presentation tends to accentuate sloppy pronunciation. If you want to say the word "environment," don't say "envirament." Say "nuclear," not "nucular." Don't drop final g's. Say "contaminating," not "contaminatin'." A related pronunciation problem concerns technical words and phrases—especially the important ones. When a speaker uses a phrase over and over, it tends to get clipped and becomes difficult to understand. Unless you articulate carefully, "Scanlon Plan" will end up as "Scanluhplah," or "total dissolved solids" will be heard as "toe-dizahved sahlds."

5. *Verbal fillers.* Avoid such meaningless fillers as "you know," "okay," "right," "uh," and "um." These phrases do not disguise the fact that you aren't saying anything; they call attention to it. A thoughtful pause is better than a nervous and annoying verbal tic.

YOUR BODY

The audience will be listening to what you say, but they will also be looking at you. Effective speakers know how to take advantage of this fact by using their bodies to help the listeners follow the presentation.

Perhaps most important in this regard are your eyes. People will be looking at you as you speak, and it is only polite to look back at them. This is called "eye contact." For small groups, look at each listener randomly; for larger groups, be sure to look at each segment of the audience during your speech. Do not stare at your notes, at the floor, at the ceiling, or out the window. Even if you are reading your presentation, you should be well enough rehearsed to allow frequent eye contact. Eye contact is not only polite; it also gives you a valuable indication of how the audience is receiving your presentation. You will be able to tell, for instance, if the listeners in the back are having trouble hearing you.

Your arms and hands also are important. Watch practiced speakers to see how they use their arms and hands to signal pauses and to emphasize important points. Gestures are particularly useful when you are referring to graphic aids. Point to direct the audience's attention to an area of the graphic.

Watch out for mannerisms—those physical gestures that serve no useful purpose. Don't play with your jewelry or the coins in your pocket. Don't tug at your beard or fix your hair. These nervous gestures can quickly distract an audience from what you are saying. Like verbal mannerisms, physical mannerisms are often unconscious. Constructive criticism from friends can help you to pinpoint them.

AFTER THE PRESENTATION

On all but the most formal occasions, an oral presentation is followed by a question-and-answer period.

When you invite questions, don't abruptly say, "Any questions?" This phrasing suggests that you don't really want any questions. Instead, say something like this: "If you have any questions, I'd be happy to try to answer them now." If asked politely, people will be much more likely to ask questions; therefore, you will be much more likely to communicate your information effectively.

In fielding a question, first make sure that everyone in the audience has heard it. If there is no moderator to do this job, you should ask if people have heard the question. If they haven't, repeat or paraphrase it yourself. Sometimes this can be done as an introduction to your response: "Your question about the efficiency of these three techniques . . ."

If you hear the question but don't understand it, ask for a clarification. After responding, ask if you have understood the question correctly.

If you understand the question but don't know the answer, tell the truth. Only novices believe that they ought to know all the answers. If you have some ideas about how to find out the answer— by checking a certain reference text, for example—share them. If the question is obviously important to the person who asked it, you might offer to meet with him or her after the question-and-answer period to discuss ways to give a more complete response, such as through the mail.

If you are unfortunate enough to have a belligerent member of the audience who is not content with your response and insists on restating his or her original point, a useful technique is to offer politely to discuss the matter further after the session. This will prevent the person from boring or annoying the rest of the audience.

If it is appropriate to stay after the session to talk individually with members of the audience, offer to do so. Don't forget to thank them for their courtesy in listening to you.

SPEAKER'S CHECKLIST

This checklist covers the steps involved in preparing to give an oral presentation.

1. Have you assessed the speaking situation—the audience and purpose of the presentation?

2. Have you determined the content of your presentation?

3. Have you shaped the content into a form appropriate to your audience and purpose?

4. Have you prepared an outline or note cards?

5. Have you prepared graphic aids that are
   a. Clear and easy to understand?
   b. Easy to see?

6. Have you made sure that the presentation room will have the necessary equipment for the graphic aids?

7. Have you rehearsed the presentation so that it flows smoothly?

8. Have you checked that the presentation will be the right length?

EXERCISES

1. Prepare a five-minute oral version of your proposal for a final report topic.

2. After the report is written, prepare a five-minute oral presentation based on your findings and conclusions.

3. Write a memo to your instructor in which you analyze a recent oral presentation of a guest speaker at your college or a politician on television.

4. Prepare a five-minute extemporaneous presentation on a term or concept in your major field of study. Address the presentation to an audience that is unfamiliar with that term or concept.

5. Prepare a five-minute extemporaneous presentation explaining how to use a piece of equipment or how to carry out a procedure in your major field of study. Address the presentation to an audience that is unfamiliar with that piece of equipment or procedure.

CHAPTER EIGHTEEN

CORRESPONDENCE

The letter is the basic means of communication between two organizations; some eighty million business letters are written each working day. Although the use of the telephone is constantly increasing, letters remain the basic communication link, because they provide documentary records. Often, phone conversations and transactions are immediately written up in the form of a letter, to become a part of the records of both organizations. The increasing use of electronic mail will not change this fact. Therefore, even as a new employee, you can expect to write letters regularly. And as you progress to positions of greater responsibility, you will write even more letters, for you will be representing your organization more often.

To a greater extent than any other kind of technical writing, letters represent the dual nature of the working world. On the one hand, a letter must be every bit as accurate and responsible as a legal contract. On the other hand, a letter is an individual communication between two people. Your reader will form an opinion of your organization on the basis of the impression that you make in your letters. Regarding your reader as an individual—while at the same time representing your organization effectively—is the challenge of writing good business letters.

THE "YOU ATTITUDE"

Like any other type of technical writing, the business letter should be clear, concise, comprehensive, accessible, correct, and accurate. It must convey information in a logical order. It should not contain small talk: the first paragraph should get directly to the point without wasting words. And to enable the reader to locate the information in the letter quickly and easily, it must make careful use of topic sentences at the start of the paragraphs. Often, letters use headings and indention to make information more accessible, just as reports do. In fact, some writers use the term *letter report* to describe a technical letter of more than, say, two or three pages. In substance, it is a report; in form, it is a letter, containing all the letter's traditional elements.

Moreover, because it is a communication from one person to another, a letter must also convey a courteous, positive tone. The key to this tone is sometimes called the "you attitude." This term means looking at the situation from your reader's point of view and adjusting the content, structure, and tone to meet that person's needs. The "you attitude" is largely common sense. If, for example, you are writing to a supplier who has failed to deliver some merchandise on the agreed-upon date, the "you attitude" dictates that you not discuss problems you are having with other suppliers—those problems don't concern your reader. Rather, you should concentrate on explaining clearly and politely to your reader that he or she has violated your agreement and that not having the merchandise is costing you money. Then you should propose ways to expedite the shipment of the merchandise.

Looking at things from the other person's point of view would be simple if both parties always saw things the same way. They don't, of course. Sometimes the context of the letter is a dispute. Nevertheless, good letter writers always maintain a polite tone. Civilized behavior is good business.

Following are examples of thoughtless sentences, each followed by an improved version that exhibits the "you attitude."

EGOTISTICAL

Only our award-winning research and development department could have devised this revolutionary new sump pump.

BETTER

Our new sump pump features significant innovations.

BLUNT

You wrote to the wrong department. We don't handle complaints.

BETTER

Your letter has been forwarded to the Customer Service Division.

ACCUSING

You must have dropped the engine. The housing is badly cracked.

BETTER

The badly cracked housing suggests that your engine must have fallen onto a hard surface from some height.

SARCASTIC

You'll need two months to deliver these parts? Who do you think you are, the Post Office?

BETTER

A two-month delay for the delivery of the parts is unacceptable.

BELLIGERENT

I'm sure you have a boss, and I doubt if he'd like to hear about how you've mishandled our account.

BETTER

I would prefer to settle our account with you rather than having to bring it to your supervisor's attention.

CONDESCENDING

Haven't you ever dealt with a major corporation before? A 60-day payment period happens to be standard.

BETTER

We had assumed that you honored the standard 60-day payment period.

OVERSTATED

Your air-filter bags are awful. They're all torn. We want our money back.

BETTER

Nineteen of the 100 air-filter bags we purchased are torn. We would therefore like you to refund the purchase price of the 19 bags: $190.00

After you have drafted a letter, look back through it. Put yourself in your reader's place. How would you react if you received it? A calm, respectful, polite tone always makes the best impression and therefore increases your chances of achieving your goal.

AVOIDING LETTER CLICHÉS

Related to the "you attitude" is the issue of letter clichés. Over the decades, a whole set of words and phrases has come to be associated with letters, phrases such as "as per your request." For some reason, many people think that these phrases are required. They're not. They make the letter sound stilted and insincere. If you would feel awkward or uncomfortable saying these clichés to a friend, avoid them in your letters.

Following is a list of some of the common clichés and their more natural equivalents.

LETTER CLICHÉS	NATURAL EQUIVALENTS
attached please find	attached is
at your earliest convenience	soon
cognizant of	aware that
enclosed please find	enclosed is
endeavor (verb)	try
herewith ("We herewith submit . . .)	(None. "Herewith" doesn't say anything. Skip it.)
hereinabove	previously, already
in receipt of ("We are in receipt of . . .")	"We have received . . ."
permit me to say	(None. Permission granted. Just say it.)
pursuant to our agreement	as we agreed
referring to your ("Referring to your letter of March 19, the shipment of pianos . . .")	"In reference to your letter of March 19, the . . ." or subordinate the reference at the end of your sentence.
same (as a pronoun: "Payment for same is requested . . .")	(Use the noun instead: "Payment for the merchandise is requested . . .")
wish to advise ("We wish to advise that . . .")	(The phrase doesn't say anything. Just say what you want to say.)
the writer ("The writer believes that . . .")	"I believe . . ."

Following are two versions of the same letter: one written in letter clichés, the other in plain language.

```
Dear Mr. Smith:

Referring to your letter regarding the problem encountered with your
new Eskimo Snowmobile. Our Customer Service Department has just ten-
dered its report.
```

It is their conclusion that the malfunction is caused by water being present in the fuel line. It is our unalterable conclusion that you must have purchased some bad gasoline. We trust you are cognizant of the fact that while we guarantee our snowmobiles for a period of not less than one year against defects in workmanship and materials, responsibility cannot be assumed for inadequate care. We wish to advise, for the reason mentioned hereinabove, that we cannot grant your request to repair the snowmobile free of charge.

Permit me to say, however, that the writer would be pleased to see that the fuel line is flushed at cost, $30. Your Eskimo would then give you many years of trouble-free service.

Enclosed please find an authorization card. Should we receive it, we shall endeavor to perform the above-mentioned repair and deliver your snowmobile forthwith.

Sincerely yours,

Dear Mr. Smith:

Thank you for writing to us about the problem with your new Eskimo Snowmobile.

Our Customer Service Department has found water in the fuel line. Apparently some of the gasoline was bad. While we guarantee our snowmobiles for one year against defects in workmanship and materials, we cannot assume responsibility for problems caused by bad gasoline. We cannot, therefore, grant your request to repair the snowmobile free of charge.

However, no serious harm was done to the snowmobile. We would be happy to flush the fuel line at cost, $30. Your Eskimo would then give you many years of trouble-free service.

If you will authorize us to do this work, we will have your snowmobile back to you within four working days. Just fill out the enclosed authorization card and drop it in the mail.

Sincerely yours,

The second version of this letter not only avoids the clichés but also shows a much better grasp of the "you attitude." Rather than building the letter around the violation of the warranty, as the first writer does, the second writer presents the message as good news: the snowmobile is not ruined, and it can be returned in less than a week for a low cost.

THE ELEMENTS OF THE LETTER

Almost every business letter has a heading, inside address, salutation, body, complimentary close, signature, and reference initials. In addition, some letters contain one or more of the following notations: attention, subject, enclosure, and copy.

For short, simple letters, it is possible to compose the elements of the letter in the sequence in which they will appear. For more complex letters, however, it is best to use the strategy discussed with all the other technical writing applications: start with the body and continue to the end. Then go back and add the preliminary elements, ending with the first paragraph.

The elements of the letter are discussed in the following paragraphs in the order in which they would ordinarily appear in a letter. Six common types of letters will be discussed in detail later in this chapter.

HEADING   The typical organization has its own stationery, with its name, address, and phone number—the letterhead—printed at the top. The letterhead and the date on which the letter will be sent (typed two lines below the letterhead) make up the heading. When typing on blank paper, use your address (without your name) as the letterhead. Note that only the first page of any letter is typed on letterhead stationery. Type the second and all subsequent pages on blank paper, with the name of the recipient, the page number, and the date in the upper left-hand corner. For example:

```
Mr. John Cummings
Page 2
July 3, 19--
```

Do not number the first page of any letter.

INSIDE ADDRESS   The inside address is your reader's name, position, organization, and business address. If your reader has a courtesy title, such as *Dr.*, *Professor*, or—for public officials—*Honorable*, use it. If not, use *Mr.* or *Ms.* (unless you know the reader prefers *Mrs.* or *Miss*). If your reader's position can be fit conveniently on the same line as his or her name, add it after a comma; otherwise, place it on the line below. Spell the name of the organization the way it does: for example, International Business Machines calls itself IBM. Include the complete mailing address: the street, city, state and zip code.

ATTENTION LINE   Sometimes you will be unable or unwilling to address the letter to a particular person. If you don't know (and cannot easily find out) the person's first name or don't know the person's name at all, use the attention line:

```
Attention: Technical Director
```

The attention line is also useful if you want to make sure that the organization you are writing to responds to your letter even if the person you would ordinarily write to is unavailable. In this case, the first line of the inside address contains the name of the organization or of one of its divisions:

```
Operations Department
Haverford Electronics
117 County Line Road
Haverford, MA 01765

Attention: Charles Fulbright, Director
```

SUBJECT LINE
The subject line contains either a project number (for example, "Subject: Project 31402") or a brief phrase defining the subject of the letter (for example, "Subject: Price Quotation for the R13 Submersible Pump").

```
Operations Department
Haverford Electronics
117 County Line Road
Haverford, MA 01765

Attention: Charles Fulbright, Director

Subject: Purchase Order #41763
```

SALUTATION
Assuming there is no attention line or subject line, the salutation is placed two lines below the inside address. The traditional salutation is *Dear* followed by the reader's courtesy title and last name. Use a colon after the name, not a comma. If you are fairly well acquainted with your reader, use *Dear* followed by the first name. When you do not know the name of the person to whom you are writing, use a general salutation:

```
Dear Technical Director:

Dear Sir or Madam:
```

When you are addressing the letter to a group of people, use one of the following salutations:

```
Ladies and Gentlemen:

Gentlemen: (if all the readers are male)

Ladies: (if all the readers are female)
```

Or you can tailor the salutation to your readers:

```
Dear Members of the Restoration Committee:
Dear Members of Theta Chi Fraternity:
```

This same strategy is useful for sales letters without individual inside addresses:

```
Dear Homeowner:
Dear Customer:
```

BODY   The body is the substance of the letter. Although in some cases it might be only a few words, generally it will be three or more paragraphs. The first paragraph introduces the subject of the letter, the second elaborates the message, and the third concludes it.

COMPLIMEN-TARY CLOSE   After the body of the letter, include one of the traditional closing expressions: *Sincerely, Sincerely yours, Yours sincerely, Yours very truly, Very truly yours.* Notice that only the first word in the complimentary close is capitalized, and that all such expressions are followed by a comma. Today, all the phrases have lost whatever particular meanings and connotations they once possessed. They can be used interchangeably.

SIGNATURE   Type your full name on the fourth line below the complimentary close. Sign the letter, in ink, above the typewritten name. Most organizations prefer that you add, beneath your typed name, your position. For example:

```
Very truly yours,
```

Chester Hall

```
Chester Hall
Personnel Manager
```

REFERENCE LINE   If someone else types your letters, the reference line identifies, usually by initials, both you and the typist. It appears a few spaces below the signature line, along the left margin. Generally, the writer's initials—which always come first—are capitalized, and the typist's initials are lowercase. For example, if Marjorie Connor wrote a letter that Alice Wagner typed, the standard reference notation would be MC/aw.

ENCLOSURE LINE If the envelope contains any documents other than the letter itself, identify the number of enclosures:

FOR ONE ENCLOSURE

```
Enclosure
```

OR

```
Enclosure (1)
```

FOR MORE THAN ONE ENCLOSURE

```
Enclosures (2)
Enclosures (3)
```

In determining the number of enclosures, count only the separate items, not the number of pages. A three-page memo and a ten-page report constitute only two enclosures. Some writers like to identify the enclosures by name:

```
Enclosure: 1986 Placement Bulletin

Enclosures (2): "This Year at Ammex"

          1986 Annual Report
```

COPY LINE If you want the reader of your letter to know that other people are receiving a copy of it, use the symbol *cc* (for "carbon copy"—even though photocopies have replaced carbon copies in most organizations) followed by the names of the other recipients (listed either alphabetically or according to organizational rank). If you do not want your reader to know about other copies, type *bcc* ("blind carbon copy") on the copies only—not on the original.

THE FORMAT OF THE LETTER

There are three popular formats used for business letters: modified block, modified block with paragraph indentions, and full block. Figures 18-1, 18-2, and 18-3 show diagrams of letters written in these three formats.

TYPES OF LETTERS

No two communication situations in the technical world are identical. There are literally dozens of kinds of letters meant for specific occasions. This chapter could not possibly cover all the types of letters that are sent routinely. It focuses, instead, on the six

FIGURE 18-1

*Modified Block
Format*

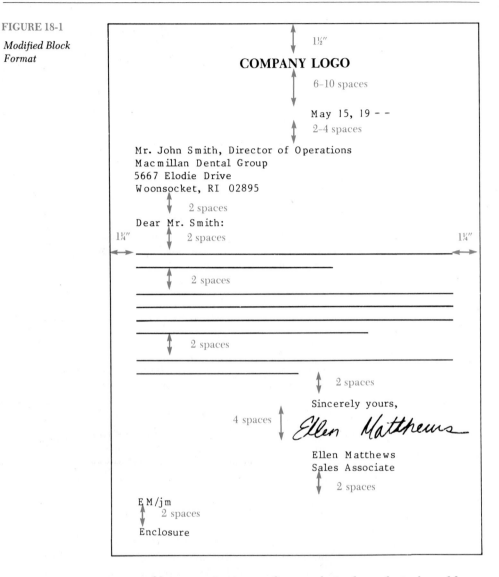

types of letters written most frequently in the technical world: order, inquiry, response to inquiry, sales, claim, and adjustment. The transmittal letter is discussed in Chapter 11. The job-application letter is discussed in Chapter 19. For a more detailed discussion of business letters, consult one of the several full-length books on the subject.

**THE ORDER
LETTER**   Perhaps the most basic form of business correspondence is the order letter, written to a manufacturer, wholesaler, or retailer. When writing an order letter, be sure to include all the informa-

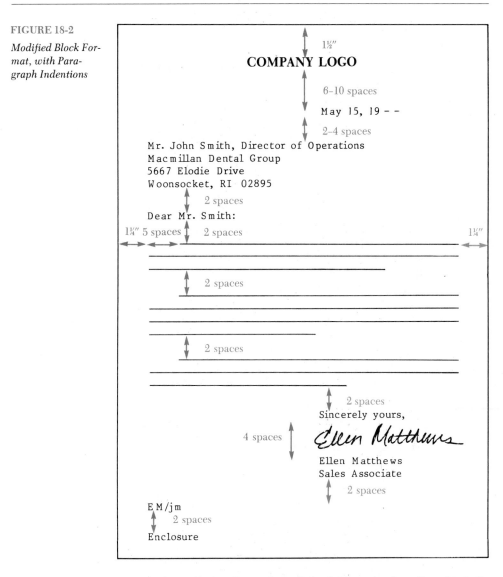

COMPANY LOGO

1½″

6–10 spaces

May 15, 19 – –

2–4 spaces

Mr. John Smith, Director of Operations
Macmillan Dental Group
5667 Elodie Drive
Woonsocket, RI 02895

2 spaces

Dear Mr. Smith:

1¼″ 5 spaces   2 spaces

1¼″

2 spaces

2 spaces

2 spaces

Sincerely yours,

4 spaces   *Ellen Matthews*

Ellen Matthews
Sales Associate

2 spaces

EM/jm

2 spaces

Enclosure

tion your reader will need to identify the merchandise. Include the quantity, model number, dimensions, capacity, material, price, and any other pertinent details. Also specify the terms of payment (if other than payment in full upon receipt of the merchandise) and method of delivery. A typical order letter is shown in Figure 18-4. (Notice that the writer of this order letter uses an informal table to describe the parts he wishes to purchase.)

Many organizations have preprinted forms, called purchase orders, for ordering products or services. A purchase order calls for the same information that appears in an order letter.

FIGURE 18-3

Full Block Format

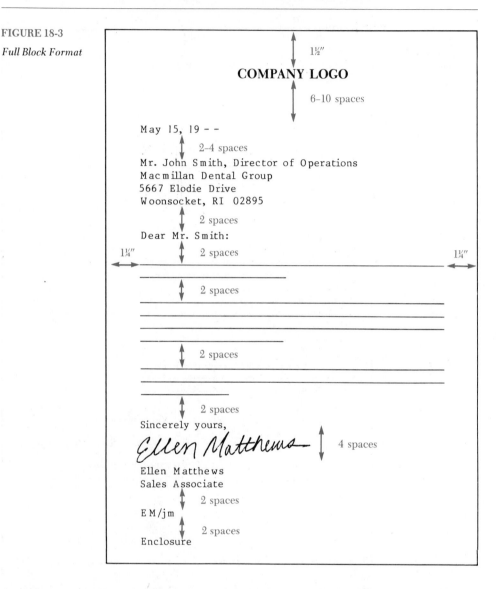

THE INQUIRY LETTER

Your purpose in writing an inquiry letter is to obtain information from your reader. The difficulty of writing an inquiry letter is determined by whether the reader is expecting your letter.

If the reader is expecting the letter, your task is easy. For example, if a company that makes institutional furniture has advertised that it will send its 48-page, full-color brochure to prospective clients, you need write merely a one-sentence letter: "Would you please send me the brochure advertised in *Higher Education To-*

FIGURE 18-4

Order Letter

WAGNER AIRCRAFT

116 North Miller Road
Akron, OH 44313

September 4, 19--

Franklin Aerospace Parts
623 Manufacturer's Blvd.
Bethpage, NY 11741

Attention: Mr. Frank DeFazio

Gentlemen:

Would you please send us the following parts by parcel post. All
page numbers refer to your 19-- catalog.

Quantity	Model No.	Catalog Page	Description	Price
2	36113-NP	42	Seal fins	$ 34.95
1	03112-Bx	12	Turbine bearing support	19.75
5	90135-QN	102	Turbine disc	47.50
1	63152-Bx	75	Turbine bearing housing	16.15
			Total Price:	$118.35

Yours very truly,

Christopher O'Hanlon

Christopher O'Hanlon
Purchasing Agent

day, May 13, 19--?" The manufacturer knows why you're writing
and, naturally, wants to receive letters such as yours, so no explanation is necessary. Similarly, a technical question, or set of questions, about any product or service for sale can be asked quickly.
For instance, an inquiry letter might begin, "We are considering
purchasing your new X-15 self-correcting typewriters for an office
staff of 35 and would like some further information. Would you
please answer the following questions?" The detail about the size
of the potential order is not necessary, but it does make the point
that the inquiry is serious and that the sale could be substantial.
An inquiry letter of this kind will get a prompt and gracious reply.

If your reader is not expecting your letter, your task is more difficult. At times, you will want information from someone who will not directly benefit from supplying it to you. You must ask your reader simply to do you a favor. This situation requires careful, persuasive writing, for you must make your reader *want* to respond despite the absence of the profit motive.

In the first paragraph of this kind of letter, state why you decided to write to *this* person or organization, rather than any other organization that could supply the same or similar information. Subtle flattery is useful at this point—for example, "I was hoping that as the leader in solid-state electronics, you might be able to furnish some information about. . . ." Then explain why you want the information. Obviously, a company will not furnish information to a competitor. You have to show that your interests are not commercial—for instance, "I will be using this information in a senior project in electrical engineering at Illinois State University. I am trying to devise a . . ." If you need the information by a certain date, this might be a good place to mention it: "The project is to be completed by April 15, 19––."

The bulk of the inquiry letter should be composed of a numbered list of the specific questions you want answered. Readers are understandably annoyed by thoughtless requests to send "everything you have" on a topic. They much prefer a set of carefully thought-out technical questions that show that the writer has already done substantial research. "Is your Model 311 compatible with Norwood's Model B12?" is obviously much easier to respond to than "Would you please tell me about your Model 311?" If your questions can be answered in a small space, it is a good idea to leave room for your reader's reply after each question or in the margin.

Because you are asking someone to do something for you, it is natural to offer to do something in return. In many cases, all you can offer are the results of your research. If possible, explicitly state that you would be happy to send your reader a copy of the report you are working on. And finally, express your appreciation. Don't say, "Thank you for sending me this information." Such a statement is presumptuous, because it assumes that the reader is both willing and able to meet your request. A statement such as the following is more effective: "I would greatly appreciate any help you could give me in answering these questions." Finally, if the answers will be brief, enclose a stamped self-addressed envelope for your reader's reply.

You should of course write a brief thank-you note to someone

FIGURE 18-5

Inquiry Letter

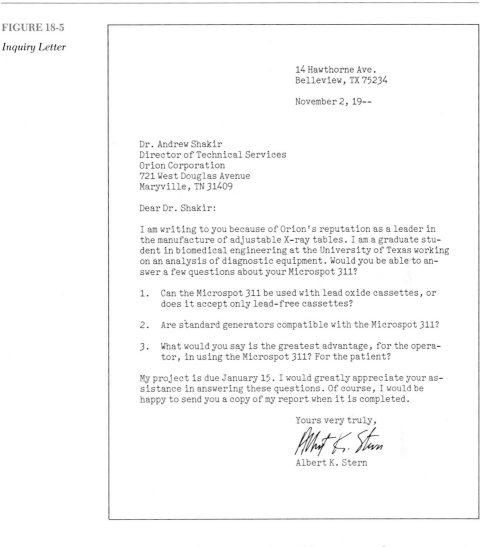

14 Hawthorne Ave.
Belleview, TX 75234

November 2, 19--

Dr. Andrew Shakir
Director of Technical Services
Orion Corporation
721 West Douglas Avenue
Maryville, TN 31409

Dear Dr. Shakir:

I am writing to you because of Orion's reputation as a leader in the manufacture of adjustable X-ray tables. I am a graduate student in biomedical engineering at the University of Texas working on an analysis of diagnostic equipment. Would you be able to answer a few questions about your Microspot 311?

1. Can the Microspot 311 be used with lead oxide cassettes, or does it accept only lead-free cassettes?

2. Are standard generators compatible with the Microspot 311?

3. What would you say is the greatest advantage, for the operator, in using the Microspot 311? For the patient?

My project is due January 15. I would greatly appreciate your assistance in answering these questions. Of course, I would be happy to send you a copy of my report when it is completed.

Yours very truly,

Albert K. Stern

who has taken the time and trouble to respond to your inquiry letter.

Figure 18-5 provides an example of a letter of inquiry.

THE RESPONSE
TO AN
INQUIRY

If you ever receive an inquiry letter, keep the following suggestions in mind. If you can provide the information the writer asks for, do so graciously. If the questions are numbered, number your responses correspondingly. If you cannot answer the questions, either because you don't know the answers or because you cannot divulge proprietary information, explain this to the writer and of-

FIGURE 18-6

*Response to an
Inquiry*

ORION

721 WEST DOUGLAS AVE.   (615) 619-8132
MARYVILLE TN 31409   TECHNICAL SERVICES

November 7, 19--

Mr. Albert K. Stern
14 Hawthorne Ave.
Belleview, TX 75234

Dear Mr. Stern:

I would be pleased to answer your questions about the Microspot
311. We think it is the best unit of its type on the market today.

1.   The 311 can handle lead oxide or lead-free cassettes.

2.   At the moment, the 311 is fully compatible only with our Dura-
matic generator. However, special wiring kits are available
to make the 311 compatible with our earlier generator
models--the Olympus and the Saturn. We are currently working
on other wiring kits.

3.   For the operator, the 311 increases the effectiveness of the
radiological procedure while at the same time cutting down
the amount of film used. For the patient, it cuts down the number of repeat exposures and therefore reduces the total dose.

I am enclosing our brochure on the Microspot 311. If you would
like copies, please let me know. I would be happy to receive a copy
of your analysis when it is complete. Good luck!

Sincerely yours,

Andrew Shakir, M.D.
Director of Technical Services

AS/le

Enclosure
cc - Robert Anderson, Executive Vice-President

---

fer to try to be of assistance with other requests. Figure 18-6 shows
a response to the inquiry letter in Figure 18-5.

THE SALES
LETTER

A large, sophisticated sales campaign costs millions of dollars—for
marketing surveys and consulting fees, printing, postage, and promotions. This kind of campaign is beyond the scope of this book.
However, it is not unusual for an employee to draft a sales letter
for a product or service.

The "you attitude" is crucial in sales letters. Your readers are
not interested in why you want to sell your product or service.
They want to know why they should buy it. You are asking your

readers to spend valuable time studying the letter. Provide clear, specific information to help them understand what you are selling and how it will help them. Be upbeat and positive in tone, but never forget that your readers are looking for facts.

A sales letter generally has four parts. In the first part, you must gain the reader's attention. Unless the opening sentence seems either interesting or important, the reader will put the letter aside. To attract the reader, use interesting facts, quotations, or questions. In particular, try to identify a problem that will be of interest to your reader. A few examples of effective openings follow.

How much have construction costs risen since your plant was built? Do you know how much it would cost to rebuild at today's prices?

The Datafix copier is better than the Xerox--and it costs less, too. We'll repeat: it's better and it costs less!

If you're like most training directors, we bet you've seen your share of empty promises. We've heard all the stories, too. And that's why we think you'll be interested in what Fortune said about us last month.

Second, describe the product or service you are trying to sell. What does it do? How does it work? What problems does it solve?

The Datafix copier features automatic loading, so your people don't waste time watching the copies come out. Datafix copies from a two-sided original--automatically! And Datafix can turn out 30 copies a minute--which is 25% faster than our fastest competitor. . . .

Third, convince your reader that your claims are accurate. Refer to users' experience, testimonials, or evaluations performed by reputable experts or testing laboratories.

In a recent evaluation conducted by Office Management Today, more than 85% of our customers said they would buy another Datafix. The next best competitor? 71%. And Datafix earned a "Highly Reliable" rating, the highest recommendation in the reliability category. All in all, Datafix scored higher than any other copier in the desk-top class. . . .

Finally, tell your reader how to find out more about your product or service. If possible, provide a postcard that the reader can use to request more information or arrange for a visit from one of your sales representatives. Make it easy to proceed to the next step in the sales process. Figure 18-7 provides an example of a sales letter.

FIGURE 18-7

Sales Letter

DAVIS TREE CARE

1300 Lancaster Avenue

Berwyn, PA 19092

May 13, 19--

Dear Homeowner:

Do you know how much your trees are worth? That's right--your trees. As a recent purchaser of a home, you know how much of an investment your house is. Your property is a big part of your total investment.

Most people don't know that even the heartiest trees need periodic care. Like shrubs, trees should be fertilized and pruned. And they should be protected against the many kinds of diseases and pests that are common in this area.

At Davis Tree Care, we have the skills and experience to keep your trees healthy and beautiful. Our diagnostic staff is made up of graduates of major agricultural and forestry universities, and all of our crews attend special workshops to keep current with the latest information in tree maintenance. Add this to our proven record of 43 years of continuous service in the Berwyn area, and you have a company you can trust.

May we stop by to give you an analysis of your trees--absolutely without cost or obligation? A few minutes with one of our diagnosticians could prove to be one of the wisest moves you've ever made. Just give us a call at 865-9187 and we'll be happy to arrange an appointment at your convenience.

Sincerely yours,

Daniel Davis III

President

**THE CLAIM
LETTER**
A claim letter is a polite and reasonable complaint. If as a private individual or a representative of an organization you purchase a defective or falsely advertised product or receive inadequate service, your first recourse is a claim letter.

The purpose of the claim letter is to convince your reader that you are a fair and honest customer who is justifiably dissatisfied. If you can do this, your chances of receiving an equitable settlement are good. Most organizations today are very accommodating toward reasonable claims, because they realize that unhappy customers are bad business. In addition, claim letters provide a valuable index of the strong and weak points of an organization's product or service.

The claim letter has a four-part structure:

1. A specific identification of the product or service. List the model numbers, serial numbers, sizes, and any other pertinent data.
2. An explanation of the problem. State explicitly the symptoms. What function does not work? What exactly is wrong with the service?
3. A proposed adjustment. Define what you want the reader to do: for example, refund the purchase price, replace or repair the item, improve the service.
4. A courteous conclusion. Say that you trust the reader, in the interest of fairness, to abide by your proposed adjustment.

Tone, the "you attitude," is just as important as content in a claim letter. You must project a calm and rational tone. A complaint such as "I'm sick and tired of being ripped off by companies like yours" will hurt your chances of an easy settlement. If, however, you write, "I am very disappointed in the performance of my new Eversharp Electric Razor," you sound like a responsible adult. There is no reason to show anger in a claim letter, even if the other party has made an unsatisfactory response to an earlier claim letter of yours. Calmly explain what you plan to do, and why. Your reader then will be much more likely to try to see the situation from your perspective. Figure 18-8 provides an example of a claim letter.

THE ADJUST-
MENT LETTER
In an adjustment letter, you respond to a customer's claim letter. The purpose of the adjustment letter is to tell the customer how you plan to handle the situation. Whether you are granting the customer everything that the claim letter proposed, part of it, or none of it, your purpose remains the same: to show the customer that your organization is fair and reasonable, and that you value his or her business.

If you are able to grant the customer's request, the letter will be simple to write. Express your regret about the situation, state the adjustment you are going to make, and end on a positive note that will encourage the customer to continue doing business with you.

If you cannot grant the customer's request, you must try to salvage as much goodwill as you can from the situation. Obviously, your reader is not going to be happy. If your letter is carefully written, however, he or she might at least believe that you have acted reasonably. In denying a request, you are attempting to explain your side of the matter and thus to educate your reader about how the problem occurred and how to prevent it in the future.

FIGURE 18-8

Claim Letter

ROBBINS CONSTRUCTION, INC.

255 Robbins Place   Centerville, MO 65101   (417) 934-1850

August 19, 19--

Mr. David Larsen
Larsen Supply Company
311 Elmerine Avenue
Anderson, MO 63501

Dear Mr. Larsen:

As steady customers of yours for over 15 years, we came to you
first when we needed a quiet pile driver for a job near a residential area. On your recommendation, we bought your Vista 500 Quiet
Driver, at $14,900. We have since found, much to our embarrass-
ment, that it is not substantially quieter than a regular pile
driver.

We received the contract to do the bridge repair here in Centerville after promising to keep the noise to under 90 db during the
day. The Vista 500 (see enclosed copy of bill of sale for particu-
lars) is rated at 85db, maximum. We began our work and, although
one of our workers said the driver didn't seem sufficiently quiet
to him, assured the people living near the job site that we were
well within the agreed sound limit. One of them, an acoustical engineer, marched out the next day and demonstrated that we were
putting out 104 db. Obviously, something is wrong with the pile
driver.

I think you will agree that we have a problem. We were able to secure other equipment, at considerable inconvenience, to finish
the job on schedule. When I telephoned your company that humili-
ating day, however, a Mr. Meredith informed me that I should have
done an acoustical reading on the driver before I accepted delivery.

This more difficult kind of adjustment letter generally has a
four-part structure:

1. An attempt to meet the customer on some neutral ground. Often, an
   expression of regret—never an apology!—is used. Never admit that
   the customer is right in this kind of claim letter. If you write, "We are
   sorry that the engine you purchased from us is defective," the customer would have a good case against you if the dispute ended up in
   court. Sometimes, the writer will even thank the customer for bringing the matter to the attention of the company.

2. An explanation of why your company is not at fault. Most often, you
   explain to the customer the steps that led to the failure of the product

Mr. David Larsen
Page 2
August 19, 19--

I would like you to send out a technician--as soon as possible--
either to repair the driver so that it performs according to specifications or to take it back for a full refund.

Yours truly,

Jack Robbins, President

JR/lr
Enclosure

or service. Do not say, "You caused this." Instead, use the less blunt passive voice: "The air pressure apparently was not monitored. . . ."

3. A clear statement that your company, for the above-mentioned reasons, is denying the request. This statement must come later in the letter. If you begin with it, most readers will not finish the letter, and therefore you will not be able to achieve your twin goals of education and goodwill.

4. An attempt to create goodwill. Often, an organization will offer a special discount if the customer buys another, similar product. A company's profit margin on any one item is almost always large enough that the company can offer the customer very attractive savings as an inducement to continue doing business with it.

FIGURE 18-9

"Good News" Adjustment Letter

Larsen Supply Company
311 Elmerine Avenue
Anderson, MO 63501

August 21, 19--

Mr. Jack Robbins, President
Robbins Construction, Inc.
255 Robbins Place
Centerville, MO 65101

Dear Mr. Robbins:

I was very unhappy to read your letter of August 19 telling me
about the failure of the Vista 500. I regretted most the treatment
you received from one of my employees when you called us.

Harry Rivers, our best technician, has already been in touch with
you to arrange a convenient time to come out to Centerville to
talk with you about the driver. We will of course repair it, replace it, or refund the price. Just let us know your wish.

I realize that I cannot undo the damage that was done on the day
that a piece of our equipment failed. To make up for some of the
extra trouble and expense you incurred, let me offer you a 10%
discount on your next purchase or service order with us, up to
$1,000 total discount.

You have indeed been a good customer for many years, and I would
hate to have this unfortunate incident spoil that relationship.
Won't you give us another chance? Just bring in this letter when
you visit us next, and we'll give you that 10% discount.

Sincerely,

Dave Larsen

Dave Larsen, President

Figures 18-9 and 18-10 show examples of "good news" and "bad news" adjustment letters. The first letter is a reply to the claim letter shown in Figure 18-8.

WRITER'S CHECKLIST

The following checklist covers the basic letter format and the six types of letters discussed in the chapter.

LETTER FORMAT

1. Is the first page of the letter typed on letterhead stationery?

2. Is the date included?

FIGURE 18-10

*"Bad News" Adjust-
ment Letter*

QUALITY VIDEO PRODUCTS

February 3, 19--

Mr. Dale Devlin
1903 Highland Avenue
Glenn Mills, NE 69032

Dear Mr. Devlin:

Thank you for writing us about the videotape
you purchased on January 11, 19--.

You used the videotape to record your daughter's
wedding. While you were playing it back last
week, the tape jammed and broke as you were
trying to remove it from your VCR. You are
asking us to reimburse you $500 because of
the sentimental value of that recording.

As you know, our videotapes carry a lifetime
guarantee covering parts and workmanship.
We will gladly replace the broken videotape.
However, the guarantee states that the manufacturer will not assume any incidental liability.
Thus we are responsible only for the retail
value of the blank tape.

However, your wedding tape can probably be
fixed. A reputable dealer can splice tape
so skillfully that you will hardly notice the
break. It's a good idea to make backup copies
of your valuable tapes.

3. Is the inside address complete and correct? Is the appropriate courtesy title used?

4. If appropriate, is an attention line included?

5. If appropriate, is a subject line included?

6. Is the salutation appropriate?

7. Is the complimentary close typed with only the first word capitalized? Is the complimentary close followed by a comma?

8. Is the signature legible and the writer's name typed beneath the signature?

FIGURE 18-10

*"Bad News" Adjust-
ment Letter (Con-
tinued)*

Mr. Dale Devlin
page 2
February 3, 19--

Attached to this letter is a list of our authorized
dealers in your area, who would be glad to do the
repairs for you. We have already sent out your new
videotape. It should arrive within the next two days.

Please contact us if we can be of any further
assistance.

Sincerely yours,

Paul R. Blackwood

Paul R. Blackwood, Manager
Customer Relations

9. If appropriate, are the reference initials included?

10. If appropriate, is an enclosure line included?

11. If appropriate, is a copy line included?

12. Is the letter typed in one of the standard formats?

TYPES OF LETTERS

1. Does the order letter
   a. Include the necessary identifying information, such as quantities
   and model numbers?

    b. Specify, if appropriate, the terms of payment?

    c. Specify the method of delivery?

2. Does the inquiry letter
    a. Explain why you chose the reader to receive the inquiry?
    b. Explain why you are requesting the information and to what use you will put it?
    c. Specify by what date you need the information?
    d. List the questions clearly and, if appropriate, provide room for the reader's response?
    e. Offer, if appropriate, the product of your research?

3. Does the response to an inquiry letter answer the reader's questions or explain why they cannot be answered?

4. Does the sales letter
    a. Gain the reader's attention?
    b. Describe the product or service?
    c. Convince the reader that the claims are accurate?
    d. Encourage the reader to find out more about the product or service?

5. Does the claim letter
    a. Identify specifically the unsatisfactory product or service?
    b. Explain the problem(s) clearly?
    c. Propose an adjustment?
    d. Conclude courteously?

6. Does the "good-news" adjustment letter
    a. Express your regret?
    b. Explain the adjustment you will make?
    c. Conclude on a positive note?

7. Does the "bad-news" adjustment letter
    a. Meet the reader on neutral ground, expressing regret but not apologizing?
    b. Explain why the company is not at fault?
    c. Clearly deny the reader's request?
    d. Attempt to create goodwill?

EXERCISES

1. Write an order letter to John Saville, general manager of White's Electrical Supply House (13 Avondale Circle, Los Angeles, CA 90014). These are the items you want: one SB11 40-ampere battery

backup kit, at $73.50; twelve SW402 red wire kits, at $2.50 each; ten SW400 white wire kits, at $2.00 each; and one SB201 mounting hardware kit, at $7.85. Invent any reasonable details about methods of payment and delivery.

2. Secure the graduate catalog of a university offering a graduate program in your field. Write an inquiry letter to the appropriate representative, asking at least three questions about the program the university offers.

3. You are the marketing director of the company that publishes this book. Draft a sales letter that might be sent to teachers of the course you are presently taking.

4. You are the marketing director of the company that makes your bicycle (calculator, stereo set, running shoes, etc.). Write a sales letter that might be sent to retailers to encourage them to sell the product.

5. You are the recruiting officer for your college. Write a letter that might be sent to juniors in local high schools to encourage them to apply to the college when they are seniors.

6. You purchased four "D"-size batteries for your cassette player, and they didn't work. The package they came in says that the manufacturer will refund the purchase price if you return the defective items. Inventing any reasonable details, write a claim letter asking for not only the purchase price, but other related expenses.

7. A thermos you just purchased for $8.95 has a serious leak. The grape drink you put in it ruined a $15.00 white tablecloth. Inventing any reasonable details, write a claim letter to the manufacturer of the thermos.

8. The gasoline you purchased from New Jersey Petroleum contained water and particulate matter; after using it, you had to have your automobile tank flushed at a cost of $50. You have a letter signed by the mechanic explaining what he found in the bottom of your tank. As a credit-card customer of New Jersey Petroleum, write a claim letter. Invent any reasonable details.

9. As the recipient of one of the claim letters described in Exercises 6 through 8, write an adjustment granting the customer's request.

10. You are the manager of a private swimming club. A member has written saying that she lost a contact lens (value $55) in your pool. She wants you to pay for a replacement. The contract that all members sign explicitly states that the management is not responsible for loss of personal possessions. Write an adjustment letter denying the request. Invent any reasonable details.

11. As manager of a stereo equipment retail store, you guarantee that you will not be undersold. If a customer who buys something from

you can prove within one month that another retailer sells the same equipment for less money, you will refund the difference in price. A customer has written to you, enclosing an ad from another store showing that it is selling the equipment he purchased for $26.50 less than he paid at your store. The advertised price at the other store was a one-week sale that began five weeks after the date of his purchase from you. He wants his $26.50 back. Inventing any reasonable details, write an adjustment letter denying his request. You are willing, however, to offer him a blank cassette tape worth $4.95 for his equipment if he would like to come pick it up.

12. The following letters could be improved in both tone and substance. Revise them to increase their effectiveness, adding any reasonable details.

a. _____

Modern Laboratories, Inc.
DEAUVILLE, IN 43504

July 2, 19--

Adams Supply Company
778 North Henson Street
Caspar, IN 43601

Gentlemen:

Would you please send us the following items:

        one dozen petri dishes
        one gross pyrex test tubes
        three bunsen burners

Please bill us.

Sincerely,

Corey Dural

Corey Dural

CD/kw

b.   _____

                                  1967 Sunset Avenue
                                  Rochester, N. Y. 06803

                                  November 13, 19--

Admissions Department
University of Pennsylvania Law School
Philadelphia, Pa. 19106

Gentlemen:

I am a senior considering going to law school. Would you please answer
the following questions about your law school:

1.   How well do your graduates do?
2.   Is the LSAT required?
3.   Do you have any electives, or are all the courses required?
4.   Are computer skills required for admission?

A swift reply would be appreciated. Thank you.

                                  Sincerely yours,

                                  Eileen Forster

                                  Eileen Forster

c.   _____

14 Wilson Avenue
Wilton, ME 04949

November 13, 19--

Union Pacific Railroad
100 Columbia Street
Seattle, WA 98104

Attention: The President

Dear Sir:

I'm a college student doing a report on the future of mass transportation in the United States. Would you please tell me how many passenger
miles your railroad traveled last year, the number of passengers,
whether this was up or down from the year before, and the socio-
economic level of your riders?

My report is due in a week and a half. If you sent your answer in the
next two days or so I'd be able to include your information in my report. Thank you in advance.

Very truly yours,

Jon Radley

Jonathan Radley

d. _____

Cutlass Vacuums, Inc.
Ridge Pike / Speonk, OR 97203

January 13,19--

Dear Service Station Owner:

I don't have to tell you that an indispensable tool for your garage is a good industrial-strength vacuum cleaner. It not only makes your garage better looking, but it makes it safer, too.

We at Cutlass are proud to introduce our brand-new Husky 450, which replaces our model 350. The 450 is bigger, so it has a greater suction power. It has a bigger receptacle, too, so you don't have to empty it as often.

I truly believe that our Husky 450 is the shop vacuum you've been looking for. If you would like further information about it, don't miss our ad in leading car magazines.

Yours truly,

Bob Wheeler

Bob Wheeler, President
Cutlass Vacuums, Inc.

_____

e. _____

GUARDSMAN PROTECTIVE EQUIPMENT, INC.

3751 PORTER STREET
NEWARK, DE 19304

April 11, 19--

Dear Smith Family:

A rose is a rose is a rose, the poet said. But not all home protection alarms are alike. In a time when burglaries are skyrocketing, can you afford the second-best alarm system?

Your home is your most valuable possession. It is worth far more than your car. And if you haven't checked your house insurance policy recently, you'll probably be shocked to see how inadequate your coverage really is.

The best kind of insurance you can buy is the Watchdog Alarm System. What makes the Watchdog unique is that it can detect intruders before they enter your home and scare them away. Scaring them away while they're still outside is certainly better than scaring them once they're inside, where your loved ones are.

At less than two hundred dollars, you can purchase real peace of mind. Isn't your family's safety worth that much?

If you answered yes to that question, just mail in the enclosed postage-paid card, and we'll send you a 12-page, fact-filled brochure that tells you why the Watchdog is the best on the market.

Very truly yours,

Jerry Wexler

Jerry Wexler, President

f. _____

19 Lowry's Lane
Morgan, TN 30610

April 13, 19--

Sea-Tasty Tuna
Route 113
Lynchburg, TN 30563

Gentlemen:

I've been buying your tuna fish for years, and up to now it's been OK.

But this morning I opened a can to make myself a sandwich. What do you
think was staring me in the face? A fly. That's right, a housefly.
That's him you see taped to the bottom of this letter.

What are you going to do about this?

Yours very truly,

Seth Reeves

g.

Handyman Hardware, Inc.

Millersville, AL 61304

December 4, 19--

Hefty Industries, Inc.
19 Central Avenue
Dover, TX 76104

Gentlemen:

I have a problem I'd like to discuss with you. I've been carrying your line of hand tools for many years.

Your 9" pipe wrench has always been a big seller. But there seems to be something wrong with its design. I have had three complaints in the last few months about the handle snapping off when pressure is exerted on it. In two cases, the user cut his hand, one seriously enough to require nineteen stitches.

Frankly, I'm hesitant to sell any more of the 9" pipe wrenches, but I still have over two dozen in inventory.

Have you had any other complaints about this product?

Sincerely yours,

Peter Arlen, Manager
Handyman Hardware

PA/sc

h. _____

Sea-Tasty Tuna
Route 113
Lynchburg, TN 30563

April 21, 19--

Mr. Seth Reeves
19 Lowry's Lane
Morgan, TN 30610

Dear Mr. Reeves:

We were very sorry to learn that you found a fly in your tuna fish.

Here at Sea-Tasty we are very careful about the hygiene of our plant. The tuna are scrubbed thoroughly as soon as we receive them. After they are processed, they are inspected visually at three different points. Before we pack them, we rinse and sterilize the cans to ensure that no foreign material is sealed in them.

Because of these stringent controls, we really don't see how you could have found a fly in the can. Nevertheless, we are enclosing coupons good for two cans of Sea-Tasty tuna.

We hope this letter restores your confidence in us.

Truly yours,

Valarie Lumaris

Valarie Lumaris
Customer Service Representative

VL/ck

Enclosures

i. _____

Hefty Industries, Inc.
19 Central Avenue
Dover, TX 76104

December 11, 19--

Mr. Peter Arlen, Manager
Handyman Hardware, Inc.
Millersville, AL 61304

Dear Mr. Arlen:

Thank you for bringing this matter to our attention.

In answer to your question--yes, we have had a few complaints about
the handle snapping on our 9" pipe wrench.

Our engineers brought the wrench back to the lab and discovered a design flaw that accounts for the problem. We have redesigned the wrench
and have not had any complaints since.

We are not selling the old model anymore because of the risk. Therefore we have no use for your two dozen. However, since you have been a
good customer, we would be willing to exchange the old ones for the new
design.

We trust this will be a satisfactory solution.

Sincerely,

Robert Panofsky, President
Hefty Industries, Inc.

_____

CHAPTER
NINETEEN

JOB-APPLICATION MATERIALS

For most of you, the first nonacademic test of your technical writing skills will come when you make up your job-application materials. These materials will provide employers with information about your academic and employment experience, personal characteristics, and reasons for applying for the position. But they provide a lot more, too. Employers have learned that one of the most important skills an employee can bring to a job is the ability to communicate effectively. Therefore, potential employers look carefully for evidence of writing skills. Job-application materials pose a double hurdle: employers want to know both what you can do and how well you can communicate.

Some students think that once they get a satisfactory job, they will never again have to worry about résumés and application letters. Statistics show that they are wrong. The typical professional changes jobs more than ten times in the course of his or her career. Although this chapter pays special attention to the student's first job hunt, the skills and materials discussed here also apply to established professionals who wish to change jobs.

FIVE WAYS TO GET A JOB

There are five traditional ways to get a professional-level position:

1. through a college or university placement office
2. through a professional placement bureau

3. through a published job ad
4. through an unsolicited letter to an organization
5. through connections

Almost all colleges and universities have placement offices, which bring hiring organizations and students together. Generally, students submit a résumé—a brief listing of credentials—to the placement office. The résumés are then made available to representatives of business and industry. After studying the résumés, these representatives use the placement office to arrange on-campus interviews with selected students. Those who do best in the campus interviews are then invited by the representatives to visit the organization for a tour and another interview. Sometimes a third interview is scheduled; sometimes an offer is made immediately or shortly thereafter. The advantage of this system is two-fold: first, it is free; second, it is easy. The student merely has to deliver a résumé to the placement office and wait to be contacted.

A professional placement bureau offers essentially the same service as a college placement office, but it charges either the employer (or, rarely, the new employee) a fee—often, 10 percent of the first year's salary—when the client accepts a position. Placement bureaus cater primarily to more advanced professionals who are looking to change jobs.

Published job ads generally offer the best opportunity for both students and professionals. Organizations advertise in three basic kinds of publications: public-relations catalogs (such as *College Placement Annual*), technical journals, and newspapers. If you are looking for a job, you should regularly check the major technical journals in your field and the large metropolitan newspapers—especially the Sunday editions. In responding to an ad, you must include with the résumé a job-application letter—one that highlights the crucial information on the résumé.

You need not wait for a published ad. You can write unsolicited letters of application to organizations you would like to work for. The disadvantage of this technique is obvious: there might not be an opening. Yet many professionals favor this technique, because there are fewer competitors for those jobs that do exist, and sometimes organizations do not advertise all available positions. And sometimes an impressive unsolicited application can prompt an organization to create a position for you. Before you write an unsolicited application, make sure you learn as much as you can about the organization (see Chapter 4): current and anticipated major projects, hiring plans, and so forth. You should know as much as you can about any organization you are applying to, of course, but when you are submitting an unsolicited application you have no other source of information on which to base your résumé and letter. The business librarian at your college or university will be able to point out sources of information, such as the Dun and Bradstreet guides, the *F&S Index of Corporations*, and the indexed newspapers such as the *New York Times* and the *Wall Street Journal*.

The use of connections to get a job is effective if you have a relative or an acquaintance in a position to exert some influence or at least point out that a position might be opening up. Other good sources of contacts include employers from your past summer jobs and faculty members in your field.

The college placement office, published ads, unsolicited letters, and connections are most useful for students. Too many students rely solely on the placement office, thereby limiting the range of possibilities to those organizations that choose to visit the college. The best system is to use the placement office and respond to published ads. If an attractive organization is not advertising, send an unsolicited application letter after telephoning them to find out the name of the appropriate person and department.

THE RÉSUMÉ

Whether it is to be submitted to a college placement office or sent along with a job-application letter, the résumé communicates in two ways: by its appearance and by its content.

APPEARANCE OF THE RÉSUMÉ
Because in almost all cases potential employers see the résumé before they see the person who wrote it, the résumé has to make a good first impression. Employers believe—often correctly—that the appearance of the résumé reflects the professionalism of the writer. When employers see a sloppy résumé, they assume that the writer would do sloppy work. A neat résumé implies that the writer would do professional work. When employers look at a résumé, therefore, they see more than a single piece of paper; they see dozens of documents they will be reading if they hire the writer.

Some colleges and universities advise students to have their résumés professionally printed. A printed résumé is attractive, and that's good—provided, of course, that the information on it is consistent with its professional appearance. Most employers agree, however, that a neatly typed or word-processed résumé photocopied on good-quality paper is just as effective.

People who photocopy a typed or word-processed résumé are more likely to tailor different versions to the needs of the organizations to which they apply—a good strategy. People who go to the trouble and expense of a professional printing job are far less likely to make up different résumés; they tend to submit the printed one to all their prospects. The résumé looks so good that they feel it is not worth tinkering with. This strategy is dangerous, for it encourages the writer to underestimate the importance of directing the content of the résumé to a specific audience.

However they are reproduced, résumés should appear neat and professional. They should have

1. *Generous margins.* Leave a one-inch margin on all four sides.
2. *Clear type.* Use a typewriter or printer with clear, sharp, unbroken letters. Avoid strikeovers and obvious corrections. Avoid dot-matrix printers.
3. *Symmetry.* Arrange the information so that the page has a balanced appearance.
4. *Clear organization.* Use adequate white space. Make sure the spacing *between* items is greater than the spacing *within* an item. That is, the spacing between your education section and your employment section should be greater than the spacing within one of those sections. You should be able to see the different sections clearly if you stand while the résumé is on the floor in front of your feet.

Use indention clearly. When you arrange items in a vertical list, indent a few spaces all second and subsequent lines of any item. Notice, for example, that the following list from the computer-skills section of a résumé can be confusing.

```
Computer Experience
Systems: IBM AT/XT pc, Apple Macintosh, Apple IIe/II-Plus,
Andover AC-256, Prime 360
Software: Lotus 1-2-3, DBase II and III, Multiplan, PFS
File/Report, Wordstar
Languages: Fortran, Pascal, BASIC
```

With the second lines indented, the arrangement is much easier to understand.

```
Computer Experience
Systems: IBM AT/XT pc, Apple Macintosh, Apple IIe/II-Plus, Andover
    AC-256, Prime 360
Software: Lotus 1-2-3, DBase II and III, Multiplan, PFS File/Report,
    Wordstar
Languages: Fortran, Pascal, BASIC
```

CONTENT OF THE RÉSUMÉ

Although different experts advocate different approaches to résumé writing, everyone agrees on two things.

First, the résumé must be completely free of errors. Grammar, punctuation, usage, and spelling errors undercut your professionalism by casting doubt on the accuracy of the information contained in the résumé. Ask for assistance after you have written the draft, and proofread the finished product at least twice.

Second, the résumé must provide clear and specific information, without generalizations or self-congratulation. Your résumé is a sales document, but you are both the salesperson and the product. You cannot say, "I am a terrific job candidate," as if the product were a toaster or an automobile. Instead, you have to provide the specific details that will lead the reader to the conclusion that you are a terrific job candidate. You must *show* the reader. Telling the reader is graceless and, worse, unconvincing.

A résumé should be long enough to include all the pertinent information but not so long as to bore or irritate the reader. Generally, a student's résumé should be kept to one page. If, however, the student has special accomplishments to describe—such as journal articles or patents—a two-page résumé is appropriate. If your information comes to just over one page, either eliminate or condense some material to make it fit onto one page, or modify the physical layout of the résumé so that it fills a substantial part of the second page.

ELEMENTS OF THE RÉSUMÉ

Almost every résumé has five basic sections:

1. identifying information
2. education
3. employment
4. personal information
5. references

But your résumé should reflect one particular person: you. Many people have some special skills or background that could be conveyed in additional sections. These sections, as discussed below, should be used where appropriate.

IDENTIFYING INFORMATION   Your full name, address, and phone number should always be placed at the top of the page. Generally, you should present your name in full capitals, centered at the top. Use your complete address, with the zip code. The two-letter state abbreviations used by the Post Office are now preferred.

Also use your complete phone number, with the area code. For the telephone exchange, use numbers rather than letters.

Students who have two addresses and phone numbers should make sure that both are listed and identified clearly. Often, an employer will try to call a student during an academic holiday to arrange an interview.

EDUCATION   The education section usually follows the identifying information on the résumé of a student or a recent graduate. People with substantial professional experience usually place the employment experience section before the education section.

Include the following information in the education section: the degree, the institution, its location, and the date of graduation. After the degree abbreviation (such as B.S., B.A., A.A., or M.S.), list the academic major (and, if you have one, the minor)—for example, "B.S. in Materials Engineering." Identify the institution by its full name: "The Pennsylvania State University," not "Penn State." Also include the city and state of the institution. If your degree has not yet been granted, write "Anticipated date of graduation" or "Degree expected in" before the month and year.

You should also list any other institutions you attended beyond the high-school level—even those at which you did not earn a degree. Students are sometimes uneasy about listing community colleges or junior colleges; they shouldn't be. Employers are generally impressed to learn that a student began at a smaller or less advanced school and was able to transfer to a four-year college or university. The listing for other institutions attended should include the same information as the main listing. Arrange the entries in reverse chronological order: that is, list first the school you attended most recently.

In addition to this basic information, many students and recent graduates like to include more details about their educational experiences. They feel, quite correctly, that the basic information alone implies that they merely endured an institution and received a degree. Of the several ways for you to fill out the education section, perhaps the easiest is to list your grade-point average (provided that it is substantially above the median for the graduating class). Or list your average in the courses in your major, if that is more impressive.

You can also expand the education section by including a list of courses that would be of particular interest to the reader. Advanced courses in an area of your major concentration might be appropriate, especially if the potential employer has mentioned that area in the job advertisement. Also useful would be a list of communications courses—technical writing, public speaking, organizational communications, and the like. Employers are looking for people who can communicate what they know. A listing of business courses on an engineer's résumé—or the reverse—also shows special knowledge and skills. The only kind of course listing that is *not* particularly helpful is one that merely names the traditional required courses for your major. Make sure to list courses by title, not by number; employers won't know what Chemistry 250 is.

Another useful way to amplify the education section of the résumé is to describe a special accomplishment. If you did a special senior design or research project, for example, a two- or three-line description would be informative. Include in this description the title and objective of the project, any special or advanced techniques or equipment you used, and—if they are known—the major results. Such a description might be phrased as follows: "A Study of Composite Substitutes for Steel—a senior design project intended to formulate a composite material that can be used to replace steel in car axles." Even a traditional academic course in which you conducted a sophisticated project and wrote a sustained report can be described profitably. A project discussion makes you look less like a student—someone who takes courses—and more like a professional—someone who designs and conducts projects.

Finally, you also can list in the education section any honors and awards you received. Scholarships, internships, and academic awards all offer evidence of an exceptional job candidate. If you have received a number of such honors, or some that were not exclusively academic, it might be more effective to list them separately (in a section called "Honors" or "Awards") rather than in a subsection of the education section. Often, some body of information could logically be placed in two or even three different locations, and you must decide where it will make the best impression.

The education section is the easiest part of the résumé to adapt for different positions. For example, a student majoring in electrical engineering who is applying for a position that calls for strong communications skills can draw up a list of communications courses in one version of his or her résumé. The same student can use a list of advanced electrical engineering courses in a résumé directed to another potential employer. As you compose the education section of your résumé, consider carefully what aspects of your background can be emphasized to address the requirements for the particular job opening.

EMPLOYMENT The employment section, like the education section, conveys at least the basic information about each job you've held: the dates of employment, the organization's name and location, and your position or title. This information is self-explanatory.

However, a skeletal listing of nothing more than these basic facts would not be very informative or impressive. As in the education section, you should provide carefully selected details.

What readers want to know, after they have learned where and when you were employed, is what you actually did. Therefore, you should provide a two- to three-line description for each position although it could be longer for particularly important or relevant jobs. Focus the description on one or more of the following factors:

1. *Reports.* What kinds of documents did you write or assist in writing? List, especially, various governmental forms and any long reports.

2. *Clients.* What kinds of, and how many, clients did you transact business with as a representative of your organization?

3. *Skills.* What kinds of technical skills did you learn or practice in doing the job?

4. *Equipment.* What kinds of technical equipment did you operate or supervise? Mention, in particular, any computer skills you demonstrated, for they can be very useful in almost every kind of position.

5. *Money.* How much money were you responsible for? Even if you considered your bookkeeping position fairly simple, the fact that the organization grossed, say, $2 million a year shows that the position involved real responsibility.

6. *Personnel.* How many personnel were under your supervision? Students sometimes supervise small groups of other students or technicians. Supervising, naturally, shows maturity and responsibility.

Make sure you use the active voice—"supervised three workers"— rather than the passive voice—"three workers were supervised by me." The active voice emphasizes the verb, which describes the action. In thinking about your functions and responsibilities in your various positions, keep in mind the strong action verbs that clearly communicate your activities. Following is a list of some strong verbs:

administered	collected	corresponded
advised	completed	created
analyzed	conducted	delivered
assembled	constructed	developed
built	coordinated	devised

directed	instituted	provided
discovered	maintained	purchased
edited	managed	recorded
evaluated	monitored	reported
examined	obtained	researched
expanded	operated	served
hired	organized	solved
identified	oversaw	supervised
implemented	performed	trained
improved	prepared	wrote
increased	produced	

Here is a sample listing:

```
June-September 19--: Millersville General Hospital, Millersville,
                    TX. Student Dietician. Gathered dietary histo-
                    ries and assisted in preparing menus for a 300-
                    bed hospital. Received "excellent" on all
                    items in evaluation by head dietician.
```

In just a few lines, you can show that you sought and accepted responsibility and that you acted professionally. Do not write, "I accepted responsibility"; rather, present facts that lead inevitably to that impression.

Naturally, not all jobs entail such professional skills and responsibilities. Many students find summer work as laborers, sales clerks, short-order cooks, and so forth. If you have not had a professional position, list the jobs you have had, even if they were completely unrelated to your career plans. If the job title is self-explanatory—such as waitress or service-station attendant—don't elaborate. Every job is valuable. You learn that you are expected to be someplace at a specific time, wear appropriate clothes, and perform some specific duties. Also, every job helps pay college expenses. If you can say that you earned, say, 50 percent of your annual expenses through a job, employers will be impressed by your self-reliance. Most of them probably started out with nonprofessional positions. And don't forget that any job you have held can yield a valuable reference.

The various jobs should be listed and described in reverse chronological order on the résumé, to highlight those positions you have held most recently.

One further word: if you have held a number of nonprofessional positions as well as several professional positions, the nonprofessional ones can be grouped together in one listing:

```
Other Employment: Cashier (summer, 1987), salesperson (part-time,
                 1987), clerk (summer, 1986).
```

This technique prevents the nonprofessional positions from drawing the reader's attention away from the more important positions.

PERSONAL INFORMATION   This section of the résumé has changed considerably in recent years. Only a decade ago, most résumés would list the writer's height, weight, date of birth, and marital status. Most résumés today include none of these items. One explanation for this change is that federal legislation now prohibits organizations from requiring this information from applicants. Perhaps a more important explanation is that most people have come to feel that such personal information is irrelevant to a person's ability to perform a job effectively.

The personal information section of the résumé *is* the appropriate place for a few items about your outside interests. Participation in community service organizations—such as Big Brothers—or volunteer work in a hospital is, of course, an extremely positive factor. Hobbies that are in some way related to your career interests—for example, amateur electronics for an engineer—are useful, too. You also should list any sports, especially those that might be socially useful in your professional career, such as tennis, racquetball, and golf. Any university-sanctioned activity—such as membership on a team, participation on the college newspaper, or election to a responsible position in an academic organization or a residence hall—should also be pointed out. In general, do not include activities that might create a negative impression in the reader: hunting, gambling, performing in a rock band. And always omit such activities as reading and meeting people—everybody reads and meets people.

REFERENCES   You may list the names of three or four referees—people who have written letters of recommendation or who have agreed to speak on your behalf—on your résumé. Or you may simply say that you will furnish the names of the referees upon request. The length of your résumé sometimes dictates which style to use. If the résumé is already long, the abbreviated form might be preferable. If it does not fill out the page, the longer form might be the one to use. However, each style has advantages and disadvantages that you should consider carefully.

Furnishing the referees' names appears open and forthright. It shows that you have already secured your referees and have nothing to hide. If one or several of the referees are prominent people in their fields, the reader is likely to be impressed. And, perhaps most important, the reader can easily phone the referees or write them a letter. Listing the referees makes it easy for the prospective employer to proceed with the hiring process. The only disadvantage of this style is that it takes up space on the résumé that might be needed for other information.

Writing "References will be furnished upon request," on the other hand, takes up only one line. In addition, it leaves you in a flexible position. You can still secure referees after you have sent out the résumé. Also, you can send selected letters of reference to prospective employers according to your analysis of what they want. Using different references for different positions is sometimes just as valuable as sending different résumés. However, some readers will interpret the nonlisting style as evasive or secretive or assume that you have not yet asked prospective referees. A greater disadvantage is that if the readers are impressed by the résumé and want to learn more about you, they cannot do so quickly and directly.

If you decide to include the listing, identify every referee by name, title, organization, mailing address, and phone number. For example:

```
Dr. Robert Ariel
Assistant Professor of Biology
Central University
Portland, OR 97202
(503) 666-6666
```

Remember that a careful choice of referees is as important as careful writing of the résumé. Solicit references only from persons who know your work best and for whom you have done the best work—for instance, a professor from whom you have received A's. It is unwise to ask prominent professors who do not know your work well, for the advantage of having a famous name on the résumé will be offset by the referee's brief and uninformative letter. Often, a young and less-well-known professor can write the most informative letter or provide the best recommendation. Try also to have at least one reference letter from an employer, even if the job was not professional.

After you have decided whom to ask, give the potential referee an opportunity to decline the request gracefully. Sometimes the person has not been as impressed with your work as you think. Also, if you simply ask, "Would you please write a reference letter for me?" the potential referee might accept and then write a lukewarm letter. It is better to follow up the first question with

"Would you be able to write an enthusiastic letter for me?" or "Do you feel you know me well enough to write a strong recommendation?" If the potential referee shows any signs of hesitation or reluctance, you can withdraw the invitation at that point. The scene is a little embarrassing, but it is better than receiving a half-hearted recommendation.

OTHER ELEMENTS   The discussion so far has concentrated on the sections that appear on virtually everyone's résumé. Other sections are either discretionary or are appropriate only for certain writers.

A statement of objectives, in the form of a brief sentence—for example, "Objective: Entry-level accounting position with medium-sized public accounting firm"—can be placed near the top of the résumé, generally right below the identifying information. A statement of objectives gives the appearance that the student has a clear sense of direction and goals, which is usually good. Sometimes, however, employers think it is unrealistic for students to state their goals with such self-assurance and certainty. Most students, in fact, have had relatively little experience doing the job for which they have been studying. As a result, their career goals often change radically soon after they leave school. An additional disadvantage of the statement of objectives is that it can sometimes eliminate you from consideration. An employer who knows that the vacant position is not the one you want might conclude that you would not be interested in the job and therefore reject your application.

If you are a veteran of the armed forces, include a military service section on the résumé. Define your military service as if it were any other job, citing the dates, locations, positions, ranks, and tasks. Often a serviceperson receives regular evaluations from a superior; these "marks" can work in your favor.

If you have a working knowledge of a foreign language, your résumé should include a *Language Skills* section. Language skills are particularly relevant if the potential employer has international interests and you could be useful in translation or foreign service.

Figures 19-1 and 19-2 provide examples of effective résumés. The job-application letters that accompany these résumés are shown in Figure 19-3 and Figure 19-4.

Notice that résumés often omit the *I* at the start of sentences. Rather than writing, "I prepared bids, . . ." many people will write, "Prepared bids. . . ." Whichever style you use, be consistent.

FIGURE 19-1

Résumé

```
KENNETH CHAING                              753 Westborn Drive
(215) 525-6881                              Ardmore, PA 19316


EDUCATION
B.S. in Civil Engineering
Eastern University, Lynwood, PA
Anticipated June 19--

Grade-Point Average: 3.35 (of 4.0)

Advanced Business Courses
Financial Accounting                    Budgeting
Legal Options in Decision Making        Advanced Accounting
Manpower Management                     Labor Law


EMPLOYMENT
May 1987-           Gilmore Construction, Redford, PA
September 1987          Prepared bids and estimates for storm and
                       sanitary piping. Revised drafting for
                       700-unit housing complex. Ordered piping
                       materials amounting to $325,000.

June 1986-          Gilmore Construction, Redford, PA
September 1986         Worked in drafting and reproductions de-
                      partment. Supervised piping sales.

June 1985-          Pertwell Construction, Salford, PA
September 1985         Laborer.


PERSONAL INFORMATION
Member, American Society of Civil Engineers; ROTC


REFERENCES
Mr. Allen Chrome      Dr. Harold Murphy     Mr. Len Lefkowitz, P.E.
President             Prof. of Physics      Chief Engineer,
Gilmore               Eastern University     Pertwell Construction
 Construction         Lynwood, PA  19314    1911 Market Street
Frazer Park           (215) 669-4300        Redford, PA   18611
Redford, PA   18611                         (306) 432-1814
(306) 912-1773
```

THE JOB-APPLICATION LETTER

The job-application letter is crucial because it is the first thing your reader sees. If the letter is ineffective, the reader probably will not bother to read the résumé.

If job candidates had infinite time and patience, they would make up a different résumé for each prospective employer, highlighting their particular appropriateness for that one job. But because most candidates don't have unlimited time and patience, they make up one or two different versions of their résumés. As a result, the typical résumé makes a candidate look only relatively

FIGURE 19-2

Résumé

ANDREA SARNO

Home Address: School Address:
   1314 Old Oaks St.   311 Hamilton Hall
   Wynnewood, WA 99123   Western University
   (206) 612-1414   Warren, WA 99314
                                     (206) 669-4136

Objective: Entry-level position in accounting.

Education: Bachelor of Science in Accounting
               Anticipated June 19--
               Western University, Warren, WA 99314

               Advanced Business Courses
               Investment Analysis
               Advanced Accounting
               Principles of Management Accounting
               Management Information Systems
               Federal Tax Law
               Collective Bargaining

Experience: Harmon Kline, Inc.   Washington, D.C.   June-
               September 1987
               Worked in the audit department. Supervised the
               auditing of seven multinational corporations.
               Wrote audit reports (average length, 18 pages)
               for each of the clients. Recommended internal
               improvements that saved over $40,000 annually.

               Franklin Warner, Inc., Spokane.   June-
               September 1985
               Worked as a clerk in the accounts receivable
               department.

Honors and Awards: Brackenbury Scholarship, 1987
               Outstanding Senior Woman, 1986-87
               President, American Accounting Society
                  Student Chapter, Western University, 1987
               Member, Lacrosse Team, 1985-87

References: Furnished upon request.

close to the ideal the employer has in mind. Thus, candidates must use job-application letters to appeal directly and specifically to the needs and desires of particular employers.

APPEARANCE OF THE JOB-APPLICATION LETTER

Like the résumé it introduces, the letter must be error-free and professional-looking. A good job-application letter has all the virtues of any business letter: adequate margins, clear and uniform type, no strikeovers or broken letters. And, of course, it must conform to one of the basic letter formats (see Chapter 18, "Correspondence").

Such advice is easy to give but hard to follow. The problem is that every job-application letter must be typed individually, because its content is unique. (Even if you have access to a word processor, the inside address and the salutation, at least, must be typed by hand.) In creating the job-application letter, therefore, you have a double burden: the one page that makes the greatest impression has to be typed separately each time you apply for a position. Compared to job-application letters, résumés are simple: you have to type them perfectly only a few times.

CONTENT OF THE JOB-APPLICATION LETTER
Like the résumé, the job-application letter is a sales document. Its purpose is to convince the reader that you are an outstanding candidate and should be called in for an interview. Of course, you accomplish this through skillful use of hard evidence, not through empty self-praise.

The job-application letter is *not* an expanded version of the résumé. In the letter, you choose from the résumé two or three points that will be of greatest interest to the potential employer and develop them into paragraphs. If one of your previous part-time positions called for specific skills that the employer is looking for, that position might be the subject of a substantial paragraph in the letter, even though the résumé devotes only a few lines to it. The key to a good application letter is selectivity. If you try to cover every point on your résumé, the letter will be fragmented. The reader then will have a hard time forming a clear impression of you, and the purpose of the letter will be thwarted.

In most cases, a job-application letter should fill up the better part of a page. Like all business correspondence, it should be single-spaced, with double spaces between paragraphs. A full page gives you space to develop a substantial argument. For more-experienced candidates, the letter may be longer, but most students find that they can adequately summarize their credentials in one page. A long letter is not necessarily a good letter. If you write at length on a minor point, you end up being boring. Worse still, you appear to have poor judgment. Employers are always seeking candidates who can say much in a limited space, not the reverse.

ELEMENTS OF THE JOB-APPLICATION LETTER
Among the mechanical elements of the job-application letter, the inside address—the name, title, organization, and address of the recipient—is most important. If you know the correct form of this information from an ad, there is no problem. However, if you are uncertain about any of the information—the recipient's name, for example, might have an unusual spelling—you should verify it by

phoning the organization. Many people are very sensitive about such matters, so you should not risk beginning the letter with a misspelling or incorrect title. When you do not know who should receive the letter, do not address it to a department of the company—unless the job ad specifically says to do so—because nobody in that department might feel responsible for dealing with it. Instead, phone the company to find out who manages the department. If you are unsure of the appropriate department or division to write to, address the letter to a high-level executive, such as the president of the organization. The letter will be directed to the right person. Also, because the application includes both a letter and a résumé, the enclosure notation (see Chapter 18) should be typed in the lower left-hand corner of the letter.

The four-paragraph letter that will be discussed here is, naturally, only a very basic model. Because every substantial job-application letter has an introductory and a concluding paragraph, the minimum number of paragraphs for the job-application letter has to be four. But there is no reason that an application letter cannot have five or six paragraphs. The four-paragraph letter, however, is a useful model for most situations. The four paragraphs can be categorized as follows:

1. introductory paragraph
2. education paragraph
3. employment paragraph
4. concluding paragraph

Because this is such an important letter, you should plan it carefully. Think about what kind of information from your background best responds to the needs of the potential employer. Draft it and then revise it. Let it sit for a while, then revise it again. Spend as much time on it as you can. Make sure each paragraph is unified and functions effectively as part of the whole letter. Check that your transitions from one paragraph to the next are clear.

THE INTRODUCTORY PARAGRAPH The introductory paragraph establishes the tone for the rest of the letter: it is your opportunity to capture the reader's attention. Specifically, the introductory paragraph has four functions:

1. It identifies your source of information.
2. It identifies the position you are interested in.
3. It states that you wish to be considered for that position.
4. It forecasts the rest of the letter.

Your source of information is usually a published advertisement or an employee already working for the organization. If it is an ad, identify specifically the publication and its date of issue. If it is an employee, identify that person by name and title. (In an unsolicited application, all you can do is ask if a position is available.) In addition to naming the source of information, you should clearly identify the position for which you are applying. Often, the reader of the letter will be unfamiliar with all the different ads that the organization has placed. Without the title of the position, he or she will not know which position you are interested in. Also, specifically state that you would like to be considered for the position. And finally, carefully choose a few phrases that forecast the body of the letter by summarizing its main points. These four aspects of the introductory paragraph need not appear in any particular order, nor need they each be covered in a single sentence. The following examples of introductory paragraphs demonstrate different ways to provide the necessary information.

I am writing in response to your notice in the May 13 New York Times. I would like to be considered for the position in systems programming. My studies at Eastern University in computer science, along with my programming experience at Airborne Instruments, would qualify me, I believe, for the position.

My academic training in hotel management and my experience with Sheraton International have given me a solid background in the hotel industry. Would you please consider me for any management trainee position that might be available?

Mr. Howard Alcott of your Research and Development Department has suggested that I write. He thinks that my organic chemistry degree and my practical experience with Brown Laboratories might be of value to you. Do you have an entry-level position in organic chemistry for which I might be considered?

As these sample paragraphs indicate, the important information can be conveyed in any number of ways. The difficult part of the introductory paragraph—and of the whole letter as well—is to achieve the proper tone. You must appear quietly self-confident. Because the letter will be read by someone who is professionally superior to you, the tone must be modest, but it should not be self-effacing or negative. Never say, for example, "I do not have a very good background in civil engineering, but I'm willing to learn." The reader will take this kind of statement at face value and probably stop reading right there. You should show pride in your education and experience, while at the same time suggesting by your tone that you have much to learn about your field.

THE EDUCATION PARAGRAPH For most students the education paragraph should come before the employment paragraph, because the content of the former will be stronger. If, however, your work experience is more pertinent than your education, discuss your work first.

In devising your education paragraph, take your cue from the job ad. What aspect of your education most directly responds to the job requirements? If the ad stresses the need for versatility, you might structure your paragraph around the range and diversity of your courses. Also, you might discuss course work in a field related to your major, such as business or communications skills. Extracurricular activities are often very valuable; if you were an officer in a student organization in your field, you could discuss the activities and programs that you coordinated. Perhaps the most popular strategy for developing the education paragraph is to discuss skills and knowledge gained from advanced course work in the major field.

Whatever information you provide, the key to the education paragraph is to develop one unified idea, rather than to toss a series of unrelated facts onto the page. Notice how each of the following education paragraphs develops a unified idea:

At Eastern University, I have taken a wide range of courses in the sciences, but my most advanced work has been in chemistry. In one laboratory course, I developed a new aseptic technique that lowered the risk of infection by over 40 percent. This new technique was the subject of an article in the Eastern Science Digest. Representatives from three national breweries have visited our laboratory to discuss the technique with me.

To broaden my education at Southern, I took eight business courses in addition to my requirements for the Civil Engineering degree. Because your ad mentions that the position will require substantial client contact, I believe that my work in marketing, in particular, would be of special value. In an advanced marketing seminar, my project culminated in a 20-page sales brochure on the different kinds of building structures for sale to industrial customers in our section of the city.

The most rewarding part of my education at Western University took place outside the classroom. My entry in a fashion-design competition sponsored by the university won second place. More important, through the competition I met the chief psychologist at Eastern Regional Hospital, who invited me to design clothing for handicapped persons. I have since completed six different outfits. This project has given me an interest in an aspect of design that has up to now received little attention. I would hope to be able to pursue this interest once I start work.

Notice that each of these paragraphs begins with a topic sentence—a forecast of the rest of the paragraph—and uses considerable detail and elaboration to develop the main idea. Notice, too, a small point: if you haven't already specified your major and college or university in the introductory paragraph, be sure to do so in the discussion of your education.

THE EMPLOYMENT PARAGRAPH Like the education paragraph, the employment paragraph should begin with a topic sentence and then elaborate a single idea. That idea might be that you have had a broad background or that a single position has given you special skills that make you particularly suitable for the available position. The following examples show effective experience paragraphs.

For the past three summers and part-time during the academic year, I have worked for Redego, Inc., a firm that specializes in designing and planning industrial complexes. I began at Redego as an assistant in the drafting room. By the second summer, I was accompanying a civil engineer on field inspections. Most recently, I have been involved with an engineer in designing and drafting the main structural supports for a 15-acre, $30 million chemical facility.

Although I have worked every summer since I was fifteen, my most recent position, as a technical editor, has been the most rewarding. I was chosen by Digital Systems, Inc., from among 30 candidates because of my dual background in computer science and writing. My job was to coordinate the editing of computer manuals. Our copy editors, who are generally not trained in computer science, need someone to help verify the technical accuracy of their revisions. When I was unable to answer their questions, I was responsible for getting the correct answer from our systems analysts and making sure the computer novice could follow it. This position gave me a good understanding of the process by which operating manuals are created.

I have worked in merchandising for three years as a part-time and summer salesperson in men's fashions and accessories. I have had experience in inventory control and helped one company to switch from a manual to an on-line system. Most recently, I assisted in clearing $200,000 in out-of-date men's fashions: I coordinated a campaign to sell half of the merchandise at cost and was able to convince the manufacturer's representative to accept the other half for full credit. For this project, I received a certificate of appreciation from the company president.

Notice how these writers carefully define their duties to give their readers a clear idea of the nature and extent of their responsibilities.

Although you will discuss your education and experience in separate paragraphs, try to link these two halves of your background. If an academic course led to an interest that you were able to follow up in a job, make that point clear in the transition from one paragraph to the other. Similarly, if a job experience helped shape your academic career, tell the reader about it.

THE CONCLUDING PARAGRAPH The concluding paragraph of the job-application letter is like the ending of any sales letter: its function is to stimulate action. In this case, you want the reader to invite you for an interview. In the preceding paragraphs of the letter, you provided the information that should have convinced your reader to give you another look. In the last paragraph, you want to make it easy for him or her to do so. The concluding paragraph contains three main elements:

1. a reference to your résumé
2. a request for an interview
3. your phone number

If you have not yet referred to the enclosed résumé, do so at this point. Then, politely but confidently request an interview—making sure to use the phrase "at your convenience." Don't make the request sound as if you're asking a personal favor. And be sure to include your phone number and the time of day you can be reached—even though the phone number is also on your résumé. Many employers will pick up a phone and call promising candidates personally, so you should make it as easy as possible for them to get in touch with you.

Following are examples of effective concluding paragraphs.

The enclosed resume fills in the details of my education and experience. Could we meet at your convenience to discuss further the skills and experience I could bring to Pentamax? A message can be left for me anytime at (333) 444-5555.

More information about my education and experience is included on the enclosed resume, but I would appreciate the opportunity to meet with you at your convenience to discuss my application. I can be reached after noon Tuesdays and Thursdays at (333) 444-5555.

Figures 19-3 and 19-4 show examples of effective job-application letters corresponding to the résumés presented in Figures 19-1 and 19-2.

THE FOLLOW-UP LETTER

The final element in a job search is a follow-up letter, which is sent to a potential employer after an interview or plant tour. Your purpose in writing a follow-up letter is to thank the representative for taking the time to see you and to remind him or her of your particular qualifications for the job. You also can take this opportunity to restate your interest in the position.

FIGURE 19-3

*Job-Application
Letter*

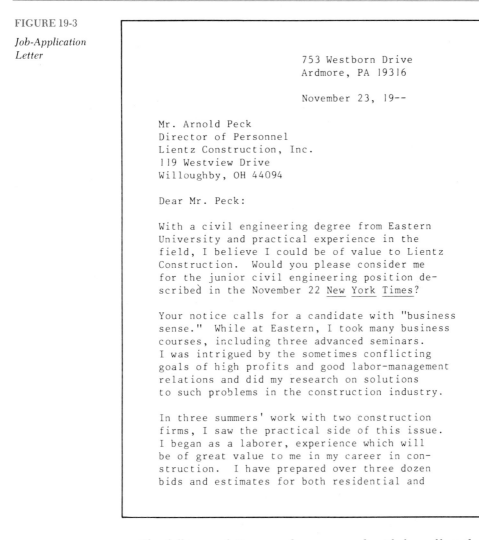

753 Westborn Drive
Ardmore, PA 19316

November 23, 19--

Mr. Arnold Peck
Director of Personnel
Lientz Construction, Inc.
119 Westview Drive
Willoughby, OH 44094

Dear Mr. Peck:

With a civil engineering degree from Eastern
University and practical experience in the
field, I believe I could be of value to Lientz
Construction. Would you please consider me
for the junior civil engineering position described in the November 22 New York Times?

Your notice calls for a candidate with "business
sense." While at Eastern, I took many business
courses, including three advanced seminars.
I was intrigued by the sometimes conflicting
goals of high profits and good labor-management
relations and did my research on solutions
to such problems in the construction industry.

In three summers' work with two construction
firms, I saw the practical side of this issue.
I began as a laborer, experience which will
be of great value to me in my career in construction. I have prepared over three dozen
bids and estimates for both residential and

The follow-up letter can do more good with less effort than any other step in the job-application procedure. The reason is simple: so few candidates take the time to write the letter. Figure 19-5 provides an example of a follow-up letter.

WRITER'S
CHECKLIST

The following checklist covers the résumé, the job-application letter, and the follow-up letter.

RÉSUMÉ

1. Does the résumé have a professional appearance, with generous margins, clear type, a symmetrical layout, adequate white space, and effective use of indention?

FIGURE 19-3

*Job-Application
Letter (Continued)*

Mr. Arnold Peck
page 2
November 23, 19--

commercial customers. In addition, I revised
the entire drafting of a 700-unit housing complex
in southern New Jersey.

The enclosed résumé provides an overview of
my skills and experience. Could I meet with
you at your convenience to discuss my qualifications for this position? You can leave a
message for me any weekday at (215) 525-6681.

Very truly yours,

Kenneth Chaing

Kenneth Chaing

Enclosure (1)

2. Is the résumé completely free of errors?

3. Does the identifying information section contain your name, address, and phone number?

4. Does the education section include your degree, your institution and its location, and your anticipated date of graduation, as well as any other information that will help your reader appreciate your qualifications?

5. Does the employment section include, for each job, the dates of em-

FIGURE 19-4

*Job-Application
Letter*

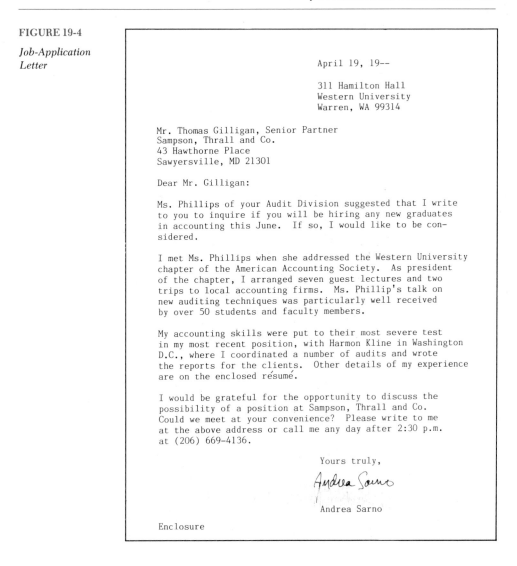

April 19, 19--

311 Hamilton Hall
Western University
Warren, WA 99314

Mr. Thomas Gilligan, Senior Partner
Sampson, Thrall and Co.
43 Hawthorne Place
Sawyersville, MD 21301

Dear Mr. Gilligan:

Ms. Phillips of your Audit Division suggested that I write
to you to inquire if you will be hiring any new graduates
in accounting this June. If so, I would like to be considered.

I met Ms. Phillips when she addressed the Western University
chapter of the American Accounting Society. As president
of the chapter, I arranged seven guest lectures and two
trips to local accounting firms. Ms. Phillip's talk on
new auditing techniques was particularly well received
by over 50 students and faculty members.

My accounting skills were put to their most severe test
in my most recent position, with Harmon Kline in Washington
D.C., where I coordinated a number of audits and wrote
the reports for the clients. Other details of my experience
are on the enclosed résumé.

I would be grateful for the opportunity to discuss the
possibility of a position at Sampson, Thrall and Co.
Could we meet at your convenience? Please write to me
at the above address or call me any day after 2:30 p.m.
at (206) 669-4136.

Yours truly,

Andrea Sarno

Andrea Sarno

Enclosure

ployment, the organization's name and location, and your position or
title, as well as a description of your duties and accomplishments?

6. Does the personal information section include relevant hobbies or activities, including extracurricular interests?

7. Does the references section include the names, job title, organization,
mailing address, and phone number of three or four referees? If you
are not listing this information, does the strength of the rest of the résumé offset the omission?

FIGURE 19-5

Follow-up Letter

1901 Chestnut Street
Phoenix, AZ 63014

July 13, 19--

Mr. Daryl Weaver
Director of Operations
Cynergo, Inc.
Spokane, WA 92003

Dear Mr. Weaver:

I would like to thank you for taking the time yesterday to show me
your facilities and to introduce me to your colleagues.

Your advances in piping design were particularly impressive. As a
person with hands-on experience in piping design, I can fully appreciate the advantages your design will have.

The vitality of your projects and the obvious good fellowship
among your employees further confirm my initial belief that
Cynergo would be a fine place to work. I would look forward to
joining your staff.

Sincerely yours,

Albert Rossman

8. Does the résumé include any other appropriate sections, such as military service, language skills, or honors?

JOB-APPLICATION LETTER

1. Does the letter have a professional appearance?

2. Does the introductory paragraph identify your source of information
and the position you are applying for, state that you wish to be considered, and forecast the rest of the letter?

3. Does the education paragraph respond to your reader's needs with a
unified idea introduced by a topic sentence?

4. Does the employment paragraph respond to your reader's needs with a unified idea introduced by a topic sentence?

5. Does the concluding paragraph include a reference to your résumé, a request for an interview, and your phone number?

6. Does the letter include an enclosure notation?

FOLLOW-UP LETTER

Does the letter thank the interviewer and remind him or her of your qualifications?

EXERCISES

1. In a newspaper or journal, find a job advertisement for a position in your field for which you might be qualified. Write a résumé and a job-application letter in response to the ad.

2. Revise the following résumés to make them more effective. Make up any reasonable details.

    a.

```
                                    Bob Jenkins
                                    2319 Fifth Avenue
                                    Waverly, CT 01603
                                    Phone: 611-3356

Personal Data:                      22 years old
                                    Height 5'11"
                                    Weight 176 lbs.

Education:                          B.S. in Electrical Engineering
                                    University of Connecticut,
                                    June 19--

Experience:        6/---9/--        Falcon Electronics
                                    Examined panels for good wiring.
                                    Also, I revised several schemat-
                                    ics.

                   6/---9/--        MacDonalds Electrical Supply Co.
                                    Worked parts counter.

                   6/---9/--        Happy Burger
                                    Made hamburgers, fries, shakes,
                                    fish sandwiches, and fried
                                    chicken.

                   6/---9/--        Town of Waverly
                                    Outdoor maintenance. In charge of
                                    cleaning up McHenry Park and Mu-
                                    nicipal Pool picnic grounds. Did
                                    repairs on some electrical equip-
                                    ment.
```

Background:	Born and raised in Waverly. Third baseman, Fisherman's Rest softball team. Hobbies: jogging, salvaging and repairing appliances, reading magazines, politics.
References:	Will be furnished upon request.

b.

Harold Perkins

1415 Ninth Street
Altoona, Pa. 16013
667-1415

319 W. Irvin Avenue
State College, Pa.
304-9316

Education:	Economics Major Penn State, 19-- Cum: 3.03
	Important courses: Econ. 412, 413, 501, 516. Fin. 353, 614 (seminar), Mgt. 502-503.
Experience:	Radio Shack, State College Bookkeeper. Maintained the books.
	Radio Shack, State College Clerk
	Holy Redeemer Hospital Volunteer
Organizations:	Scuba Club Alpha Sigma Sigma--Pledge Master Treasurer, Penn State Economics Association

References:

George Williams Economics Dept. McKee Hall State College, Pa.	Arthur Lawson Radio Shack State College, Pa.	Dr. John Tepper Holy Redeemer Hospital Boalsburg, Pa.

3. Revise the following job-application letters to make them more effective. Make up any reasonable details.

a.

April 13, 19--

Wayne Grissert
Best Department Store
113 Hawthorn
Atlanta, Georgia

Dear Mr. Grissert:

As I was reading the Sunday Examiner, I came upon your ad for a buyer. I have always been interested in learning about the South, so would you consider my application?

I will receive my degree in fashion design in one month. I have taken many courses in fashion design, so I feel I have a strong background in the field.

Also, I have had extensive experience in retail work. For two summers I sold women's accessories at a local clothing store. In addition, I was a temporary department head for two weeks.

I have enclosed a resume and would like to interview you at your convenience. I hope to see you in the near future. My phone number is 436-6103.

Sincerely,

Brenda Adamson

Brenda Adamson

b.

113 Holloway Drive
Nanuet, Oklahoma 61403

May 11, 19--

Miss Betty Richard
Manager of Technical Employment
Scott Paper Co.
Philadelphia, Pa. 19113

Dear Miss Betty,

I am writing this letter in response to your ad in the Philadelphia Inquirer, March 23, regarding a job opening in research and development.

I am expecting a B.S. degree in chemical engineering this June. I am particularly interested in design and have a strong background in it. I am capable of working independently and am able to make certain important decisions pertaining to certain problems encountered in chemical processes. I am including my résumé with this letter.

If you think I can be of any help to you at your company, please contact me at your convenience. Thank you.

Sincerely,

Glenn Corwin

Glenn Corwin

4. Revise this follow-up letter to make it more effective. Make up any reasonable details.

914 Imperial Boulevard
Durham, NC 27708

November 13, 19--

Mr. Ronald O'Shea
Division Engineering
Safeway Electronics, Inc.
Holland, MI 49423

Dear Mr. O'Shea:

Thanks very much for showing me around your plant. I hope I was able to convince you that I'm the best person for the job.

Sincerely yours,

Robert Wilcox

Robert Wilcox

APPENDIX A

HANDBOOK

This handbook concentrates on style, punctuation, and mechanics. Where appropriate, common errors are defined directly after the correct usage is discussed.

Many of the usage recommendations made here are only suggestions. If your organization or professional field has a style guide that makes different recommendations, you should of course follow it.

Also, note that this is a selective handbook. It cannot replace full-length treatments, such as the handbooks often used in composition courses.

STYLE

USE MODIFIERS EFFECTIVELY Technical writing is full of modifiers—phrases and clauses that describe other elements in the sentence. To make your meaning clear, you must use modifiers effectively. You must clearly communicate to your readers whether a modifier provides necessary information about its referent (the word or phrase it refers to) or simply provides additional information. Further, you must make sure that the referent itself is always clearly identified.

RESTRICTIVE AND NONRESTRICTIVE MODIFIERS A **restrictive modifier**, as the term implies, restricts the meaning of its referent: that is, it provides information necessary to identify the referent. In the following example, the restrictive modifiers are italicized.

The aircraft *used in the exhibitions* are slightly modified.
Please disregard the notice *you just received from us.*

In most cases, the restrictive modifier doesn't require a pronoun, such as *that* or *which*. If you choose to use a pronoun, however, use *that*: "The aircraft that are used in the exhibits are slightly modified." (If the pronoun refers to a person or persons, use *who*.) Notice that restrictive modifiers are not set off by commas.

A **nonrestrictive modifier** does not restrict the meaning of its referent: it provides information that is not necessary to identify the referent. In the following examples, the nonrestrictive modifiers are italicized.

The space shuttle, *described by NASA as a space bus*, made its first flight in 1980.
When you arrive, go to the Registration Area, *which is located on the second floor.*

Like the restrictive modifier, the nonrestrictive modifier usually does not require a pronoun. If you use one, however, choose *which* (*who* or *whom* when referring to a person). Note that nonrestrictive modifiers are separated from the rest of the sentence by commas.

MISPLACED MODIFIERS The placement of the modifier often determines the meaning of the sentence. Notice, for instance, how the placement of *only* changes the meaning in the following sentences.

Only Turner received a cost-of-living increase last year.
(*Meaning:* Nobody else received one.)

Turner received only a cost-of-living increase last year.
(*Meaning:* He didn't receive a merit increase.)

Turner received a cost-of-living increase only last year.
(*Meaning:* He received a cost-of-living increase as recently as last year.)

Turner received a cost-of-living increase last year only.
(*Meaning:* He received a cost-of-living increase in no other year.)

Misplaced modifiers—those that appear to modify the wrong referent—pose a common problem in technical writing. The solution is, in general, to make sure the modifier is placed as close as possible to its intended referent. Frequently, the misplaced modifier is a phrase or a clause:

MISPLACED

The subject of the meeting is the future of geothermal energy in the downtown Webster Hotel.

CORRECT

The subject of the meeting in the downtown Webster Hotel is the future of geothermal energy.

MISPLACED

Jumping around nervously in their cages, the researchers speculated on the health of the mice.

CORRECT

The researchers speculated on the health of the mice jumping around nervously in their cages.

A special kind of misplaced modifier is called a **squinting modifier**—one that is placed ambiguously between two potential referents, so that the reader cannot tell which one is being modified:

UNCLEAR

We decided immediately to purchase the new system.

CLEAR

We immediately decided to purchase the new system.

CLEAR

We decided to purchase the new system immediately.

UNCLEAR

The men who worked on the assembly line reluctantly picked up their last paychecks.

CLEAR

The men who worked reluctantly on the assembly line picked up their last paychecks.

CLEAR

The men who worked on the assembly line picked up their last paychecks reluctantly.

DANGLING MODIFIERS A **dangling modifier** is one that has no referent in the sentence:

Searching for the correct answer to the problem, the instructions seemed unclear.

In this sentence, the person doing the searching has not been identified. To correct the problem, rewrite the sentence to put the clarifying information either *within* the modifier or *next to* the modifier:

As I was searching for the correct answer to the problem, the instructions seemed unclear.

Searching for the correct answer to the problem, I thought the instructions seemed unclear.

A writer sometimes can correct a dangling modifier by switching from the indicative mood (a statement of fact) to the imperative mood (a request or command):

DANGLING

To initiate the procedure, the BEGIN button should be pushed.

CORRECT

To initiate the procedure, push the BEGIN button.

In the imperative, the referent—in this case, *you*—is understood.

KEEP
PARALLEL
ELEMENTS
PARALLEL

A sentence is **parallel** if its coordinate elements are expressed in the same grammatical form: that is, all its clauses are either passive or active, all its verbs are either infinitives or participles, and so forth. By creating and sustaining a recognizable pattern for the reader, parallelism makes the sentence easier to follow.

Notice how faulty parallelism weakens the following sentences.

NONPARALLEL

Our present system is costing us profits and reduces our productivity. (*nonparallel verbs*)

PARALLEL

Our present system is costing us profits and reducing our productivity.

NONPARALLEL

The dignitaries watched the launch, and the crew was applauded. (*nonparallel voice*)

PARALLEL

The dignitaries watched the launch and applauded the crew.

NONPARALLEL

The typist should follow the printed directions; do not change the originator's work. (*nonparallel mood*)

PARALLEL

The typist should follow the printed directions and not change the originator's work.

A subtle form of faulty parallelism often occurs with the correlative constructions, such as *either . . . or, neither . . . nor,* and *not only. . . but also*:

NONPARALLEL

The new refrigerant not only decreases energy costs but also spoilage losses.

PARALLEL

The new refrigerant decreases not only energy costs but also spoilage losses.

In this example, "decreases" applies to both "energy costs" and "spoilage losses." Therefore, the first half of the correlative construction should follow "decreases." Note that if the sentence contains two different verbs, the first half of the correlative construction should precede the verb:

The new refrigerant not only decreases energy costs but also prolongs product freshness.

When creating parallel constructions, make sure that parallel items in a series do not overlap, thus changing or confusing the meaning of the sentence:

CONFUSING

The speakers will include partners of law firms, businesspeople, and civic leaders.

CLEAR

The speakers will include businesspeople, civic leaders, and partners of law firms.

The problem with the original sentence is that "partners" appears to apply to "businesspeople" and "civic leaders." The revision solves the problem by rearranging the items so that "partners" cannot apply to the other two groups in the series.

Parallelism should be maintained not only within individual sentences but also among sentences in paragraphs. When you establish a pattern in a paragraph, follow it through. The pattern may be as simple as numbering the steps in a process:

PARALLEL

Correlating the two results is a three-part procedure. First, ...
. . . . Second, ...
And third, . . .

Keep the voice consistent—either active or passive:

PARALLEL

The sample is placed in the petri dish. Two drops of culture are added.
. . . After two days, the growth rate is examined by. . . . Finally, the sample is weighed.

AVOID
AMBIGUOUS
PRONOUN
REFERENCE

Pronouns must refer clearly to the words or phrases they replace. **Ambiguous pronoun references** can lurk in even the most innocent-looking sentences:

UNCLEAR

Remove the cell cluster from the medium and analyze it. (*Analyze what, the cell cluster or the medium?*)

CLEAR

Analyze the cell cluster after removing it from the medium.

CLEAR

Analyze the medium after removing the cell cluster from it.

CLEAR

Remove the cell cluster from the medium. Then analyze the cell cluster.

CLEAR

Remove the cell cluster from the medium. Then analyze the medium.

Ambiguous references can also occur when a relative pronoun such as *which*, or a subordinating conjunction such as *where*, is used to introduce a dependent clause:

UNCLEAR

She decided to evaluate the program, which would take five months. (*What would take five months, the program or the evaluation?*)

CLEAR

She decided to evaluate the program, a process that would take five months. (*By replacing "which" with "a process that," the writer clearly indicates that the evaluation will take five months.*)

CLEAR

She decided to evaluate the five-month program. (*By using the adjective "five-month," the writer clearly indicates that the program will take five months.*)

UNCLEAR

This procedure will increase the handling of toxic materials outside the plant, where adequate safety measures can be taken. (*Where can adequate safety measures be taken, inside the plant or outside?*)

CLEAR

This procedure will increase the handling of toxic materials outside the plant. Because adequate safety measures can be taken only in the plant, the procedure poses risks.

CLEAR

This procedure will increase the handling of toxic materials outside the plant. Because adequate safety measures can be taken only outside the plant, the procedure will decrease safety risks.

As the last example shows, sometimes the best way to clarify an unclear pronoun is to split the sentence in two, eliminate the problem, and add clarifying information. Clarity is always the primary characteristic of good technical writing. If more words will make your writing clearer, use them.

Ambiguity can also occur at the beginning of a sentence:

UNCLEAR

Allophanate linkages are among the most important structural components of polyurethane elastomers. They act as cross-linking sites. (*What act as cross-linking sites, allophanate linkages or polyurethane elastomers?*)

CLEAR

Allophanate linkages, which are among the most important structural components of polyurethane elastomers, act as cross-linking sites. (*The writer has changed the second sentence into a clear nonrestrictive modifier.*)

Your job is to use whichever means—restructuring the sentence or dividing it in two—will best ensure that the reader will know exactly which word or phrase the pronoun is replacing.

COMPARE
ITEMS
CLEARLY

When comparing or contrasting items, make sure your sentence clearly communicates the relationship. A simple comparison between two items often causes no problems: "The X3000 has more storage than the X2500." However, don't let your reader confuse a comparison and a simple statement of fact. For example, in the sentence "Trout eat more than herring minnows," does the writer mean that trout don't restrict their diet to minnows or that trout eat more than herring minnows eat? If a comparison is intended, a second verb should be used: "Trout eat more than minnows do." And if three items are introduced, make sure that the reader can tell which two are being compared:

AMBIGUOUS

Trout eat more algae than herring minnows.

CLEAR

Trout eat more algae than they do herring minnows.

CLEAR

Trout eat more algae than herring minnows do.

Beware of comparisons in which different aspects of the two items are compared:

ILLOGICAL

The resistance of the copper wiring is lower than the tin wiring.

LOGICAL

The resistance of the copper wiring is lower than that of the tin wiring.

In the illogical construction, the writer contrasts "resistance" with "tin wiring" rather than the resistance of copper with the resistance of tin. In the revision, the pronoun *that* is used to substitute for the repetition of "resistance."

USE
ADJECTIVES
CLEARLY

In general, adjectives are placed before the nouns they modify: "the plastic washer." Technical writing, however, often requires clusters of adjectives. To prevent confusion, use commas to separate coordinate adjectives, and use hyphens to link compound adjectives.

Adjectives that describe different aspects of the same noun are known as coordinate adjectives:

In this case, the comma replaces the word *and*.

Note that sometimes an adjective is considered part of the noun it describes: "electric drill." When one adjective is added to "electric drill," no comma is required: "a reversible electric drill." The addition of two or more adjectives, however, creates the traditional coordinate construction: "a two-speed, reversible electric drill."

The phrase "two-speed" is an example of a compound adjective—one made up of two or more words. Use hyphens to link the elements in compound adjectives that precede nouns:

a variable-angle accessory

increased cost-of-living raises

The hyphens in the second example prevent the reader from momentarily misinterpreting "increased" as an adjective modifying "cost" and "living" as a participle modifying "raises."

A long string of compound adjectives can be confusing even if hyphens are used appropriately. To ensure clarity in such a case, put the adjectives into a clause or phrase following the noun:

UNCLEAR

an operator-initiated, default-prevention technique

CLEAR

a technique initiated by the operator for preventing default

In turning a string of adjectives into a phrase or clause, make sure the adjectives cannot be misread as verbs. Use the pronouns *that* and *which* to prevent confusion:

CONFUSING

The good experience provides is often hard to measure.

CLEAR

The good that experience provides is often hard to measure.

MAINTAIN NUMBER AGREEMENT   Number disagreement commonly takes one of two forms in technical writing: (1) the verb disagrees in number with the subject when a prepositional phrase intervenes; (2) the pronoun disagrees in number with its referent when the latter is a collective noun.

SUBJECT-VERB DISAGREEMENT A prepositional phrase does not affect the number of the subject and the verb. The following examples show that the object of the preposition can be plural in a singular sentence, or singular in a plural sentence. (The subjects and verbs are italicized.)

INCORRECT

The *result* of the tests *are* promising.

CORRECT

The *result* of the tests *is* promising.

INCORRECT

The *results* of the test *is* promising.

CORRECT

The *results* of the test *are* promising.

Don't be misled by the fact that the object of the preposition and the verb don't sound natural together, as in *tests is* or *test are*. Grammatical agreement of subject and verb is the primary consideration.

PRONOUN-ANTECEDENT DISAGREEMENT The problem of pronoun-antecedent disagreement crops up most often when the antecedent, or referent, is a collective noun—one that can be interpreted as either singular or plural, depending on its usage:

INCORRECT

The *company* is proud to announce a new stock option plan for *their* employees.

CORRECT

The *company* is proud to announce a new stock option plan for *its* employees.

In this example, "the company" acts as a single unit; therefore, the singular verb, followed by a singular pronoun, is appropriate. When the individual members of a collective noun are stressed, however, plural pronouns and verbs are appropriate: "The inspection team have prepared their reports."

PUNCTUATION

THE PERIOD   Periods are used in the following instances.

1. At the end of sentences that do not ask questions or express strong emotion:

The lateral stress still needs to be calculated.

2. After some abbreviations:

M.D.

U.S.A.

etc.

(For a further discussion of abbreviations, see pp. 572–573.)

3. With decimal fractions:

4.056

$6.75

75.6%

THE EXCLA-
MATION POINT

The exclamation point is used at the end of a sentence that ex-
presses strong emotion, such as surprise or doubt:

The nuclear plant, which was originally expected to cost $1.6 billion,
eventually cost more than $4 billion!

Because technical writing requires objectivity and a calm, understated tone, technical writers rarely use exclamation points.

THE
QUESTION
MARK

The question mark is used at the end of a sentence that asks a direct question:

What did the commission say about effluents?

Do not use a question mark at the end of a sentence that asks an
indirect question:

He wanted to know whether the procedure had been approved for use.

When a question mark is used within quotation marks, the quoted
material needs no other end punctuation:

"What did the commission say about effluents?" she asked.

THE COMMA

The comma is the most frequently used punctuation mark, as well
as the one about whose usage many writers most often disagree.
Following are the basic uses of the comma.

1. To separate the clauses of a compound sentence (one composed of two or more independent clauses) linked by a coordinating conjunction (*and, or, nor, but, so, for, yet*):

Both methods are acceptable, but we have found that the Simpson procedure gives better results.

In many compound sentences, the comma is needed to prevent the reader from mistaking the subject of the second clause for an object of the verb in the first clause:

The RESET command affects the field access, and the SEARCH command affects the filing arrangement.

Without the comma, the reader is likely to interpret the coordinating conjunction "and" as a simple conjunction linking "field access" and "SEARCH command."

2. To separate items in a series composed of three or more elements:

The manager of spare parts is responsible for ordering, stocking, and disbursing all spare parts for the entire plant.

The comma following the second-to-last item is required by most technical writing style manuals, despite the presence of the conjunction "and." The comma clarifies the separation and prevents misreading. For example, sometimes in technical writing the second-to-last item will be a compound noun containing an "and."

The report will be distributed to Operations, Research and Development, and Accounting.

3. To separate introductory words, phrases, and clauses from the main clause of the sentence:

However, we will have to calculate the effect of the wind.
To facilitate trade, the government holds a yearly international conference.
Whether the workers like it or not, the managers have decided not to try the flextime plan.

In each of these three examples, the comma helps the reader follow the sentence. Notice in the following example how the comma actually prevents misreading:

Just as we finished eating, the rats discovered the treadmill.

The comma is optional if the introductory text is brief and cannot be misread.

CORRECT

First, let's take care of the introductions.

CORRECT

First let's take care of the introductions.

4. To separate the main clause from a dependent clause:

The advertising campaign was canceled, although most of the executive council saw nothing wrong with it.
Most accountants wear suits, whereas few engineers do.

5. To separate nonrestrictive modifiers (parenthetical clarifications) from the rest of the sentence:

Jones, the temporary chairman, called the meeting to order.

6. To separate interjections and transitional elements from the rest of the sentence:

Yes, I admit your findings are correct.
Their plans, however, have great potential.

7. To separate coordinate adjectives:

The finished product was a sleek, comfortable cruiser.
The heavy, awkward trains are still being used.

The comma here takes the place of the conjunction "and." If the adjectives are not coordinate—that is, if one of the adjectives modifies the combination of the adjective and the noun—do not use a comma:

They decided to go to the first general meeting.

8. To signal that a word or phrase has been omitted from an elliptical expression:

Smithers is in charge of the accounting; Harlen, the data management; Demarest, the publicity.

In this example, the commas after "Harlen" and "Demarest" show that the phrase "is in charge of" has been omitted.

9. To separate a proper noun from the rest of the sentence in direct address:

John, have you seen the purchase order from United?
What I'd like to know, Betty, is why we didn't see this problem coming.

10. To introduce most quotations:

He asked, "What time were they expected?"

11. To separate towns, states, and countries:

Bethlehem, Pennsylvania, is the home of Lehigh University.
He attended Lehigh University in Bethlehem, Pennsylvania, and the University of California at Berkeley.

Note the use of the comma after "Pennsylvania."

12. To set off the year in dates:

August 1, 1989, is the anticipated completion date.

Note the use of the comma after "1989." If the month separates the date from the year, the commas are not used, because the numbers are not next to each other:

The anticipated completion date is 1 August 1989.

13. To clarify numbers:

12,013,104

(European practice is to reverse the use of commas and periods in writing numbers: periods are used to signify thousands, and commas to signify decimals.)

14. To separate names from professional or academic titles:

Harold Clayton, Ph.D.
Marion Fewick, CLU
Joyce Carnone, P.E.

Note that the comma also follows the title in a sentence:

Harold Clayton, Ph.D., is the featured speaker.

COMMON ERRORS

1. No comma between the clauses of a compound sentence:

INCORRECT

The mixture was prepared from the two premixes and the remaining ingredients were then combined.

CORRECT

The mixture was prepared from the two premixes, and the remaining ingredients were then combined.

2. No comma (or just one comma) to set off a nonrestrictive modifier:

INCORRECT

The phone line, which was installed two weeks ago had to be disconnected.

CORRECT

The phone line, which was installed two weeks ago, had to be disconnected.

3. No comma separating introductory words, phrases, or clauses from the main clause, when misreading can occur:

INCORRECT

As President Canfield has been a great success.

CORRECT

As President, Canfield has been a great success.

4. No comma (or just one comma) to set off an interjection or a transitional element:

INCORRECT

Our new statistician, however used to work for Konaire, Inc.

CORRECT

Our new statistician, however, used to work for Konaire, Inc.

5. Comma splice (a comma used to "splice together" independent clauses not linked by a coordinating conjunction):

INCORRECT

All the motors were cleaned and dried after the water had entered, had they not been, additional damage would have occurred.

CORRECT

All the motors were cleaned and dried after the water had entered; had they not been, additional damage would have occurred.

CORRECT

All the motors were cleaned and dried after the water had entered. Had they not been, additional damage would have occurred.

6. Superfluous commas:

INCORRECT

Another of the many possibilities, is to use a "first in, first out" sequence. (*In this sentence, the comma separates the subject, "Another," from the verb, "is."*)

CORRECT

Another of the many possibilities is to use a "first in, first out" sequence.

INCORRECT

The schedules that have to be updated every month are, 14, 16, 21, 22, 27, and 31. (*In this sentence, the comma separates the verb from its complement.*)

CORRECT

The schedules that have to be updated every month are 14, 16, 21, 22, 27, and 31.

INCORRECT

The company has grown so big, that an informal evaluation procedure is no longer effective. (*In this sentence, the comma separates the predicate adjective "big" from the clause that modifies it.*)

CORRECT

The company has grown so big that an informal evaluation procedure is no longer effective.

INCORRECT

Recent studies, and reports by other firms confirm our experience. (*In this sentence, the comma separates the two elements in the compound subject.*)

CORRECT

Recent studies and reports by other firms confirm our experience.

INCORRECT

New and old employees who use the processed order form, do not completely understand the basis of the system. (*In this sentence, a comma separates the subject and its restrictive modifier from the verb.*)

CORRECT

New and old employees who use the processed order form do not completely understand the basis of the system.

THE SEMICOLON

Semicolons are used in the following instances.

1. To separate independent clauses not linked by a coordinating conjunction:

The second edition of the handbook is more up-to-date; however, it is more expensive.

2. To separate items in a series that already contains commas:

The members elected three officers: Jack Resnick, president; Carol Wayshum, vice-president; Ahmed Jamoogian, recording secretary.

In this example, the semicolon acts as a "supercomma," keeping the names and titles clear.

COMMON ERROR Use of a semicolon when a colon is called for:

INCORRECT

We still need one ingredient; luck.

CORRECT

We still need one ingredient: luck.

THE COLON

Colons are used in the following instances.

1. To introduce a word, phrase, or clause that amplifies or explains a general statement:

The project team lacked one crucial member: a project leader.

Here is the client's request: we are to provide the preliminary proposal by November 13.

We found three substances in excessive quantities: potassium, cyanide, and asbestos.

The week had been productive: fourteen projects had been completed and another dozen had been initiated.

Note that the text preceding a colon should be able to stand on its own as a main clause:

INCORRECT

We found: potassium, cyanide, and asbestos.

CORRECT

We found potassium, cyanide, and asbestos.

2. To introduce items in a vertical list, if the introductory text would be incomplete without the list:

We found the following:

   potassium
   cyanide
   asbestos

3. To introduce long or formal quotations:

The president began: "In the last year . . ."

COMMON ERROR   Use of a colon to separate a verb from its complement:

INCORRECT

The tools we need are: a plane, a level, and a T-square.

CORRECT

The tools we need are a plane, a level, and a T-square.

CORRECT

We need three tools: a plane, a level, and a T-square.

THE DASH   Dashes are used in the following instances.

1. To set off a sudden change in thought or tone:

The committee found—can you believe this?—that the company bore *full* responsibility for the accident.

That's what she said—if I remember correctly.

2. To emphasize a parenthetical element:

The managers' reports—all ten of them—recommend production cutbacks for the coming year.

Arlene Kregman—the first woman elected to the board of directors—is the next scheduled speaker.

3. To set off an introductory series from its explanation:

Wetsuits, weight belts, tanks—everything will have to be shipped in.

When a series *follows* the general statement, a colon replaces the dash:

Everything will have to be shipped in: wetsuits, weight belts, and tanks.

Note that typewriters and most word processors do not have a key for the dash. In typewritten or word-processed text, a dash is represented by two uninterrupted hyphens. No space precedes or follows the dash.

COMMON ERROR Use of a dash as a "lazy" substitute for other punctuation marks:

INCORRECT

The regulations—which were issued yesterday—had been anticipated for months.

CORRECT

The regulations, which were issued yesterday, had been anticipated for months.

INCORRECT

Many candidates applied—however, only one was chosen.

CORRECT

Many candidates applied; however, only one was chosen.

PARENTHESES   Parentheses are used in the following instances.

1. To set off incidental information:

Please call me (x3104) when you get the information.

Galileo (1564–1642) is often considered the father of modern astronomy.

H. W. Fowler's *Modern English Usage* (New York: Oxford University Press, 2nd ed., 1965) is still the final arbiter.

2. To enclose numbers and letters that label items listed in a sentence:

To transfer a call within the office, (1) place the party on HOLD, (2) press TRANSFER, (3) press the extension number, and (4) hang up.

Use both a left and a right parenthesis—not just a right parenthesis—in this situation.

COMMON ERROR Use of parentheses instead of brackets to enclose the writer's interruption of a quotation (see the discussion of brackets):

INCORRECT

He said, "The new manager (Farnham) is due in next week."

CORRECT

He said, "The new manager [Farnham] is due in next week."

THE HYPHEN   Hyphens are used in the following instances.

1. In general, to form compound adjectives that precede nouns:

general-purpose register
meat-eating dinosaur
chain-driven saw

Note that hyphens are not used after words that end in -ly:

newly acquired terminal

Also note that hyphens are not used when the compound adjective follows the noun:

The Woodchuck saw is chain driven.

Many organizations have their own preferences about hyphenating compound adjectives. Check to see if your organization has a preference.

2. To form some compound nouns:

vice-president
editor-in-chief

3. To form fractions and compound numbers:

one-half
fifty-six

4. To attach some prefixes and suffixes:

post-1945
president-elect

5. To divide a word at the end of a line:

We will meet in the pavil-
ion in one hour.

Whenever possible, avoid such breaks; they annoy some readers. When you do use them, check the dictionary to make sure you have divided the word *between* syllables.

**THE
APOSTROPHE**

Apostrophes are used in the following instances.

1. To indicate the possessive case:

the manager's goals
the foremen's lounge
the employees' credit union
Charles's T-square

For joint possession, add the apostrophe and the *s* to only the last noun or proper noun:

Watson and Crick's discovery

For separate possession, add an apostrophe and an *s* to each of the nouns or pronouns:

Newton's and Galileo's ideas

Make sure you do not add an apostrophe or an *s* to possessive pronouns: *his, hers, its, ours, yours, theirs.*

2. To form contractions:

I've

can't

shouldn't

it's

The apostrophe usually indicates an omitted letter or letters. For example, *can't* is *can(no)t*, *it's* is *it(i)s*.

Some organizations discourage the use of contractions; others have no preference. Find out the policy your organization follows.

3. To indicate special plurals:

three 9's

two different JCL's

the why's and how's of the problem

As in the case of contractions, it is a good idea to learn the stylistic preferences of your organization. Usage varies considerably.

COMMON ERROR Use of the contraction *it's* in place of the possessive pronoun *its*.

INCORRECT

The company does not feel that the problem is it's responsibility.

CORRECT

The company does not feel that the problem is its responsibility.

QUOTATION MARKS Quotation marks are used in the following instances.

1. To indicate titles of short works, such as articles, essays, or chapters:

Smith's essay "Solar Heating Alternatives"

2. To call attention to a word or phrase that is being used in an unusual way or in an unusual context:

A proposal is "wired" if the sponsoring agency has already decided who will be granted the contract.

Don't use quotation marks as a means of excusing poor word choice:

The new director has been a real "pain."

3. To indicate direct quotation: that is, the words a person has said or written:

"In the future," he said, "check with me before authorizing any large purchases."
As Breyer wrote, "Morale *is* productivity."

Do not use quotation marks to indicate indirect quotation:

INCORRECT

He said that "third-quarter profits would be up."

CORRECT

He said that third-quarter profits would be up.

CORRECT

He said, "Third-quarter profits will be up."

RELATED PUNCTUATION Note that if the sentence contains a "tag"—a phrase identifying the speaker or writer—a comma is used to separate it from the quotation:

John replied, "I'll try to fly out there tomorrow."
"I'll try to fly out there tomorrow," John replied.

Informal and brief quotations require no punctuation before the quotation marks:

She said "Why?"

In the United States (but not in most other English-speaking nations), commas and periods at the end of quotations are placed within the quotation marks:

The project engineer reported, "A new factor has been added."
"A new factor has been added," the project engineer reported.

Question marks, dashes, and exclamation points, on the other hand, are placed inside the quotation marks when they apply only to the quotation and outside the quotation marks when they apply to the whole sentence:

He asked, "Did the shipment come in yet?"
Did he say, "This is the limit"?

Note that only one punctuation mark is used at the end of a set of quotation marks:

INCORRECT

Did she say, "What time is it?"?

CORRECT

Did she say, "What time is it?"

BLOCK QUOTATIONS

When quotations reach a certain length—generally, more than four lines—writers tend to switch to a block format. In typewritten manuscript, a block quotation is usually

1. indented ten spaces from the left-hand margin
2. single-spaced
3. typed without quotation marks
4. introduced by a colon

(Different organizations observe their own variations on these basic rules.)

McFarland writes:

> The extent to which organisms adapt to their environment is still being charted. Many animals, we have recently learned, respond to a dry winter with an automatic birth control chemical that limits the number of young to be born that spring. This prevents mass starvation among the species in that locale.

Hollins concurs. She writes, "Biological adaptation will be a major research area during the next decade."

MECHANICS

ELLIPSES

Ellipses (three spaced periods) indicate the omission of some material from a quotation. A fourth period with no space before it precedes ellipses when the sentence in the source has ended and you are omitting material that follows or when the omission follows a portion of the source's sentence that is in itself a grammatically complete sentence:

"Send the updated report . . . as soon as you can."

Larkin refers to the project as "an attempt . . . to clarify the issue of compulsory arbitration. . . . We do not foresee an end to the legal wrangling . . . but perhaps the report can serve as a definition of the areas of contention."

In the second example, the writer has omitted words after "attempt" and after "wrangling." In addition, she has used a sentence period plus three spaced periods after "arbitration," which ends the original writer's sentence; and she has omitted the following sentence.

BRACKETS   Brackets are used in the following instances.

1. To indicate words added to a quotation:

The minutes of the meeting note that "He [Pearson] spoke out against the proposal."

A better approach would be to shorten the quotation:

Pearson "spoke out against the proposal."

2. To indicate parentheses within parentheses:

(For further information, see Charles Houghton's *Civil Engineering Today* [New York: Arch Press, 1982].)

ITALICS   Italics (or underlining) are used in the following instances.

1. For words used as words:

In this report, the word *operator* will refer to any individual who is actually in charge of the equipment, regardless of that individual's certification.

2. To indicate titles of long works (books, manuals, etc.), periodicals and newspapers, films, plays, and long musical works:

See Houghton's *Civil Engineering Today.*
We subscribe to the *New York Times.*

3. To indicate the names of ships, trains, and airplanes:

The shipment is expected to arrive next week on the *Penguin*.

4. To set off foreign expressions that have not become fully assimilated into English:

The speaker was guilty of *ad hominem* arguments.

5. To emphasize words or phrases:

Do not press the ERASE key.

If your typewriter or word processor does not have italic type, indicate italics by underlining.

Darwin's <u>Origin of Species</u> is still read today.

NUMBERS
The use of numbers varies considerably. Therefore, you should find out what guidelines your organization or research area follows in choosing between words and numerals. Many organizations use the following guidelines.

1. Use numerals for technical quantities, especially if a unit of measurement is included:

3 feet
12 grams
43,219 square miles
36 hectares

2. Use numerals for nontechnical quantities of 10 or more:

300 persons
12 whales
35% increase

3. Use words for nontechnical quantities of fewer than 10:

three persons
six whales

4. Use both words and numerals

  a. For back-to-back numbers:

six 3-inch screws
fourteen 12-foot ladders
3,012 five-piece starter units

In general, use the numeral for the technical unit. If the non-technical quantity would be cumbersome in words, use the numeral.

  b. For round numbers over 999,999:

14 million light-years
$64 billion

  c. For numbers in legal contracts or in documents intended for international readers:

thirty-seven thousand dollars ($37,000)
five (5) relays

  d. For addresses:

3801 Fifteenth Street

SPECIAL CASES

1. If a number begins a sentence, use words, not numerals:

Thirty-seven acres was the agreed-upon size of the lot.

Many writers would revise the sentence to avoid this problem:

The agreed-upon size of the lot was 37 acres.

2. Don't use both numerals and words in the same sentence to refer to the same unit:

On Tuesday the attendance was 13; on Wednesday, 8.

3. Write out fractions, except if they are linked to technical units:

two-thirds of the members
3½ hp.

4. Write out approximations:

approximately ten thousand people
about two million trees

5. Use numerals for titles of figures and tables and for page numbers:

Figure 1
Table 13
page 261

6. Use numerals for decimals:

3.14
1,013.065

Add a zero before decimals of less than one:

0.146
0.006

7. Avoid expressing months as numbers, as in "3/7/84": in the United States, this means March 7, 1984; in many other countries, it means July 3, 1984. Use one of the following forms:

March 7, 1984
7 March 1984

8. Use numerals for times if A.M. or P.M. are used:

6:10 A.M.
six o'clock

ABBREVIATIONS   Abbreviations provide a useful way to save time and space, but you must use them carefully; you can never be sure that your readers will understand them. Many companies and professional organizations have lists of approved abbreviations.

Analyze your audience in determining whether and how to abbreviate. If your readers include nontechnical people unfamiliar with your field, either write out the technical terms or attach a list of abbreviations. If you are new in an organization or are writing for publication for the first time in a certain field, find out what abbreviations are commonly used. If for any reason you are unsure about whether or how to abbreviate, write out the word.

The following are general guidelines about abbreviations.

1. You may make up your own abbreviations. For the first reference to the term, write it out and include, parenthetically, the abbreviation. In subsequent references, use the abbreviation. For long works, you might want to write out the term at the start of major units, such as chapters.

The heart of the new system is the self-loading cartridge (slc).

This technique is also useful, of course, in referring to existing abbreviations that your readers might not know:

The cathode-ray tube (CRT) is your control center.

2. Most abbreviations do not take plurals:

1 lb
3 lb

3. Most abbreviations in scientific writing are not followed by periods:

lb
cos
dc

If the abbreviation can be confused with another word, however, use a period:

in.
Fig.

4. Spell out the unit if the number preceding it is spelled out or if no number precedes it:

How many square meters is the site?

CAPITAL-
IZATION   For the most part, the conventions of capitalization in general writing apply in technical writing.

1. Capitalize proper nouns, titles, trade names, places, languages, religions, and organizations:

William Rusham
Director of Personnel
Quick Fix Erasers
Bethesda, Maryland
Methodism
Italian
Society for Technical Communication

In some organizations, job titles are not capitalized unless they refer to specific persons:

Alfred Loggins, Director of Personnel, is interested in being considered for vice-president of marketing.

2. Capitalize headings and labels:

A Proposal to Implement the Wilkins Conversion System
Section One
The Problem
Figure 6
Mitosis
Table 3
Rate of Inflation, 1960–1980

APPENDIX B

AUTHOR-DATE CITATIONS

NUMBERED CITATIONS

DOCUMENTATION OF SOURCES

Documentation is the explicit identification of the sources of the ideas and quotations used in your report.

For you as the writer, complete and accurate documentation is primarily a professional obligation—a matter of ethics. But it also serves to substantiate the report. Effective documentation helps to place the report within the general context of continuing research and to define it as a responsible contribution to knowledge in the field. Failure to document a source—whether intentionally or unintentionally—is plagiarism. At most universities and colleges, plagiarism means automatic failure of the course and, in some instances, suspension or dismissal. In many companies, it is grounds for immediate dismissal.

For your readers, complete and accurate documentation is an invaluable tool. It enables them to find the source you have relied on, should they want to read more about a particular subject.

What kind of material should be documented? Any quotation from a printed source or an interview, even if it is only a few words, should be documented. In addition, a paraphrased idea, concept, or opinion gathered from your reading should be documented. There is one exception to this rule: if the idea or concept is so well known that it has become, in effect, general knowledge, it need not be documented. Examples of knowledge that is within the public domain would include Einstein's theory of relativity and the Laffer curve. If you are unsure whether an item is within the public domain, document it anyway, just to be safe.

Many organizations have their own preferences for documentation style; others use published style guides, such as the *U.S. Government Printing Office Style Manual*, the American Chemical Society's *Handbook for Authors*, or *The Chicago Manual of Style*. You should find out what your organization's style is and abide by it.

The system of documentation you are probably most familiar with—the notes and bibliography—is used rarely today, especially outside the liberal arts. Even the Modern Language Association, a major organization in the humanities, no longer recommends the use of the notes-and-bibliography style because of the duplication involved in creating two different lists.

Two basic systems of documentation have become standard:

1. author/date citations and bibliography
2. numbered citations and numbered bibliography.

Variations on these systems abound.

AUTHOR/DATE CITATIONS

In the author/date style, a parenthetical notation that includes the name of the source's author and the date of its publication immediately follows the quoted or paraphrased material:

```
This phenomenon was identified as early as forty years ago (Wilkinson
1943).
```

Sometimes, particularly if the reference is to a specific fact or idea, the page (or pages) from the source is also listed:

```
(Wilkinson 1943: 36-37)
```

A citation can also be integrated into the sentence:

```
Wilkinson (1943: 36-37) identified this phenomenon as early as forty
years ago.
```

If two or more sources by the same author in the same year are listed in the bibliography, the notation may include an abbreviated title, to prevent confusion:

(Wilkinson, "Cornea Research," 1943: 36-37)

Or the citation for the first source written that year can be identified with a lowercase letter:

(Wilkinson 1943a: 36-37)

The second source would be identified similarly:

(Wilkinson 1943b: 19-21)

For sources written by two authors, cite both names:

(Smith and Jones 1987)

For sources written by three or more authors, use the first name followed by "*et al.*"

(Smith et al. 1987)

The simplicity and flexibility of the author/date system make it highly attractive. Of course, because the citations are minimal and because their form is dictated more by common sense than by a style sheet, a conventional bibliography that contains complete publication information must be used in conjunction with them, following the text. Your obligation as a writer when you are using this system is to leave your reader no doubt about which of your many sources you are citing in any particular instance. *The Chicago Manual of Style*, 13th edition (Chicago: University of Chicago Press, 1982), advocates using the author/date system throughout the natural and social sciences and recommends the following bibliographic conventions for use with author/date citations.

The order of basic information included in this style of bibliography is as follows:

FOR BOOKS

1. author (last name followed by initials)

2. date of publication

3. title
4. place of publication
5. publisher

1. author
2. date of publication
3. article title
4. journal or anthology title
5. volume (of a journal) or place of publication and publisher (of an anthology)
6. inclusive pages

    The individual entries are arranged alphabetically by author (and then by date if two or more works by the same author are listed) within the bibliography. Anonymous works are integrated into the alphabetical listing by title. Where several works by the same author are included, they are arranged by date under the author's name. In such cases, a long dash (10 hyphens on a typewriter) takes the place of the author's name in all entries after the first. The first line of a bibliographic entry is flushed left with the margin; each succeeding line is indented (five spaces on a typewriter).

Chapman, D. L. 1987. The closed frontier: Why Detroit can't make cars that people will buy. Motorist's Metronome 12 (June):17-26.
----------. 1985. The driver's guide to evaluating compact automobiles. Athens, Ga.: Consumer Press.
Courting credibility: Detroit and its mpg figures. 1986. Countercultural Car & Driver (July):19.

Following are the standard bibliographic forms used with various types of sources.

A BOOK

Cunningham, W. 1980. Crisis at Three Mile Island. New York: Madison.

The author's surname is followed by first initials only. The date of publication comes next, followed by the title of the book, underlined (italicized in print). In the natural and social sciences, the style is generally to capitalize only the first letter of the title, the subtitle (if there is one), and all proper nouns. The last items are the location and name of the publisher.

For a book by two or more authors, all the authors are named.

Cunningham, W., and A. Breyer . . .

Only the name of the first author is inverted.

A BOOK ISSUED BY AN ORGANIZATION

Department of Energy. 1979. The energy situation in the eighties.
    Washington, D.C.: U.S. Government Printing Office, Technical
    Report 11346-53.

A BOOK COMPILED BY AN EDITOR OR ISSUED UNDER AN EDITOR'S NAME

Morgan, K. E., ed. 1987. Readings in alternative energies. Boston:
    Smith-Howell.

AN EDITION OTHER THAN THE FIRST

Schonberg, N. 1987. Solid state physics. 3d ed. London: Paragon.

AN ARTICLE INCLUDED IN A BOOK

May, B., and J. Deacon. 1986. Amplification systems. In Third Annual
    Conference of the American Electronics Association, ed. A.
    Kooper, 101-14. Miami: Valley Press.

A JOURNAL ARTICLE

Hastings, W. 1983. The Skylab debate. The Modern Inquirer 13: 311-18.

The title of the journal is underlined (italicized in print). The first letters of the first and last words and all nouns, adjectives, verbs (except the infinitive *to*), adverbs, and subordinate conjunctions are capitalized. After the journal title come the volume and page numbers of the article.

AN ANONYMOUS JOURNAL ARTICLE

The state of the art in microcomputers. 1985. Newscene 56: 406-21.

Anonymous journal articles are arranged alphabetically by title. If the title begins with a grammatical article such as *the* or *a*, alphabetize it under the first word following the article (in this case, *state*).

A NEWSPAPER ARTICLE

Eberstadt, A. 1983. Why not a Rabbit, why not a Fox? Morristown Mirror
and Telegraph. Sunday 31 July: Business and Finance Section.

A PERSONAL INTERVIEW

Riccio, Dr. Louis, Professor of Operations Research, Tulane University. Interview with author. New York City, 13 July 1986.

The name and professional title of the interviewee should be cited
first, then the fact that the source was an interview, and finally
the place and date of the interview.

A QUESTIONNAIRE CONDUCTED BY SOMEONE OTHER THAN THE AUTHOR

Recycled Resources Corp. Data derived from questionnaire adminis-
tered to 33 foremen in the Bethlehem plant, June 3-4, 1987.

An individual, an outside firm, or a department might also
serve as the questionnaire's "author" for bibliographic purposes. If
you are citing data derived from your own questionnaire, a bibliographic entry is unnecessary: you include the questionnaire and
appropriate background information in an appendix. When citing
your own questionnaire data in the text of your report, refer your
readers to the appendix: "Foremen at the Bethlehem plant over-
whelmingly favored staggered shifts that would keep the plant in
operation 16 hours per day (Appendix D)."

COMPUTER SOFTWARE

Block, K. 1987. Planner. Computer software. Global Software. IBM-PC.

Tools for Drafting. 1986. Computer software. Software international. CP/M 2.3.

Begin the citation with the author of the software if an author is
identified. Underline (or italicize) the name of the program, then
identify the program as computer software. List the name of the
publisher. Finally, add any identifying information, such as the
kind of system or the brand and model of hardware the software
runs on.

INFORMATION RETRIEVED FROM A DATA-BASE SERVCE

Crayton, H. 1987. "A New Method for Insulin Implants." Biotechnology
6: 31-42. DIALOG file 17, item 230043 867745.

Information retrieved from a data-base service is treated like a printed source. At the end of the citation, however, you should list the name of the data-base service and the necessary identifying information to enable a reader to find the item.

Following is a sample bibliography for use with the author/ date citation system:

BIBLIOGRAPHY

Daly, P. H. 1984. Selecting and designing a group insurance plan. Personnel Journal 54:322-23.

Flanders, A. 1984. Measured daywork and collective bargaining. British Journal of Industrial Relations 9:368-92.

Goodman, R. K., J. H. Wakely, and R. H. Ruh. 1983. What employees think of the Scanlon Plan. Personnel 6:22-29.

Trencher, P. 1984. Recent trends in labor-management relations. New York: Madison.

Zwicker, D. Professor of Industrial Relations, Hewlett College. Interview with author. Philadelphia, 19 March 1985.

NUMBERED CITATIONS

You will occasionally encounter a variation on the author/date notation system, the numbered citation and numbered bibliography system. In this system, the items in the bibliography are arranged either alphabetically or in order of the first appearance of each source in the text and then assigned a sequential number. The citation is the number of the source, enclosed in brackets or parentheses:

According to Hodge [3], "There is always the danger that the soil will shift." However, no shifts have been noted.

As in author/date citations, page numbers may be added:

According to Hodge [3:26], "There is always . . . "

In this example, 3 means that Hodge is the third item in the numbered bibliography; 26 is the page number.

Except that they are numbered, the individual bibliography entries in this system are identical to those used with author/date.

APPENDIX C

TECHNICAL WRITING

USAGE AND GENERAL WRITING

STYLE MANUALS

TECHNICAL MANUALS

GRAPHIC AIDS

PROPOSALS

ORAL PRESENTATIONS

JOURNAL ARTICLES

WORD PROCESSING AND COMPUTER GRAPHICS

SELECTED
BIBLIOGRAPHY

TECHNICAL WRITING

Currently, more than 100 technical writing texts and guides are on the market. Following is a representative selection.

Blicq, R. S. 1986. *Technically—write! Communicating in a technological era.* 3d ed. Englewood Cliffs, N.J.: Prentice-Hall.

Brusaw, C. T., G. J. Alred, and W. E. Oliu. 1982. *Handbook of technical writing.* 2d ed. New York: St. Martin's.

Kolin, D. C., and J. Kolin. 1985. *Models for Technical Writing.* New York: St. Martin's.

Mathes, J. C., and D. W. Stevenson. 1976. *Designing technical reports.* Indianapolis: Bobbs-Merrill.

Pickett, N. A., and A. A. Laster. 1984. *Technical English.* 4th ed. New York: Harper & Row.

Sherman, T. A., and S. S. Johnson. 1983. *Modern technical writing.* 4th ed. Englewood Cliffs, N.J.: Prentice-Hall.

Wiseman, H. M. 1985. *Basic technical writing.* 5th ed. Columbus, Ohio: Charles E. Merrill.

Also see the following journals:

IEEE Transactions on Professional Communication
Journal of Business Communication
Journal of Technical Writing and Communication
Technical Communication
The Technical Writing Teacher

USAGE AND GENERAL WRITING

Barzun, J. 1985. *Simple and direct: A rhetoric for writers*. New York: Harper & Row.

Bernstein, T. M. 1965. *The careful writer*. New York: Atheneum.

Corbett, E. P. J. 1971. *Classical rhetoric for the modern student*. 2d ed. New York: Oxford University Press.

Flesch, R. 1980. *The ABC of style—A guide to plain English*. New York: Harper & Row.

Fowler, H. W. 1965. *A dictionary of modern English usage*. 2d ed., rev. by Sir E. Gowers. New York: Oxford University Press.

Hayakawa, S. I. 1978. *Language in thought and action*. 4th ed. New York: Harcourt Brace Jovanovich.

Miller, K., and K. Swift. 1980. *The handbook of nonsexist writing*. New York: Lippincott & Crowell.

Sorrels, B. D. 1983. *The nonsexist communicator: Solving the problems of gender and awkwardness in modern English*. Englewood Cliffs, N.J.: Prentice-Hall.

Strunk, W., and E. B. White. 1979. *The elements of style*. 3d ed. New York: Macmillan.

Trimble, J. R. 1975. *Writing with style: Conversations on the art of writing*. Englewood Cliffs, N.J.: Prentice-Hall.

STYLE MANUALS

American Chemical Society. 1978. *Handbook for authors*. Washington, D.C.: American Chemical Society.

American National Standards, Inc. 1979. *American national standard for the preparation of scientific papers for written or oral presentation*. ANSI Z39.16—1972. New York: American National Standards Institute.

CBE Style Manual Committee. 1983. *Council of biology editors style manual: A guide for authors, editors, and publishers in the biological sciences.* 5th ed. Washington, D.C.: Council of Biology Editors.

The Chicago manual of style. 1982. 13th ed., rev. Chicago: University of Chicago Press.

ORNL style guide. 1974. Oak Ridge, Tenn.: Oak Ridge National Laboratory.

Pollack, G. 1977. *Handbook for ASM editors.* Washington, D.C.: American Society for Microbiology.

Publications manual of the American Psychological Association. 1983. 3d ed. Washington, D.C.: American Psychological Association.

Skillin, M., R. Gay, et al. 1974. *Words into type.* 3d ed. Englewood Cliffs, N.J.: Prentice-Hall.

U.S. Government Printing Office style manual. 1967. Rev. ed. Washington, D.C.: Government Printing Office.

Also, many private corporations, such as John Deere, DuPont, Ford Motor Company, General Electric, and Westinghouse, have their own style manuals.

TECHNICAL MANUALS

Cohen, G., and D. H. Cunningham. 1984. *Creating technical manuals: A step-by-step approach to writing user-friendly instructions.* New York: McGraw-Hill.

McGehee, B. M. 1984. *The complete guide to writing software manuals.* Cincinnati, Ohio: Writers Digest.

Weiss, E. H. 1985. *How to write a usable user manual.* Philadelphia: ISI Press.

GRAPHIC AIDS

Beakley, G. C., Jr., and D. D. Autore. 1973. *Graphics for design and visualization.* New York: Macmillan.

Lefferts, R. 1982. *How to prepare charts and graphs for effective reports.* New York: Harper & Row.

MacGregor, A. J. 1979. *Graphics simplified: How to plan and prepare effective charts, graphs, illustrations, and other visual aids.* Toronto: University of Toronto Press.

Morris, G. E. 1975. *Technical illustrating.* Englewood Cliffs, N.J.: Prentice-Hall.

Thomas, T. A. 1978. *Technical illustration.* 3d ed. New York: McGraw-Hill.

Tufte, E. R. 1983. *The visual display of quantitative information.* Cheshire, Conn.: Graphics Press.

Turnbull, A. T., and R. N. Baird. 1980. *The graphics of communication.* 4th ed. New York: Holt, Rinehart and Winston.

Also see the following journals:

Graphic Arts Monthly
Graphics: USA

PROPOSALS

Lefferts, R. 1983. *The basic handbook of grants management.* New York: Basic Books.

Society for Technical Communication. 1973. *Proposals and their preparation.* Vol. 1. Washington, D.C.: Society for Technical Communication.

ORAL PRESENTATIONS

Anastasi, T. E. Jr. 1972. *Communicating for results.* Menlo Park, Calif.: Cummings.

Howell, W. S., and E. G. Barmann. 1971. *Presentational speaking for business and the professions.* New York: Harper & Row.

Weiss, H., and J. B. McGrath, Jr. 1963. *Technically speaking: Oral communication for engineers, scientists, and technical personnel.* New York: McGraw-Hill.

JOURNAL ARTICLES

Day, R. A. 1983. *How to write and publish a scientific paper.* 2d ed. Philadelphia: ISI.

Graham, B. P. 1980. *Magazine article writing: Substance and style.* New York: Holt, Rinehart and Winston.

Michaelson, H. B. 1986. *How to write and publish engineering papers and reports.* 2d ed. Philadelphia: ISI.

Mitchell, J. H. 1968. *Writing for professional and technical journals.* New York: Wiley.

Zinsser, W. 1980. *On writing well.* 2d ed. New York: Harper & Row.

Also see the following journals:

The Writer

Writer's Digest

WORD PROCESSING AND COMPUTER GRAPHICS

Bernstein, S., and L. McGarry. 1986. *Making art on your computer.* New York: Watson-Guptill.

Fluegelman, A., and J. J. Hewes. 1983. *Writing in the computer age: Word processing skills and style for every writer.* Garden City, N.Y.: Anchor Press/Doubleday.

Hyman, M. I. 1985. *Advanced IBM PC graphics: State of the Art.* New York: Brady Communications.

McCunn, D. 1982. *Write, edit, & print: Word processing with personal computers.* San Francisco: Design Enterprises of San Francisco.

Price, J., and Pinneau Urban, L. 1984. *The definitive word-processing book.* New York: Viking Penguin.

Schwartz, H. J. 1985. *Interactive writing: Composing with a word processor.* New York: Holt, Rinehart and Winston.

Stern, F. 1983. *Word processing and beyond.* Santa Fe, N.M.: John Muir.

Stone, M. D. 1984. *Getting on-line: A guide to accessing computer information services.* Englewood Cliffs, N.J.: Prentice-Hall.

Zinsser, W. 1983. *Writing with a word processor.* New York: Harper & Row.

Flow Chart reprinted with the permission of the Institute of Electrical and Electronics Engineers, Inc. © 1985 IEEE. By Neng F. Yao, "The Computer-Aided Programming System—A Friendly Programming Environment," in IEEE MICRO, vol. 5, no. 2 (April 1985), 12.

STOP discussion and illustration reprinted with the permission of the Institute of Electrical and Electronics Engineers, Inc. © 1983 IEEE. By James R. Tracey, "The Theory and Lessons of STOP Discourse," in IEEE TRANSACTIONS ON PROFESSIONAL COMMUNICATION, vol. PC-26, no. 2 (June 1983), 69.

Instructions to authors from JOURNAL OF GERONTOLOGY, V. 41, N. 2, page 305 (March 1986). Reprinted with the permission of the publisher.

Excerpts from report on rat food. Reprinted with the permission of Edwin Korody.

Proposal, progress report, and completion report on personal computers. Reprinted with the permission of the author, Cynthia C. Adams.

Excerpts from a report on a department store heating system. Reprinted with the permission of the author, Mark S. MacBride.

Excerpts from a report on a radio-based system for distribution automation. Reprinted with the permission of the author, Brian D. Crowe.

Excerpts from student-center manual. Reprinted with the permission of the author, Sandra D. Plum.

Excerpts from "Use of a Fluorescent Stain for Visualization of Nuclear Material in Living Oocytes and Early Embryos," by Elizabeth S. Crister and N.L. First. Reprinted from *Stain Technology* 61, 1:1–5 with permission of the authors.

INDEX